T0213642

Communications in Computer and Information Science 619

Commenced Publication in 2007
Founding and Former Series Editors:
Alfredo Cuzzocrea, Dominik Ślęzak, and Xiaokang Yang

More information about this series at http://www.springer.com/series/7899

Fernando Santos Osório
Rogério Sales Gonçalves (Eds.)

Robotics

12th Latin American Robotics Symposium and
Third Brazilian Symposium on Robotics, LARS 2015/SBR 2015
Uberlândia, Brazil, October 28 – November 1, 2015
Revised Selected Papers

Editors
Fernando Santos Osório
SSC – Depto. de Sistemas de Computação
USP – University of São Paulo (São Carlos)
São Carlos, SP
Brazil

Rogério Sales Gonçalves
UFU – Universidade Federal de Uberlândia
Uberlândia, MG
Brazil

ISSN 1865-0929 ISSN 1865-0937 (electronic)
Communications in Computer and Information Science
ISBN 978-3-319-47246-1 ISBN 978-3-319-47247-8 (eBook)
DOI 10.1007/978-3-319-47247-8

Library of Congress Control Number: 2016953217

Printed on acid-free paper

This Springer imprint is published by Springer Nature
The registered company is Springer International Publishing AG
The registered company address is: Gewerbestrasse 11, 6330 Cham, Switzerland

Preface

This volume on "Intelligent Robotics and Automation Systems" from the Springer *Communications in Computer and Information Science* series (Springer CCIS) consists of the best papers selected from LARS 2015 (the 12th Latin American Robotics Symposium) and SBR 2015 (the Third Brazilian Symposium on Robotics).

LARS 2015 and SBR 2015 comprised a comprehensive scientific meeting in the area of intelligent robotics, bringing together researchers as well as, undergraduate and graduate students of computer science, electrical engineering, mechatronics engineering, mechanical engineering and related areas. The LARS/SBR 2015 event can be considered as the most important Latin American symposium on robotic.

LARS/SBR 2015 were the major scientific/academic events of the Robotics Conferences 2015. These conferences also covered the 14th Latin American Robotics Competition (LARC 2015), the 13th Brazilian Robotics Competition (CBR 2015, 13ª Competição Brasileira de Robótica), the Brazilian Robotics Olympiad (OBR 2015, Olimpíada Brasileira de Robótica), and the National Robotics Shows (MNR 2015, Mostra Nacional de Robótica).

The 2015 Robotics Conferences 2015 were held from October 28 to November 1, 2015, on the campus of the Universidade Federal de Uberlândia (UFU) in the city of Uberlândia, Minas Gerais, Brazil.

These events were the result of a joint action of the Brazilian Computer Society (SBC), the IEEE Latin American Robotics Council and IEEE Robotics, the Automation Society Chapter of Brazil, and RoboCup Brazil, with support from FAPEMIG, CAPES, and CNPq.

We received 80 submissions from ten countries, from which 65 papers were accepted. Each submitted paper was reviewed by at least two experts. The acceptance/rejection decision used the following criteria: All papers with two positive reviews were accepted, and those with two negative reviews were rejected. Borderline papers (with one positive and one negative review) were analyzed carefully by the conference chairs in order to evaluate the reasons given for acceptance or rejection. Our final decision on these submissions took into account mainly the potential of each paper to foster fruitful discussions and the future development of robotics in Latin America and Brazil, including papers at initial stages. In some cases, an additional review was made in order for a decision to be taken. The best papers were selected considering the highest scores obtained according to the reviewers. We selected the 17 best papers from LARS and SBR 2015. Extended and improved versions of the best papers were requested from the authors of these 17 papers. The submitted articles were reviewed again by selected Program Committee members of the conferences and, after corrections, the articles were accepted for publication in this volume of the CCIS series.

We would like to thank all the authors, whose work and dedication made it possible to put together an exciting book. We are also enormously grateful to all Program Committee members and reviewers for their goodwill in cooperating for the success of this book.

We hope that you will find these papers interesting and consider them a helpful reference in the future.

July 2016

Fernando Santos Osório
Rogério Sales Gonçalves

Organization

LARS SBR 2015 – *Robotics Conferences 2015*

General Chair

Rogério Sales Gonçalves, UFU

LARS SBR 2015 – *12th Latin American Robotics Symposium and Third Brazilian Conference on Robotics*

Program Chairs

Douglas Guimaraes Macharet, UFMG
Fernando Santos Osório, USP
João Carlos Mendes Carvalho, UFU
Mário Fernando Montenegro Campos, UFMG

LARC/CBR 2015 – *XIV Latin American Robotics Competition and XIII Brazilian Robotics Competition*

Program Chairs

Esther Luna Colombini, FEI
Carmen R. Faria Santos, UFES
João Alberto Fabro, UTFPR
Ricardo H. de Oliveira Filho, UFTM

MNR – *National Robotics Shows*

Program Chairs

Alexandre da Silva Simões, UNESP
Flávio Tonidandel, FEI

OBR – *Brazilian Robotics Olympiad*

Program Chair

Esther Luna Colombini, FEI
João Olegário de Oliveira de Souza, UNISINOS
Rafael Vidal Aroca, UFSCar
Tatiana de Figueiredo P.A.T. Pazelli, UFSCar

Program Committee

Andre Barczak	Massey University, New Zealand
Bruno Fernandes	University of Pernambuco, Brazil
Cairo Nascimento Jr.	Instituto Tecnológico de Aeronáutica, Brazil
Carlos De Marqui Jr.	USP-EESC, Brazil
Cristiane Tonetto	Universidade Federal de Santa Catarina, Brazil
Damian Lyons	Fordham Univesrity, USA
Danilo Perico	Centro Universitário da FEI, Brazil
Denis Wolf	University of Sao Paulo, Brazil
Dennis Barrios-Aranibar	Universidad Católica San Pablo, Peru
Douglas Macharet	Universidade Federal De Minas Gerais, Brazil
Eduardo Freire	Universidade Federal de Sergipe, Brazil
Eduardo Todt	UFPR, Brazil
Erickson Nascimento	Universidade Federal de Minas Gerais, Brazil
Fabricio Junqueira	Escola Politécnica da USP, Brazil
Fernando Osorio	ICMC-USP – Universidade de Sao Paulo, Brazil
Fernando Augusto de Noronha Castro Pinto	Universidade Federal do Rio de Janeiro – UFRJ – COPPE, Brazil
Flavio Tonidandel	Centro Universitário da FEI, Brazil
Flávio Gonçalves	Universidade Estadual Paulista, Brazil
Glauco A.P. Caurin	Universidade de São Paulo (USP), Brazil
Gustavo Pessin	USP and UFPA, Brazil
Hugo Vieira Neto	Universidade Tecnológica Federal do Paraná, Brazil
Humberto Ferasoli Filho	Universidade Estadual Paulista – UNESP, Brazil
Jefferson Souza	Federal University of Uberlândia (UFU), Brazil
Jes Cerqueira	Universidade Federal da Bahia, Brazil
João Carlos Mendes Carvalho	Universidade Federal de Uberlândia, Brazil
Jochen Steil	Bielefeld University, Germany
John Archila	University of Sao Paulo, Brazil
José Martins Junior	Universidade de São Paulo, Brazil
José Tavares	Universidade Federal de Uberlândia, Brazil
José Tenreiro Machado	Institute of Engineering, Polytechnic of Porto, Portugal
Josue Junior Guimarães Ramos	CTI – Centro de Tecnologia da Informação Renato Archer, Brazil
Jun Okamoto	USP Kalinka Castelo Branco, ICMC – USP, Brazil
Leonardo Pedro	Federal University of São Carlos, Brazil
Luiz Chaimowicz	Universidade Federal de Minas Gerais, Brazil
Manfred Huber	University of Texas at Arlington, USA
Marcelo Becker	Escola de Engenharia de São Carlos – USP, Brazil
Mario Tronco	Universidade de São Paulo (USP), Brazil
Márcio José da Cunha	Federal University of Uberlândia, Brazil
Milton Heinen	Universidade Federal do Pampa (UNIPAMPA), Brazil
Paulo Farias	Universidade Federal da Bahia, Brazil

Priscila Martins	UFMS Rafael Aroca, Universidade Federal de São Carlos, Brazil
Reinaldo Bianchi	FEI, Brazil
Renato Tinos	USP, Brazil
Rodrigo Calvo	Universidade Estadual de Maringá, Brazil
Rogério Sales Gonçalves	Universidade Federal de Uberlândia, Brazil
Ronald Arkin	Georgia Tech., USA
Roseli Francelin Romero	USP-SC, Brazil
Rubens Tabile	FZEA – Universidade de São Paulo, Brazil
Sergio Campello Oliveira	Universidade de Pernambuco, Brazil
Silvia Botelho	FURG, Brazil
Tatiana Pazelli	Federal University of Sao Carlos, Brazil
Valéria Santos	Universidade de São Paulo, Brazil
Valdir Grassi Junior	University of São Paulo, Brazil
Vera Franco	Universidade Federal de Uberlândia, Brazil
Vinicius Marques	Universidade Federal do Triângulo Mineiro, Brazil
Vitor Romano	UFRJ, Brazil
Whisner Mamede	Instituto Federal de São Paulo, Brazil

Contents

Evaluating the Performance of Two Computer Vision Techniques for a Mobile Humanoid Agent Acting at RoboCup KidSized Soccer League

Claudio O Vilão Jr.[1(✉)], Vinicius Nicassio Ferreira[1], Luiz Antonio Celiberto Jr.[2], and Reinaldo A.C. Bianchi[1]

[1] Electrical Engineering Department, FEI University Center, Sao Bernardo do Campo, Sao Paulo, Brazil
{cvilao,rbianchi}@fei.edu.br

[2] Brazil Center of Engineering, Modelling and Applied Social Sciences, Federal University of ABC, Santo Andre, Sao Paulo, Brazil
luiz.celiberto@ufabc.edu.br

http://www.fei.edu.br, http://www.ufabc.edu.br

Abstract. A humanoid robot capable of playing soccer needs to identify several objects in the soccer field in order to play soccer. The robot has to be able to recognize the ball, teammates and opponents, inferring information such as their distance and estimated location. In order to achieve this key requisite, this paper analyzes two descriptor algorithms, HAAR and HOG, so that one of them can be used for recognizing humanoid robots with less false positives alarms and with best frame per second rate. They were used with their respective classical classifiers, AdaBoost and SVM. As many different robots are available in RoboCup domain, the descriptor needs to describe features in a way that they can be distinguished from the background at the same time the classification has to have a good generalization capability. Although some limitations appeared in tests, the results were beyond expectations. Given the results, the chosen descriptor should be able to identify a mainly white-ball, which is clearly a simpler object. The results for ball detection were also quite interesting.

Keywords: Humanoid robots · Monocular vision · Object tracking

1 Introduction

Object detection is one of the key challenges in computer vision and its a necessary precondition for robotic soccer players. Identifying objects, such as other robots, can determine which team will win the game. However, in this context, selecting an appropriate visual descriptor for object representation to get a high detection rate, in real time, is quite challenging.

Aiming to develop a vision system for a humanoid robot, capable of playing soccer inserted in the RoboCup domain, more specifically in KidSize league,

F. Santos Osório and R. Sales Gonçalves (Eds.): LARS 2015/SBR 2015, CCIS 619, pp. 1–19, 2016.
DOI: 10.1007/978-3-319-47247-8_1

(a) B2 Robot

Fig. 1. A Darwin-OP [2] based robot developed by the University Center of FEI

two well known techniques for visual descriptors will have their performance compared in order to achieve the goal of identify other robots. This two techniques are the HOG - Histogram of Oriented Gradients with SVM classifier and the cascade HAAR Wavelets with boosting classifier. Figure 1 shows the robots developed by the FEI University Center [1] having as basis the Darwin-OP [2] platform and the ball used in Robocup 2015 competition.

Mainly two methods are used for object detection. Sliding window techniques scan the input image several times over different positions and scales classifying each sliding window. Other methods generate hypotheses by evidence aggregation. Parts of the objects are generally used for this purpose.

A discussion on some of the most successful sliding window techniques are proposed so that can be possible to use one of those methods for robot and ball identification. Both techniques fit in appearance-based methods, which learn the characteristics of the classes robot and not robot from a set of training images. Each training image is represented by a set of local or global descriptors [features]. Those features are used by the classification algorithm to create a decision boundary between those two classes. Features are commonly used instead of pixels values as they can code a high level object information. It is possible to see in Fig. 2 the mainly white-ball used on Robocup soccer competition has few gray details.

1.1 Domain

RoboCup is an international research and education initiative, it is also a competition, which goal is to increase robotic research by providing problems where a huge range of technologies can be integrated. It is compound of competitions aiming several types of robots.

This project is based on the main soccer humanoid competition for robots, the RoboCup and in its agent performance, more specifically, in KidSize category.

(a) Nike England Size 1 Ball

Fig. 2. The ball used in Robocup humanoid KidSize league. Source: [3]

The KidSize category comprehend all humanoid robots that has up to 90 cm height.

The playing field consists of a flat, leveled surface covered with artificial grass. Ten centimeter wide segments are used for penalty marks and for the ball starting position. The field is limited by two larger lines, called side lines, and two smaller ones, defined as base lines, where the goals are positioned. All areas located outside these four lines are considered unknown and undefined.

The lighting conditions provided by the organization are considered enough but it always depends on the site of the competition. It is defined around the field a place where only a human judge, his human assistants and two human handlers can act. No participant can use below the waist, similar colors to those set for the field, ball or goals.

Only human equivalent sensors are allowed, but are not mandatory. Humanoid robots must be in the form of their bodies similar to the human being, should have legs and arms. A series of size restrictions are applied to the agents body measures, in a way that the robots don't have too long arms or too long legs. Those limitations are necessary to keep the look of robots similar to the humanoid form.

Four humanoid robots compete in RoboCup Humanoid KidSize environment each one has a single full-hd camera, and has to rely on its provided image, as it is its main perception sensor.

The kidSize environment has its own rules regarding limits for robot, which has to have only human features, but there is no standard when it comes to robot shape. This means that as long as the robot has two arms, and their size is within the rules, these arms doesn't have to be equal in terms of construction. The same analogy can be made for the torso and legs. This states that the diversity of robot can be enormous and still be in accordance with the rules.

The robots field of view is limited to 180°, this means that the maximum angle between any two points in the overlay of all the cameras mounted on the robot's field of vision must necessarily be less than 180°. The mechanism to rotate (pan) the camera is limited to 270°, 135° to either side as from the position to look forward. Since the mechanism to lift (tilt) the camera is limited to 90° measured from the horizontal line. These restrictions aims to simulate the same limitations of the joints of the human neck.

Following this line of human limitations, is to set the robot in any position and from any angle of the camera or robot, one may not be able to view both goals at once. The number of cameras is limited to two, however, the monocular vision is also allowed. Either way it has to be placed in the head of the robot.

Robots participants of the humanoid league competitions must have bodies, predominantly black, gray, silver or white, with the limitations of being non-reflective. Only exception are the legs which must be black. Any color of the field, goal, ball or opponent must be avoided in the robot. Arms, legs and bodies must have solid appearance and must be marked with the color magenta (for agents of the red team) and the cyan (for agents of the blue team), both brands must be at least 5 cm and be visible in both sides of the legs and arms.

This paper is organized as follows: in Sect. 2, related work explains how HOG and HAAR descriptors works and how they have been used by other researchers, in Sect. 3 show some methodology for testing both descriptors, Sect. 4 a few experimental results for robot detection and for ball detection, finally Sect. 5 concludes this work.

2 Related Work

Many researchers proposed a huge number of techniques to describe objects [4,5], to track pedestrians [6–9], vehicle identification [10] and face recognition [11].

In [12], the most famous visual descriptor Multi-scale Block Local Binary Pattern MB-LBP [13], HOG [14] and HAAR [15] were adapted for use with AdaBoost [16].

In our previous work [17], due to the variety of robots in the KidSize category, the lack of training images, a classifier such as the support vector machine [18] was trained using the information of several human images and a small dataset of robot images. Another worth citing work, much closer to the proposed domain, was done by [19] who used a hog descriptor to identify NAO robots [20].

2.1 HAAR-Boosting

In mathematics, the Haar wavelet is a sequence of rescaled square-shaped functions which together form a wavelet family or basis [21]. Wavelet analysis allows a target function over an interval to be represented in terms of an orthonormal function basis. The Haar wavelet is not differentiable as it is a continuous function, even though this could be a technical disadvantage this function can determine sudden transitions, specially in images [21].

Basically, when every pixel of an image has to be computed the feature calculation tends to be expensive, [22] proposed an alternative feature set that is based on Haar wavelets, instead of using pixel intensities. The idea of using Haar wavelets, was next improved by Viola and Jones [15], and originated the

Haar-like features, which are digital image features used in object recognition. One of the first applications of this technique was a real-time face detector [15].

For computing Haar like features one has to consider adjacent rectangular regions at a specific location in an image, compute the sum of pixels intensities in that region and then calculate the difference between the sums. This difference is then used to categorize subsections of an image. The position of these rectangles is defined relative to a detection window that acts like a bounding box to the target object. Figures 3 and 4 shows some of the commonly used features.

The AdaBoost algorithm has been introduced by [16], and is a classification algorithm which approach consists on creating a highly accurate prediction rule by combining relatively weak and inaccurate rules.

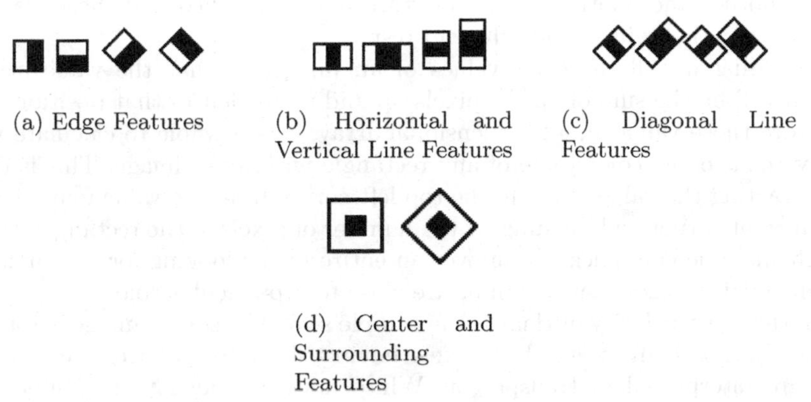

(a) Edge Features

(b) Horizontal and Vertical Line Features

(c) Diagonal Line Features

(d) Center and Surrounding Features

Fig. 3. Some simple HAAR-like features used for describing objects. Source: [23]

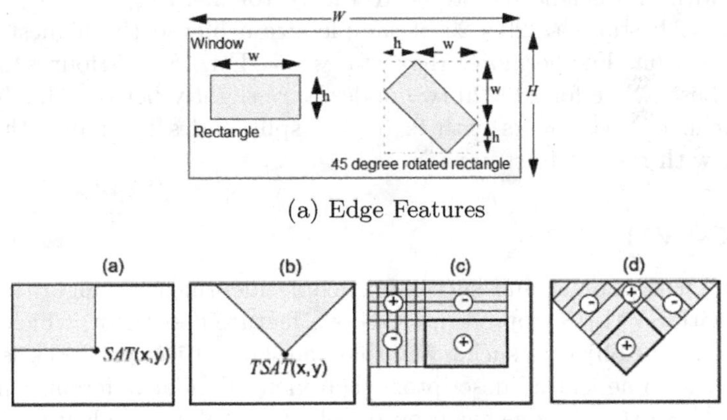

(a) Edge Features

(b) Horizontal and Vertical Line Features

Fig. 4. Some operations applied to simple HAAR-like features used for describing objects. Source: [23]

The AdaBoost calls a weak classifier repeatedly using iterations t = 1, . . . , T. For each call, distribution weights D is updated to indicate the importance of the example data set used for classification. At each iteration the weights of each example mis-categorized is increased, or alternatively, the weights correctly classified are decreased.

The AdaBoost is adaptive in sense that subsequent classifications made are adjusted in favor of instances adversely classified by previous classifications. AdaBoost uses a number of training sample images to pick a number of good features. For object recognition a feature is typically just a rectangle of pixels that has a certain average color value and a relative size. AdaBoost will look at a number of features and find out which one is the best predictor of the object based on the sample images. After it has chosen the best classifier it will continue to find another and another until some threshold is reached and those features combined together will provide the end result.

By adding up all the color values of an image, so that the value in any position will be the sum of all the pixels up and to the left of that position, and then store those values in a 2 dimensional array, it is possible to calculate very quickly the average color value of any rectangle within the image. This is done by subtracting the value found in the top left corner from the value found in the bottom right corner and dividing by the number of pixels in the rectangle. Using this schema, one can quickly scan over an entire image looking for rectangles of different relative sizes that match or are close to a particular color.

Another particularity of Haar is the sample size. The source image is rotated randomly around all axes. Then pixels having the intensity from an specific range are interpreted as transparent. White noise is added to the intensities of the foreground. The obtained image is placed onto an arbitrary background from the background description file, resized to the desired size specified by the provided width and height and stored to the vector file.

Kuranov [23] states as 20×20 of sample size achieved the highest hit rate for face detection. Furthermore, they states as "For 18×18 four split nodes performed best, while for 20×20 two nodes were slightly better. The difference between weak tree classifiers with 2, 3 or 4 split nodes is smaller than their superiority with respect to stumps."

2.2 HOG-SVM

One of the most popular and successful people identification algorithm is the HOG [14] with SVM [18] approach. HOG is a feature descriptor, which concept is to generalize the object in such a way that the same object produces as close as possible to the same feature descriptor when viewed under different conditions.

The HOG-SVM technique has proved to be very robust to identifying robots images as done in our previous work [17]. Even when using a completely different class of images (pedestrians), it was possible to detect robots. In that occasion human images, instead of robot images, were used as a way to identify robots. Computing gradient magnitude and orientation of a sliding window through the image, HOG creates vectors that are fed into SVM classifier.

Three stages are required: training, testing and color segmentation inside the window in order to determine to which team color the identified robot belongs, as done before by Castro and Radev [19] in a team of NAO [20] robots.

By using a mask or a kernel the HOG algorithm compute the gradient of every pixel of the image giving the direction of that gradient. The creators of this approach trained a Support Vector Machine, or SVM [18], to recognize HOG descriptors of people, classifying them as "person" or "not a person". It uses a global feature to describe a person rather than a collection of local features. First of all, a gradient computation is required. Gradient vectors (or image gradients) are one of the most fundamental concepts in computer vision. The calculation includes a gradient magnitude and gradient vector angle. The primary use for this technique is the edge detection, the higher the variance of the pixels, the greater the gradient magnitude will be. The entire image is represented by a histogram or, in other words, a single feature gradient vector.

This set of images, already converted in a set of gradient oriented vectors by running the above algorithm, becomes the training set of the SVM classifier. A SVM constructs a set of hyperplanes in a high-dimensional space, which can be used for classification. Intuitively, a good separation is achieved by the hyperplane that has the largest distance to the nearest training data point of any class.

A support vector machine - SVM is a concept created by [18] refers to a set of supervised learning methods that recognizes and analyzes data patterns used for classification and analysis regression. The standard SVM takes as input a set of data and predicts, for each given input, which of two possible classes the entry belongs, which makes the SVM a non probabilistic binary linear classifier.

In the literature, one predictor variable is called a processed attribute and an attribute that is used to define the hyperplane is called characteristic. The task of choosing the most appropriate representation is known as feature selection. A set of features that describes a case (i.e. a line forecast values) is called a vector. Thus the objective of shaping the SVM is to find the optimal hyperplane that separates the sets of vectors in such a manner that data of a target category are on one side of the hyperplane and the other category is the other side. The vectors which are close to the hyperplane are called support vectors.

In order to allow some flexibility in the classification of SVM, an error factor called cost parameter (C) is inserted on the sidelines, allowing some data to be accepted as belonging to another class so they would be misclassified. Reduce the value of this parameter will require a more stringent classification, but also reduce the power of generalization of the classifier. Applying the concept of reverse manner, namely, increasing the value of the cost parameter, the effect would be no classification, therefore, the choice of this parameter determines the effectiveness of this classifier.

The application of this technique results are comparable to those obtained by other learning algorithms such as Artificial Neural Network (ANN) [24] some tasks have been shown to be superior as to detect faces in images made by

[25], and categorization of texts, as proposed by [26]. According to [27], as SVM advantages, one can mention:

1. **Good generalization:** classifiers generated by an SVM generally achieve good generalization results. The ability to generalize a classifier is measured by its efficiency in data classification that do not belong to the group used in their training. In generating predictors for SVMs is therefore avoided over-fitting, in which case the predictor becomes very specialized in the training set, getting poor performance when faced with new standards;
2. **Robustness in large dimensions:** The SVMs are robust when facing large dimensions such as, for example, images. Commonly there is the occurrence of over-fitting in the classifiers generated by other intelligent methods on these types of data;
3. **Convexity of the objective function:** The application of SVMs involves the optimization of a quadratic function, which has only one global minimal. This is an advantage over, for example, Artificial Neural Networks, where there is the presence of local minimal in the objective function to be minimized;

3 Proposal

Both classical approaches were implemented, the HOG descriptor with SVM classifier and HAAR-like features with boosting classifier. The objective is to see which one would be better for robot identification.

As HOG works with pictures with limited dimensions of 64×128 pixels, all the images containing robots had to be cropped and normalized. An implementation of SVM call SVM-light were used with Radial Basis Function Kernel which was chosen, in essence, to increase the number of dimensions, in order to find the best hyperplane capable of maximize the margin between classes.

For Haar-Like features a cascade of weak classifiers was implemented using AdaBoost (Adaptive Boosting) a variation of the boosting algorithm. Most of the image region is non-object region. So a fast checking if a window is not a object region will provide a faster computation result by focusing on region where it is possible to find a object.

So instead of applying all features on a window, it is possible to group the features into different stages of classifiers. Consider the first stage, if a window does not contain an object, it is unnecessary to consider remaining features on it. If the opposite occurs apply the second stage and so on. To be considered a window that contains an object, it has to pass through all stages.

All the objects were manually selected using a rectangular bounding box from the positive images. The utility create samples, from opencv library [28], was used to create training samples from every image while feeding into a vector file. Some errors can occur and the most possible reason is insufficient count of samples in a given vec-file, to solve that, Dimashova [29] proposed the following:

Suppose a positive samples subset of size N_{Pos} was selected to train current i-stage (i starting in zero) and Min_{HitRate} is the training constraint for each stage. After the training of current stage, at least Min_{HitRate} * N_{Pos} samples from this subset have to pass this stage, i.e. current cascade classifier with i+1 stages has to recognize this part Min_{HitRate} of the selected samples as positive. Some already used samples can be filtered by each previous stage (i.e. recognized as background), but no more than ((1 - Min_{HitRate}) * N_{Pos}) on each stage. See Eq. 1 where N_{Positive} is the amount of positive images.

The Eq. 2 was used to determine the N_{Pos} parameter.

$$Samples = (N_{\text{Stages}} - 1) * (1 - Min_{\text{HitRate}}) \tag{1}$$

$$N_{\text{Positive}} = [N_{\text{Pos}} + Samples * N_{\text{Pos}}] + N_{\text{Neg}} \tag{2}$$

The count of samples from vec-file that can be recognized as background right away is the N_{Neg} in the Eq. 2. The major problem on the robot recognition algorithm is running it in real time, because the number of dimensions, e.g. features, are enormous, as its processing costs. Both algorithms should be fed with the same images and the choosing criteria will be the ROC curve obtained from both techniques and which one gets the better frame rate with greater resolution.

4 Experimental Results

There are many aspects in constructing a reliable vision system for identifying robots, but most importantly, it must be compatible to a fast acting robot. The experiments were made in a Dell Inspiron equipped with the 5th generation Intel Core i7-5500U 2.4 GHz processor with 16 GB DDR3L SODIMMs of ram memory. It also features a video card AMD Radeon HD R7 M265 with dedicated memory of 2 GB DDR3 with 4K display capabilities. In order to acquire a image from the environment a Logitech C920 full-hd camera was used.

In order to proceed, a set of positive and negative images were collected. Here, a negative dataset was gathered aiming diversity, it contains images from different locations including buildings, soccer fields, crowds and streets, no robots involved. Given the diversity of robot in the proposed domain, the positive dataset was obtained from internet videos and images which have at least one robot present and has a minimum resolution of 640 × 480 pixels, so the robots features can be easily identified and properly processed. For this purpose, it was gathered 2135 negative images and 1192 positive ones.

4.1 Robot Identification

As mentioned before, two performance parameters are used to compare the two proposed algorithms. For HOG descriptor part, a standard 64 × 128 HOG descriptor were used, a window stride of 16 × 16 pixels, a padding of 32 × 32 pixels and 0.5 as a hit threshold. For the SVM classifier, the cost parameter C was set to 10 and $\Gamma = 0{,}001$ chosen by using a grid search. For HAAR Descriptor, the chosen window had width of 100 pixels and height of 100 pixels as the size of the object features in positive images will never be less than 100×100 pixels.

For the Haar training stage, the maximum number of stages was set to 20, the minimum hit rate was set to 0.999, the N_{Pos} parameter after calculation using Eqs. 1 and 2, was set to 920, and finally the sample window was of 100×100 pixels. The image training set and the image testing set used were the same for both technique. The frame rate results are in the Fig. 11.

In Figs. 5 and 6 both descriptors were used to identify the robots of the university of Hertfordshire, as a way to identifying opponents in the soccer game. In Figs. 7 and 8 both decriptors were used to identify the robots developed by the FEI university center as a way to identifying teammates in the soccer game.

In Fig. 9 a more realistic possibility, showing the system identifying several robots from two different teams in a game held in RoboCup. Here it is possible to see the difference of two used techniques, while HOG proved to be very reliable in terms of detection showing no false positives, but it has missed one robot or more precisely half of a robot.

In Fig. 9(b) HAAR technique presented some false positives, but one detection of the camera held by the cameraman behind the goal lines was surprisingly

Fig. 5. Bold hearts robot [30] identification from the university of Hertfordshire using HAAR. Source: Frame extracted from bold hearts - RoboCup 2015 qualification video. [31]

Fig. 6. Bold hearts robot [30] identification from the university of Hertfordshire using HOG. Source: Frame extracted from bold hearts - RoboCup 2015 qualification video. [31]

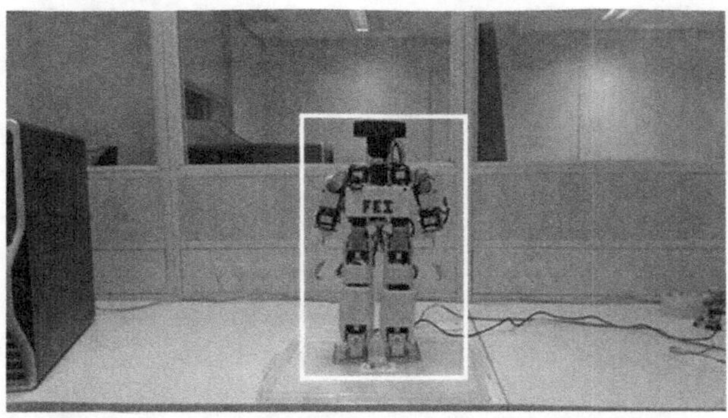

Fig. 7. RoboFEI b2 robot [17] identification using HOG.

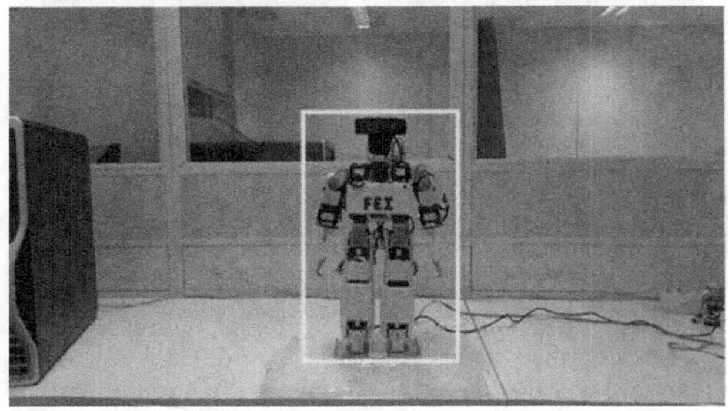

Fig. 8. RoboFEI b2 robot [17] identification using HAAR.

good, since, even though, there is no robot there, the system was capable to identify the flash light as a camera, that are most commonly present on the top of all humanoid robots.

In Fig. 10a sequence of frames of robot b2 tracking done by another b2 robot in a real artificial grass field, similar to the competition environment.

HAAR outperformed better than HOG in terms of speed specially in greater resolutions. When the robot was taken to the real environment, the illumination proved to be a real problem, however, this descriptor, even in this circumstances seems to do its job (Fig. 12).

In Fig. 10 some false positives are shown in right corner, probably due to the illumination changes. In the real competition the training set will be made taking in account the local illumination as it changes by shadows from other objects.

4.2 Ball Identification

Given the performance of HAAR Descriptor, it was possible to use this technique for ball detection. In this case the chosen window had width of 80 pixels and

(a) HOG - SVM

(b) HAAR - Boosting

Fig. 9. Several robot identified using HOG-SVM and HAAR-Boosting algorithms. Source: frame extracted from bold hearts - RoboCup 2015 qualification video. [31]

height of 80 pixels as the size of the ball features in positive images will never be less than 80×80 pixels. Following the same train of thought of the robot detection, for the Haar training stage the maximum number of stages was set to 20, the minimum hit rate was set to 0.999, the N_{Pos} parameter after computation was set to 870, and finally the sample window was of 80×80 pixels (Fig. 13).

Other teams used white color segmentations over a green background in order to identifying the ball, that solution comes with a setback the size of the ball detection has to be limited, as its detection distance. This kind of limitation occurs mostly because a great range of objects reflects white, so the minimum count of pixels that has to be considered in order to check if the region is indeed a ball has to be greater than the noise produced by those other objects presents in the soccer field.

There is two balls in the frames showed in Fig. 14 one was placed distanced from the robot up to 5 m (b) and another one placed at 7 m away from the robot (h), given the overstated environment with several mainly white regions, it is possible to see the system identifying only the two balls placed in the soccer field. Even in (c) frame circumstances where the ball is near two white shoes, the system was capable of distinguish both regions.

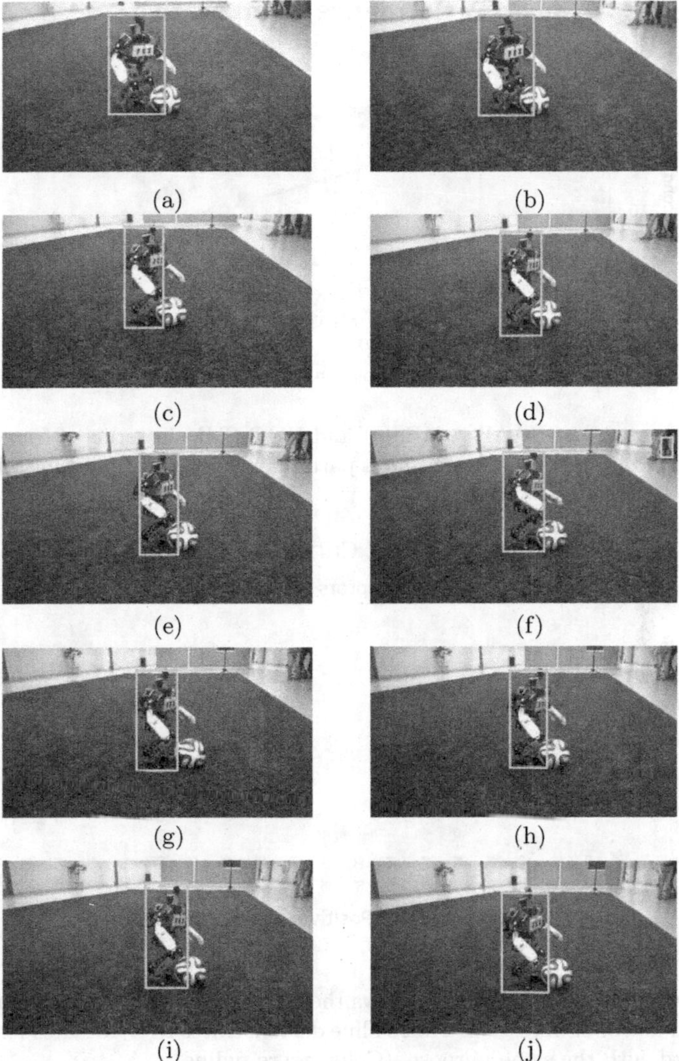

Fig. 10. Darwin tracking using HAAR-like features in an artificial grass soccer field.

A pre-processing is performed, in order to maximize the performance of HAAR. It basically consists in using temporal information to determine in what position the object was detected in previous frames, this information is particularly important to create a framework that predicts where the object is in the current position, and thus, applying the HAAR only in this context.

The cropped frame is 10 % bigger than the original object detection window, the reason why this is done is because of the moving behavior of the ball which can be inside the detection window or approaching to the edge of it. This value

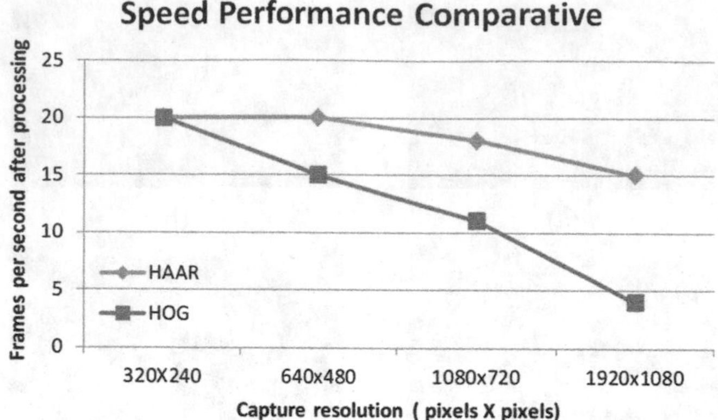

Fig. 11. Performance of HAAR-AdaBoost and HOG-SVM in terms of frames per second after computation. This information is particularly vital when dealing with a robot which has to take actions quickly

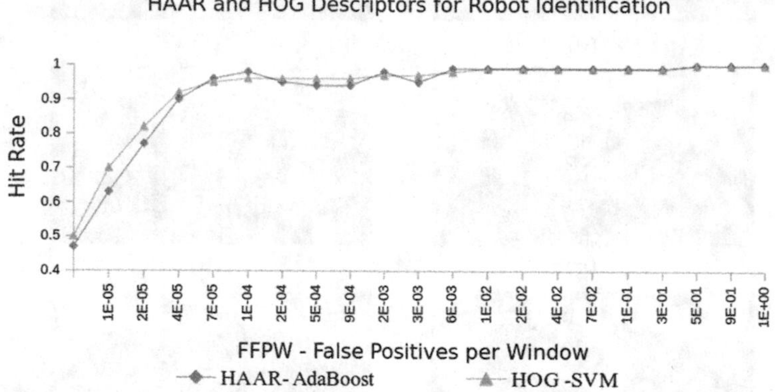

Fig. 12. ROC curves. The green line shows the classical HOG-SVM algorithm trained classifier used for robot detection. The red line demonstrates HAAR-AdaBoost classifier results trained with the same purpose. (Color figure online)

was acquired empirically, with no especial method, until the robot follows the ball without losing it.

As the ball is mainly in a green region, i.e. the field, a color segmentation was done in order to reduce the region of action of the Haar technique. For this experiment the equipment used were is a Intel Core i5-2410U processor with 4 GB of ram memory with shared onboard video card.

Here, it is seen that HAAR processing the whole frame using a fixed resolution of 1024×576 shows the performance of an average of 0.17 seconds per frame, equivalent to 5.83 frames per second. When using the cropping frame the performance was improved to 0.067 seconds per frame, equivalent to 14.92

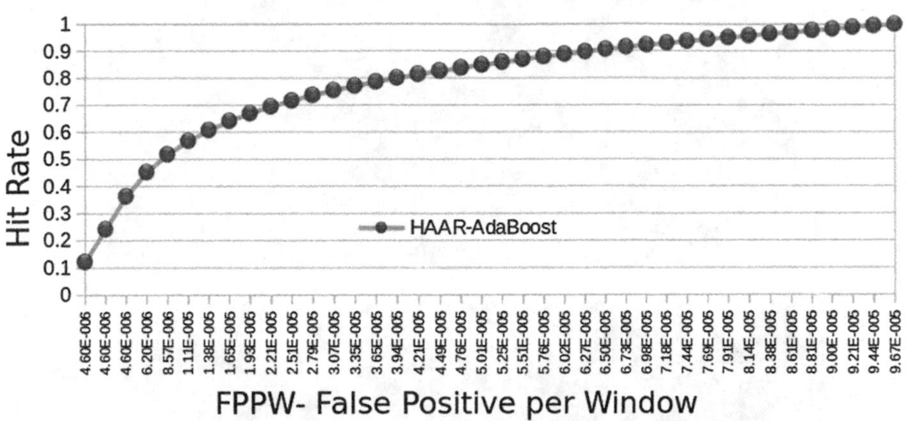

Fig. 13. ROC curve for ball identification using HAAR-Adaboost.

frames per second, increasing the performance in 9.3 frames per second even in a equipment with lower computational power.

Another process is also performed consisting in making a range of capture resolutions to be in accordance to the size of the object i.e. the ball, this means that greater resolutions are used to far away objects, where less pixels represents the object, and smaller resolutions are used for close objects, aiming computation speed performance, as there are less pixels to be computed.

For this we used a geometric progress function where its elements are the maximum and minimum limits for the resolution exchange. It was possible to go beyond increase performance, increasing its detection capacity, as objects placed in longer distances are processed with higher resolutions and closer objects with lower resolutions.

5 Conclusion

Both ROC curves suggests that the trained classifiers, seems to be equivalent. Due to the robots diversity, both classifiers shows a great generalization capacity, that could recognize a huge amount of humanoid robots, given enough normalized, standard, training images.

Giving the limitation of real time application, HAAR outperformed better taking in consideration the computation speed. However it has some problems with illumination and therefore have a high false alarm rate specially in real and complex environments.

The results show some advantages for the Haar technique in frames per second rate which could be run in full hd but with some false positives since it is a

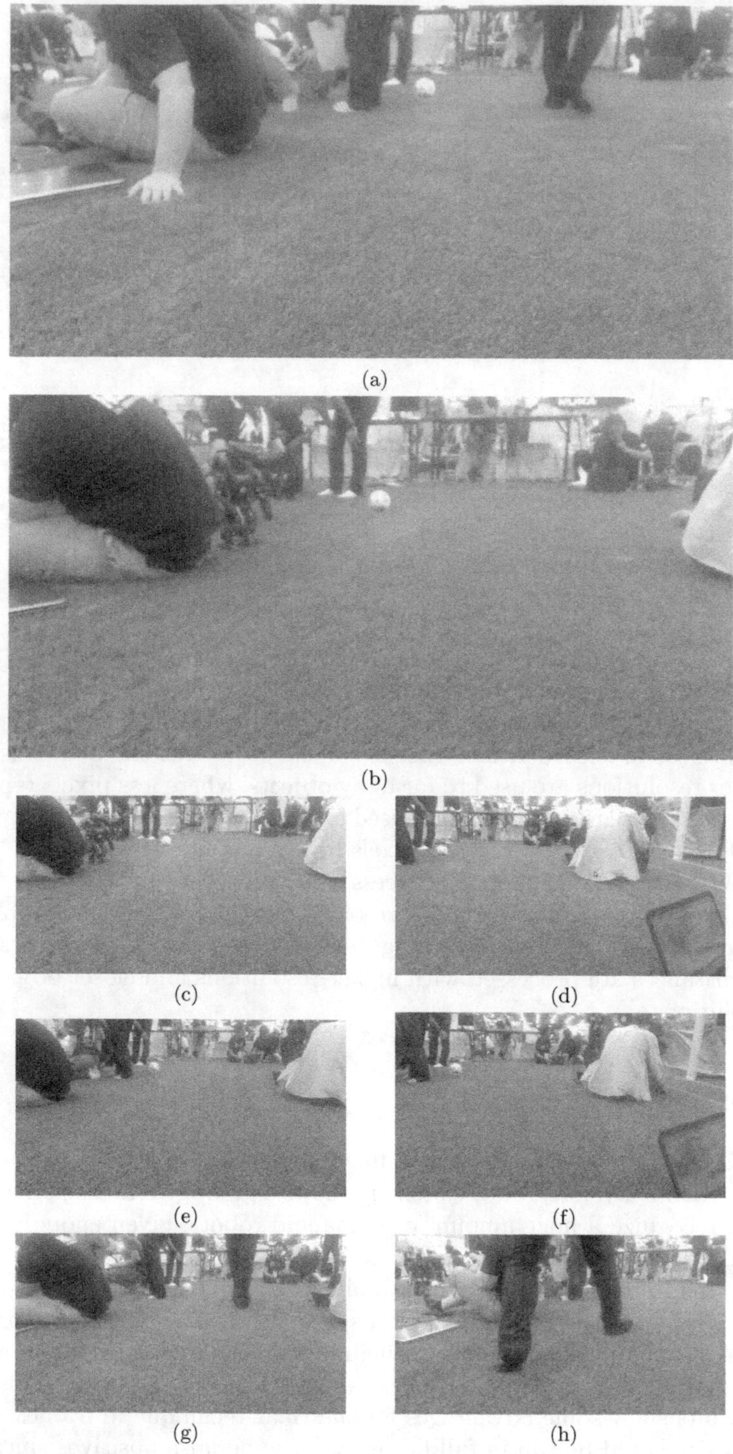

Fig. 14. Robot tracking ball using HAAR-like features in an overstated artificial grass soccer field.

technique very sensible to lightning changes and although HOG was more accurate in terms of detection, it could have a limitation of speed rate and maximum resolution of 640×480 pixels.

Here a caveat comes in order, when it comes to evaluate the descriptors the chosen parameters and the hardware used were greatly responsible for both algorithms overall speed.

The spent time in training was also a complicated issue. While Hog determine all the features by itself and take just a few minutes for training (only problem was to crop and normalize the images), HAAR needed a person to determine where the objects are, taking from 5 to 10 h to be concluded.

The next step is increase the number of images in different scenarios with diverse illumination conditions so HAAR-like features will become more robust to overstated background. Giving the context, HAAR seems to be more fitting for the proposed domain, mainly because of its processing speed.

Another possibility is the use of deep learning for object detection as done before in 2012, Jeff Dean e Andrew Y. Ng from Google Co. presented a new system of image recognition [32] using Deep Neural networks in order to identifying objects.

Even though the hardware used for training and testing DNN's were much bigger than the embedded system used in the robot, the number of classes is smaller.

So a study to identify the processing levels of the deep neural networks to identifying robots and other objects in the soccer field using a moving camera can also be glimpsed.

This paper presented a performance benchmarking of robots recognizing system for the RoboCup humanoid KidSized league environment, using well known vision descriptors, HAAR-like features and HOG descriptor.

Acknowledgment. The authors would like to thank the University Center of FEI for the available resources, and the total cooperation of the researchers involved in this project, including the Robotics and Artificial Intelligence Laboratory that supports this project. The authors would also like to thank to Brazilian research fomentation agencies, FAPESP, CNPQ and CAPES, for the scholarships provided.

References

1. Perico, D., Silva, I., Vilao, C., Homem, T., Destro, R., Tonidandel, F., Bianchi, R.: Hardware, software aspects of the design, assembly of a new humanoid robot for robocup soccer. In: 2014 Joint Conference on Robotics: SBR-LARS Robotics Symposium and Robocontrol (SBR LARS Robocontrol), pp. 73–78, October 2014. doi:10.1109/SBR.LARS.Robocontrol.2014.39
2. Ha, I., Tamura, Y., Asama, H., Han, J., Hong, D.W.: Development of open humanoid platform DARwin-OP, pp. 2178–2181 (2011)
3. Nike Website (2015). http://store.nike.com/gb/en_gb/pw/football-balls/896Zof3. Acessed July 2015

4. Lienhart, R., Maydt, J.: An extended set of Haar like features for rapid object detection. In: International Conference on Image Processing (2002). doi:10.1109/ICIP.2002.1038171
5. Gerónimo, D., López, A., Ponsa, D., Sappa, A.D.: Haar wavelets and edge orientation histograms for on-board pedestrian detection. In: Proceedings of the 3rd Iberian Conference on Pattern Recognition and Image Analysis, Part I, pp. 418–425 (2007). doi:10.1007/978-3-540-72847-4_54
6. Viola, P., Jones, M., Snow, D.: Detecting pedestrians using patterns of motion and appearance. In: Ninth IEEE International Conference on Computer Vision (2003). doi:10.1109/ICCV.2003.1238422
7. Wei, Y., Tian, Q., Guo, T.: An improved pedestrian detection algorithm integrating Haar-like features and HOG descriptors. Adv. Mech. Eng. (2013). doi:10.1155/546206
8. Schiele, B., Andriluka, M., Majer, N., Roth, S., Wojek, C.: Visual people detection - diferent models, comparison and discussion. In: Proceedings of the IEEE ICRA Workshop on People Detection and Tracking (2009)
9. Brehar, R., Nedevschi, S.: A comparative study of pedestrian detection methods using classical Haar and HOG features versus bag of words model computed from Haar and HOG features. Intell. Comput. Commun. Process. (ICCP) (2011). doi:10.1109/ICCP.2011.6047884
10. Negri, P., Clady, X., Prevost, L.: Benchmarking Haar and histograms of oriented gradients features applied to vehicle detection. In: Proceedings of the Fourth International Conference on Informatics in Control, Automation and Robotics, Robotics and Automation, ICINCO, Angers, France, 9–12 May 2007, vol. 1, pp. 359–364 (2007)
11. Paisitkriangkrai, S., Shen, C., Zhang, J.: Face detection with effective feature extraction. CoRR, abs/1009.5758 (2010)
12. Ju, Y., Zhang, H., Xue, Y.: Research of feature selection and comparision in adaboost based object detection system. J. Comput. Inf. Syst. **9**(22), 8947–8954 (2013)
13. Zhang, L., Chu, R., Xiang, S., Liao, C., Li, S.Z.: Face detection based on multi-block LBP representation. In: Lee, S.-W., Li, S.Z. (eds.) ICB 2007. LNCS, vol. 4642, pp. 11–18. Springer, Heidelberg (2007). doi:10.1007/978-3-540-74549-5_2
14. Dalal, N., Triggs, B.: Histograms of oriented gradients for human detection. In: Conference on Computer Vision and Pattern Recognition, vol. 1, pp. 886–893 (2005)
15. Viola, P., Jones, M.: Rapid object detection using a boosted cascade of simple features. In: Conference on Computer Vision and Pattern Recognition (CVPR), pp. 511–518 (2001)
16. Freund, Y., Schapire, R.E.: A decision-theoretic generalization of on-line learning and an application to boosting (1997). http://portal.acm.org/citation.cfm?id=261540.261549
17. Vilão, C.J., Silva, I.J., Perico, D.H., Homem, T.P.D., Tonidandel, F., da Costa Bianchi, R.A.: A single camera vision system for a humanoid robot. In: Joint Conference on Robotics: SBR-LARS Robotics Symposium and Robocontrol, pp. 181–186 (2014). doi:10.1109/SBR.LARS.Robocontrol.2014.51
18. Vapnik, V.N., Chervonenkis, A.Y.: Support-vector networks. Mach. Learn. **20**(3), 273 (1995). doi:10.1007/BF00994018
19. Estivill-Castro, V., Radev, J.: Humanoids learning who are teammates and who are opponents. In: The 8th Workshop on Humanoid Soccer Robots 13th at IEEE-RAS International Conference on Humanoid Robots (2013)

20. Nao: Aldebaran Robotics. SAS (Limited Company) (2015). http://www.active8robots.com/wp-content/uploads/File-1400771269.pdf. Accessed 2 June 2015
21. Kim, J., Shin, H.: Algorithm & SoC Design for Automotive Vision Systems: For Smart Safe Driving System. Springer, Dordrecht (2014)
22. Papageorgiou, C.P., Oren, M., Poggio, T.: A general framework for object detection. In: International Conference on Computer Vision (1998)
23. Lienhart, R., Kuranov, A., Pisarevsky, V.: An empirical analysis of boosting algorithms for rapid objects with an extended set of Haar-like features. Intel Technical report MRL-TR-July02-01 (1998). http://www.lienhart.de/Publications/DAGM2003.pdf
24. Haykin, S.: Neural Networks: A Comprehensive Fundation, 2nd edn. Prentice Hall, Englewood Cliffs (1999)
25. Osuna, E., Hearst, M.A., Dumais, S.T., Platt, J., Schlkopf, B.: Applying SVMS to face detection. IEEE Computer Society (1999)
26. Dumais, S.T., Hearst, M.A., Osuna, E., Platt, J., Schlkopf, B.: Using SVMS for text categorization. IEEE Computer Society (1999)
27. Lorena, A.C., Carvalho, A.C.P.d.L.F.d.C.: Introdução as máquinas de vetores suporte. Relatórios Técnicos do ICMC: USP, São Carlos, ISSN 0103-2569 (2003)
28. OpenCV: Open source computer vision library (2014). http://www.opencv.org. Accessed 20 May 2014
29. Dimashova, M.: Can not get new positive sample. http://answers.opencv.org/question/4368/traincascade-error-bad-argument-can-not-get-new-positive-sample-the-most-possible-reason-is-insufficient-count-of-samples-in-given-vec-file/#4474. Accessed 10 Apr 2014
30. van Dijk, S., Noakes, D., Barry, D., Polani, D.: Bold hearts team description robocup 2014 kid size, April 2014. http://homepages.stca.herts.ac.uk/epics/boldhearts/
31. Bold hearts - Robocup 2015 qualification - kid size humanoid league (2015). https://youtu.be/pzYHAp7b7sY. Acessed Dec 2015
32. Le, Q.V.: Building high-level features using large scale unsupervised learning. In: IEEE International Conference on Acoustics, Speech and Signal Processing (ICASSP), pp. 8595–8598. IEEE (2013)

Dense Tracking with Range Cameras Using Key Frames

Andrés Díaz[1(✉)], Lina Paz[2], Eduardo Caicedo[1], and Pedro Piniés[2]

[1] School of Electrical and Electronic Engineering, Universidad del Valle, Cali, Colombia
{andres.a.diaz,eduardo.caicedo}@correounivalle.edu.co
[2] Oxford Mobile Robotics Group, University of Oxford, Oxford, UK
{linapaz,ppinies}@robots.ox.ac.uk

Abstract. We present a low cost localization system that exploits dense image information to continuously track the position of a range camera in 6DOF. This work has two main contributions: First, the localization of the camera is performed with respect to a set of keyframes selected according to a spatial criteria producing a less populated and more uniform distribution of keyframes in space. This allows us to avoid the computational overload caused by having to estimate a depthmap at the frame rate of the camera as it is common in other dense sequential methods. Second, we propose a two-stage approach to compute the current location of the camera with respect to its closest keyframe. During the first stage, our system calculates an initial relative pose estimate from a sparse set of 3D to 2D point correspondences. This estimate is then refined during the second stage using a dense image alignment. The refinement step is stated as a Non Linear Least Squares (NLQs) optimisation embedded in a coarse to fine approach that minimizes the photo-consistency error between the current image and a warped version of the image associated to the closest keyframe and its depth map.

To validate the accuracy of our system, we conducted experiments using datasets with perfectly known trajectory and with both, perfect ray-traced images and images with noise and blur. We also evaluate the accuracy of the system using datasets with RGBD images taken at different frame-rates, and the improvements in convergence due to our coarse-to-find approach. Our assessment shows that our system is able to achieve millimeter accuracy. Most of the expensive calculations are carried out by exploiting parallel computation on a GPU.

Keywords: Localization using dense techniques · Range camera · 6 DOF · Keyframes · Non linear squares optimization · Parallel computing · Noise and blur · Different frame rates · Coarse-to-fine scheme

1 Introduction

The aim of (*Simultaneous Localization and Mapping*, SLAM) is to estimate the position of a mobile platform moving in an unknown environment and, at the

F. Santos Osório and R. Sales Gonçalves (Eds.): LARS 2015/SBR 2015, CCIS 619, pp. 20–38, 2016.
DOI: 10.1007/978-3-319-47247-8_2

same time, build a map of this environment in an incremental way. Many solutions have been proposed using different sensor modalities; sonar rings, laser scanners, perspective cameras, omnidirectional cameras, stereo cameras, inertial sensors, and combination of them. Due to its versatility and low cost, vision-based SLAM systems have shown great potential in the fields of mobile robotics, augmented reality, wearable devices, user interfaces, surgical procedures, among others. Popular visual SLAM systems use sparse features such as corners [6], edges [12], or planes [18], and have as principal goal to localize the camera. More recently, dense techniques for localization and mapping have gained attention [14]. They do not use sparse features but all pixels in the image, showing better accuracy for the localization and providing high quality maps. For achieving real time performance these dense methods are based on optimization algorithms implemented on GPUs. Moreover, the mapping and localization algorithms run in parallel, on two or more CPU cores or GPUs, without the need of propagating a joint probability of the camera-features estimates. With these techniques more accurate estimates of trajectories (even with fast motions) and more robust augmented reality applications are obtained.

We can find three main frameworks in the literature about tracking the position of a camera using dense information: frame-to-frame [16] (when the pose of the camera is computed with respect to a locally generated depth map), frame-to-model-vertices [15], frame-to-model-TSDF [2]:

In frame-to-frame approaches using the current image I_c^{RGB} and a reference image I_r^{RGB} with its corresponding depth map I_r^{depth} we can compute the relative transformation between both as follows: using an initial estimate of the relative position between the current and the reference image we can use the I_r^{depth} and I_c^{RGB} to create a warp image I_{warp}^{RGB} that should resemble the reference image I_r^{RGB}. We can then calculate the differences and correct accordingly the transformation in order to reduce them.

In a frame-to-model-vertices approach the idea is to estimate the pose of the current camera with respect a continuously built 3D dense model. Given the current depth map at time k a set of vertices and normals can be calculated from it (V_k, N_k) respectively. From the model and the current estimate of the pose of the camera with respect to the model we can make a prediction of these elements $(\hat{V}_k, \hat{N}_k)$. Again, comparing the observations with the predictions we can correct the current pose estimate.

Finally in frame-to-model-TSDF approaches the camera is also localized with respect to the model but instead of predicting vertices and normals what is compared are the values of the truncated signed distance function (TSDF) used to represent the model.

The advantage of the model based approaches is that drift in the camera pose does not accumulate if the model is continuously observed. In the case of the frame-to-frame approach, the advantage is that a model is not required to compute the trajectory of the camera. In that sense, model based approaches are more similar to Bundle Adjustment techniques whereas frame-to-frame ones to Visual Odometry algorithms.

In this paper, we focus on the problem of continuously localizing an RGBD camera that moves freely in 6DOF following a frame-to-frame approach. Our contribution is in the use a set of depth maps available only for a set of selected frames or "keyframes". In this way, we can overcome the computational payload that commonly whips other visual localization systems. These techniques will be used in a posterior work together with a dense mapping module for making localization and mapping using only a monocular camera.

In Sect. 2 we analyse the most important sparse and dense localization systems. In Sect. 3 we focus on the proposed technique for dense localization of a range camera with keyframes. We provide details about the criteria for keyframe selection, the 3D-2D visual odometry technique used for getting an initial pose estimate and the refinement of the camera pose with NLQs. Our experiments and the results of the performance of the algorithms implemented on graphic processor are explained in Sect. 4. Finally, we draw our conclusions and future work in Sect. 5.

2 Related Work

Cameras are inexpensive sensors that provide rich information from the observed scene, therefore becoming an ideal sensor for many robotics applications with embedded visual navigation systems. In addition, increase in computing power and new developments in computer vision techniques (e.g. robust feature detectors, outlier rejection and matching algorithms) allow their use in real time applications. In robotics, the first visual SLAM systems proposed were based on the Extended Kalman Filter (EKF) estimator [5,6]. Other systems have excelled for introducing novel techniques such as patch association and planes-based map building [18], inverse depth parametrization [4] and large scale SLAM [17]. Unfortunately, these approaches provide a sparse map representation, only hundreds of points can be estimated at a given instant, which limits the accuracy of the camera localization.

The system presented in [11] radically changed the filtering paradigm used until that moment to non-linear optimisation techniques using keyframes. This system is pioneer in balancing the computations between two CPU cores, one for localization and the other for mapping respectively. It uses a Five-Point algorithm [20] to initialise the system. As we mentioned before, keyframes are selected to reduce the computational complexity of the optimization. Camera tracking is still carried out on the intermediate camera frames by re-projecting the estimated 3D points and minimizing a photoconsistency error. Despite this approach is able to create very accurate maps in real time, it represents the environment sparsely. The map can only contain up to 6000 points and 120 keyframes. Nonetheless, this work represents a transition between sparse and dense systems. In our approach, we also use keyframes to track the position of the camera.

Recently, [21] proposed a mapping approach able to compute dense depth maps from monocular images. In this context, dense means that every pixel in the image contains an estimate of the depth to the corresponding object in the scene. Depth maps are computed by minimizing an energy function composed of a regularisation term and a photometric term (data term). A generalized threshold scheme is used for minimizing the photometric error of the energy function. The localization is based on bundle adjusting a set of keyframes [11] on a CPU. The system can reach 24 fps for a depth map resolution of 480×360 when a GPU is used to calculate the depthmaps. Similarly, the work in [16] builds high quality dense maps of small environments using a GPU and a monocular camera in real time. Again, depth maps are obtained by minimising an energy function. The main difference with respect to the previous approach is that a decoupling scheme is implemented during the optimization. This way, the problem is decomposed into two tied minimization problems. In the first problem, a convex energy function composed of a regularisation and a quadratic terms is minimized using a primal-dual algorithm [3]. For the second problem, an exhaustive search is carried out over a finite range of discrete inverse depth values. The camera is localized by aligning the current observed frame to a synthetic one created using a nearby image and its depthmap. The camera motion corresponds then to the relative transformation that produces the best alignment between the actual and the synthetic images. A pyramid of images is built to accelerate convergence allowing faster motions of the camera. The initialization of the system is similar to [11].

The large scale direct monocular SLAM system LSD-SLAM [7] uses direct image alignment and filtering-based estimation of semi-dense depth maps. In addition to the photometric residual, the authors incorporate a depth residual which penalizes deviations in inverse depth between keyframes, allowing to directly estimate the scaled transformation between them, operating on Sim(3). Moreover, the effect of noise depth values is included probabilistically into tracking. The global map is represented as a pose graph consisting of keyframes as vertices with 3D similarity transforms as edges. The system detects loop closures and scale-drift, and runs on real-time on a CPU.

In [19] an energy-based approach to estimate the rigid body motion of a handheld RGB-D camera for a static scene is presented. The reprojection error is defined such that the warped second image matches the first image, using consecutive images of the trajectory. The authors minimize this non-convex reprojection error by a sequence of convex optimization problems obtained by linearizing the data term and solving the resulting normal equations in a coarse to fine manner. This system is better than the iterative closest point ICP algorithm for continuous camera tracking and is able to perform at 12.5 Hz on CPU.

3 Dense Localization Algorithm

In this section we are going to explain the main processes developed for estimating the camera pose using a dense technique. Without loss of generality, consider a sequence of images I_{c_i}, for $i = 1 : m$, gathered by the camera. Our aim is to estimate the relative transformation $T_{r_j c_i}$, for $j = 1 : n$, at each frame, where

r_j is the closet keyframe. Algorithm 1 presents the order in which the processes are carried out.

Since a coarse-to-fine scheme is used, the T_{cr} obtained after convergence in a level is used in the following lower level as initial estimate, improving convergence, specially in lower levels where the computational cost is higher. The Levenberg-Marquardt iterates until the stopping criteria is fulfilled: the norm of variations in the motion vector is small (less than a threshold ε) or the number of iteration reaches the maximum value $iter_{max}$.

Algorithm 1. Relative transformation using dense information

```
Data: Current Intensity Image I_c, Reference Intensity Image I_r and Reference
      Depth Image D_r.
Result: Relative Transformation Matrix T_cr.
Compute the initial T_cr
Build a four level pyramid for I_c, I_r and D_r
Define the maximum number of iterations iter_max for each level
for level l = 4 : 1 do
    Compute the warped Image Iw, the residual r, the cost C and the Jacobian
    J for level l
    iter=1
    found=false
    while (found==false) and (iter < iter_max) do
        Compute the variation in the motion vector dξ = (J'J + αI)^{-1} J'r
        where ξ is the minimal parametrization of the motion
        Update the new Relative Transformation T_cr_new using dξ:
        T_cr_new = exp_SE3([dξ]_x) * T_cr
        Compute the new warped Image I_w_new and the new cost C_new
        if |dξ| < ε then
            found=true
        else
            if C_new < C then
                Update the cost C, the relative transformation T_cr and the
                warped Image I_w with the new values
                Compute the Jacobian J
                Update α, α ← α/10
            else
                Update α, α ← α * 10
        iter++
    Use the estimated T_cr as initial value in the next lower level
```

The camera trajectory is obtained concatenating relative transformations, e.g. $T_{wr1}T_{r_1r_2}T_{r_2r_3}...T_{r_{j-1}T_{r_j}}T_{r_jc_i}$, where the world coordinate system corresponds to the coordinate system of the first keyframe, therefore $T_{wr_1} = [I\,|\,0]$. A dense model of the environment can be generated using the poses and depth maps from keyframes T_{wr_j}, D_{r_j}.

In the remaining of this section we will describe the criteria for keyframe selection, the calculation of an initial pose estimate, how to refine this estimate using a photometric error obtained by aligning the warped and observed images, and finally, its minimization using Levenberg-Marquardt.

3.1 Keyframe Selection

The mapping and global optimization is performed only for certain frames of reference in order to achieve higher performance. These frames are known as keyframes and are selected according to certain temporal and spatial criteria [11].

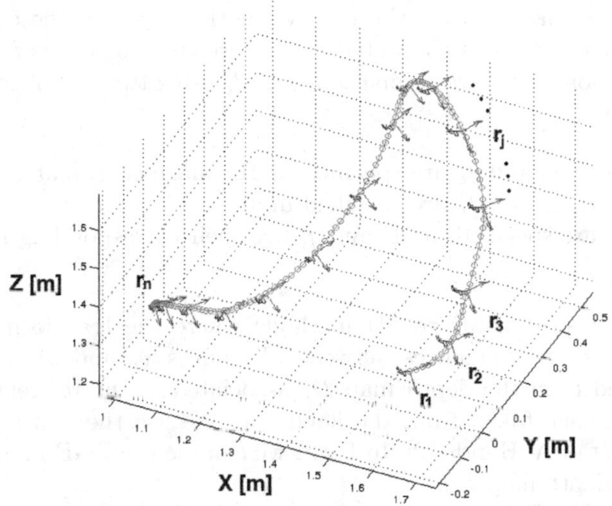

Fig. 1. The camera trajectory is drawn in cyan. Keyframes r_j are drawn with their local coordinate systems. They are selected based on a time criteria, e.g. for every nine images a new keyframe is initialized.

For the sake of simplicity, Fig. 1 shows an example of keyframes selected based just on a time criteria. Each keyframe r_j has a transformation matrix T_{wr}, an RGB image I_r and a depth map D_r. The system saves all keyframes with its corresponding information, which allows it to initialize new keyframes, estimate the relative pose of the current image I_c with respect to the nearest keyframe and perform optimization for getting a consistent map and trajectory.

We set thresholds to the displacement and to the change in orientation (no time criteria): our system initializes a new keyframe when the displacement with respect to any other keyframe is greater than 0.13 m or when the sum of changes in orientation with respect to the last keyframe (around X, Y and Z axis) is greater than 30°. Thus, if the displacement and change in orientation of the camera is smaller than these thresholds, no new keyframes will be initialized. In this way, the map is compact in the sense that it does not grow arbitrarily with time due to little or no movement, even with repeated movements in restricted areas as we will see in the results section.

The frame of the camera when the system is initialized is taken as the first keyframe, with transformation matrix $T_{wr_1} = [I \mid 0]$. For the following images I_{c_i}, the position of the previous image is used to define the nearest keyframe. The estimate of relative motion is performed considering the correspondence between the RGB image associated to the nearest keyframe and the current image. The pose of the new image is defined as $T_{wr_{near}} T_{r_{near}c_i}$.

3.2 Initial Pose Estimate

In the motion estimate process the relative motion between the current image and the nearest keyframe is computed. There are three methods for computing the relative motion T_{cr} between images I_r and I_c using two sets of corresponding features f_r and f_c:

- 2D to 2D. Both f_r and f_c are specified in 2D image coordinates.
- 3D to 3D. Both f_r and f_c are specified in 3D.
- 3D to 2D. f_r are specified in 3D and f_c are their corresponding reprojections on the image I_c.

We are interested in 3D to 2D methods where the transformation T_{cr} is computed from correspondences between 3D points χ_r and 2D points p_c. χ_r can be obtained from the depth map D_r associated to the reference image I_r, it means, the image taken from the keyframe. We use the depth map loaded from the dataset of A. Handa [9]. In Fig. 2 we can see the RGB image I_r and its corresponding depth map D_r.

We apply a SURF feature detector and a SURF descriptor extractor to I_r and to I_c which belongs to the set of images associated to I_r, that is I_{c_i} for $i = 1, 2, 3, ..., m$. Then a matcher based on the FLANN algorithm computes the matches between SURF descriptors of both images. The matches are validated if they fulfil three conditions. The first one is that the distance between the two matched descriptors is smaller than a threshold. The second one is that the correlation index between a patch around the descriptor in I_r and the descriptor in I_c is greater than a threshold. The final one is that the matches are inliers when the fundamental matrix is computed using RANSAC. This means that the descriptor in I_c must lie on the corresponding epipolar line.

We set the threshold in distance to five times the minimum distance between all matches and the correlation index threshold to 0.7. Finally, as the goal is to find T_{cr} that minimizes the image reprojection error, we apply a solver based on PNP algorithm that uses RANSAC for outlier rejection

$$\arg\min_{T_{cr}} \sum_k ||p_{c_i}^k - \hat{p}_{r_j}^k||^2, \tag{1}$$

(a) RGB image loaded from the dataset. This is a synthetic environment with small camera motions around the desk.

(b) Depth map loaded from the dataset. It is used just for keyframes

Fig. 2. Perfect ray-traced RGB Image (a) and Depth map (b) of the dataset of A. Handa [9]. They are used for computing the initial pose estimate and for the posterior refinement.

where $\hat{p}^k_{r_j}$ is the reprojection of the kth 3D point $\chi^k_{r_j}$ into image I_{c_i} according to the transformation T_{cr}. The transformation of a feature $p^k_{r_j}$ with 2D coordinates μ to a spacial coordinate χ^k_r and the transformation of a spacial point $\chi^k_c = T_{cr}\chi^k_r$ to a point in the image plane c_i, $\hat{p}^k_{r_j}$, are described in Subsect. 3.3.

3.3 Pose Refinement

The goal of the localization algorithm is to estimate the relative transformation matrix between two camera poses T_{cr} (Fig. 3) using all the depth data D_r of the associated keyframe and all the intensity data of I_r and I_c. The camera pose of the current Image C_c is given by $C_c = T_{cr}C_r$, where C_r is the camera pose of the reference image. The current image is warped with the transformation T_{cr} in order to compare it with the reference image. In case the transformation T_{cr} is the true 3D transformation between the two cameras, the images I_c and I_r should be identical and the residual should be zero [1].

The warping consists in projecting the reference depthmap D_r into 3D space, transforming by T_{cr} and projecting back into the current image Ic. Algorithm 2 summarizes the steps of this process. K is the intrinsic calibration matrix, π represents the function that computes normalized homogeneous coordinates and $\hat{\mu}$ is the homogeneous version of the coordinate in the image plane. A residual function is defined as the difference between $I_r(\mu)$ and $I_w(\mu)$,

$$r(\mu, T_{cr}) = I_r(\mu) - I_c(K\pi(T_{cr}D_r(\mu)K^{-1}\hat{\mu})). \tag{2}$$

Since the depth $D_r(\mu)$ is known, T_{cr} is defined as the argument that minimizes the residual function. T_{cr} is a homogeneous transformation matrix defined by a rotation matrix R_{cr} and a translation vector t,

$$T_{cr} = \begin{bmatrix} R_{cr} & t \\ 0 & 1 \end{bmatrix}. \tag{3}$$

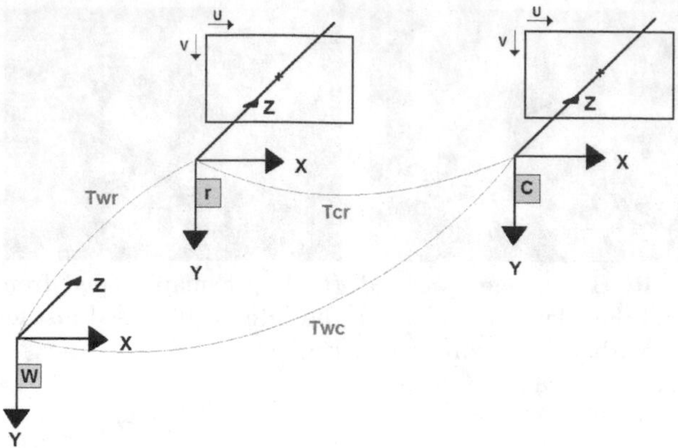

Fig. 3. Current image I_c and reference image I_r with their corresponding local and global coordinate systems. The transformations that relate coordinate systems are drawn in red. (Color figure online)

Algorithm 2. Warping an Image

Data: Current Intensity Image I_c, Reference Depth Image D_r, Transformation T_{cr} and Intrinsic Matrix K.

Result: Warped Image I_w.

Transform coordinates in the image plane I_r to spacial coordinates:
$\chi_r(\mu) = D_r(\mu)K^{-1}\hat{\mu}$
Change of coordinate system: $\chi_c(\mu) = T_{c_r}\chi_r(\mu)$
Transform spatial coordinates to coordinates in the image plane: $\acute{\mu} = K\pi(\chi_c)$,
where $\pi(\chi_c) = (X_n, Y_n, 1)^T = \left(\frac{X_c}{Z_c}, \frac{Y_c}{Z_c}, 1\right)^T$
Compute the warped Image: $I_w(\mu) = I_c(K\pi(T_{cr}D_r(\mu)K^{-1}\hat{\mu})) = I_c(\acute{\mu})$

The vector $\xi = (\boldsymbol{w}, \boldsymbol{v})$ represents the relative transformation matrix T_{cr} with the smallest possible number of parameters. The orientation of the vector $\boldsymbol{w} = (w_1, w_2, w_3)$ defines the axis of rotation and the magnitude of the vector defines the amount of rotation. The vector $\boldsymbol{v} = (v_1, v_2, v_3)$ corresponds to the translation t rotated over $\boldsymbol{w}$.

An exponential mapping between the 6-element vector ξ in the Lie algebra $\mathfrak{se}(3)$ [23] to members of the group $SE(3)$ is made, as in [11] with a motion model and feature association (sparse techniques) or in [16] using a dense estimate of rotation [13] and image alignment. Computing the exponential mapping involves the operator $[.]_x$, which maps a vector to its *skew* matrix,

$$exp_{SE3}([\xi]_x) = T_{cr}, \tag{4}$$

where $[\xi]_x$ is

$$[\xi]_x = \begin{bmatrix} 0 & -w_z & w_y & v_x \\ w_z & 0 & -w_x & v_y \\ -w_y & w_x & 0 & v_z \\ 0 & 0 & 0 & 0 \end{bmatrix}. \tag{5}$$

The updated transformation T_{cr} is expressed as the product of an initial transformation T_0 and an update in the motion vector $d\xi = \xi - \xi_0$, that is,

$$T_{cr} = dT * T = exp_{SE3}([d\xi]_x) * T_0. \tag{6}$$

This way, the singularities are avoided since the incremental update is calculated in the tangent space around the identity (i.e. it is always very small) and mapped back onto the manifold SE(3). With this transformation, the residual defined in (2) is

$$r(\mu, \xi) = I_r(\mu) - I_c(K\pi(exp_{SE3}([d\xi]_x)T_0 D_r(\mu)K^{-1}\hat{\mu})). \tag{7}$$

Finally, the localization problem consist in finding ξ that minimizes the following cost function:

$$\min_{\xi} \frac{1}{2} \int [r(\mu, \xi)]^2 d\mu. \tag{8}$$

This is a non-linear minimization problem that we solve with Levenberg Marquardt [10]. Since r is non-linear and non-convex in ξ we linealize it using a first order Taylor expansion:

$$r(\xi) \approx r(\xi_0) + \left.\frac{dr(\xi)}{\xi}\right|_{\xi_0} (\xi - \xi_0). \tag{9}$$

The quantity $\frac{dr(\xi)}{\xi}$ is the Jacobian of r and is given by

$$J = \begin{pmatrix} \nabla r_1(\xi)^T \\ \nabla r_2(\xi)^T \\ \vdots \\ \nabla r_i(\xi)^T \end{pmatrix}, \tag{10}$$

where $\nabla r_i(\xi)$ is the derivative of the ith component of the residual vector.

Computation of the Jacobian. Lets define a function W for the warping process

$$W(\mu, \xi) = \pi(\overbrace{K exp_{SE(3)}([d\xi]_x) \underbrace{T_0 D_r(\mu) K^{-1}\hat{\mu}}_{p_1}}^{p_{1n}}), \tag{11}$$

where $\pi(X) = \left(\frac{x}{z}, \frac{y}{z}, 1\right)^T$, p_1 denotes the transformed point and p_{1n} is the transformed back-projected (normalized) point. The residual can be written as

$$r(\mu, \xi) = I_c(W(\mu, \xi)) - I_r(\mu). \tag{12}$$

The derivative of the residual can be written using the chain rule,

$$\frac{dr(\xi)}{d\xi} = \frac{dI_c(W(\xi,\mu))}{dW(\xi,\mu)}\bigg|_{W(\xi_0,\mu)} \frac{dW(\xi,x)}{d\pi}\bigg|_{p_{1n}} \frac{d\pi(\xi,x)}{d\xi}\bigg|_{\xi_0}. \tag{13}$$

The first term $\frac{dI_c(W(\xi,\mu))}{dW(\xi,\mu)}\Big|_{W(\xi_0,\mu)}$ is just the image gradient of the warped image $\nabla I_w(\mu)$. The second term involves the projection $\pi(\cdot)$ and therefore

$$\frac{dW(\xi,x)}{d\pi}\bigg|_{p_{1n}} = \begin{bmatrix} \frac{1}{z} & 0 & \frac{-x}{z^2} \\ 0 & \frac{1}{z} & \frac{-y}{z^2} \end{bmatrix}. \tag{14}$$

The third term $\frac{d\pi(\xi,x)}{d\xi}\Big|_{\xi_0} = \begin{bmatrix} \frac{d\pi}{dw_x} & \frac{d\pi}{dw_y} & \frac{d\pi}{dw_z} & \frac{d\pi}{dv_x} & \frac{d\pi}{dv_y} & \frac{d\pi}{dv_z} \end{bmatrix}$. Putting together all these derivatives, it becomes

$$\frac{d\pi(\xi,x)}{d\xi}\bigg|_{\xi_0} = K \begin{bmatrix} 0 & p_z & -p_y & 1 & 0 & 0 \\ -p_z & 0 & p_x & 0 & 1 & 0 \\ p_y & -p_x & 0 & 0 & 0 & 1 \end{bmatrix}. \tag{15}$$

The first 3×3 submatrix is the negative skew matrix of the vector p_1 and the second 3×3 submatrix is the identity matrix. Equation (15) simplifies to

$$\frac{d\pi(\xi,x)}{d\xi}\bigg|_{\xi_0} = K\left[-[p_1]_x \mid I_3\right]. \tag{16}$$

Now that the Jacobian was defined, the variation in the motion vector $d\xi$ can be obtained by solving (17).

$$J^T J d\xi = -J^T r. \tag{17}$$

For the Levenberg-Marquardt method the Hessian can be approximated by $J^T J + \alpha I$. For a large α the method approaches the Steepest Descent method, while for a small α the method approaches the Gauss-Newton method. Generally, the Steepest Descent works better further away from the local minima and the Gauss-Newton approximation works better close the local minima where the quadratic approximation is good [1]. We start with a small initial value of α, $\alpha = 1e-3$. After each iteration, the new cost is compared with the previous cost (see Algorithm 1). If the cost has decreased, α is reduced by a factor of 10 and the parameters are updated. If the cost has increased, the parameters are not updated and α is increased by a factor of 10. Solving (17) is efficient because we just need invert a 6×6 matrix and use $d\xi$ in (6).

4 Results

The dense localization algorithm runs on a laptop Hewlett Packard with a processor Intel Core *i*7 - 2.2 GHz, 11.7 GB of RAM memory, a graphic processor NVIDIA GEFORCE 755M and Ubuntu 14.04 as operating system. OpenCV is employed for processing images, OpenGL for drawing the camera trajectory and the keyframes, and Cuda 6.0 for getting a better performance using graphic processor. The following analysis is made using the synthetic RGB-D dataset of A. Handa [9] and is based on three criteria; time, convergence and accuracy.

4.1 Time

We have implemented the most important algorithms of the dense localization module on GPU. The Jacobian is computed in 11.1 ms in the lower level (the full image, 640×480 pixels). The warped image and the residual is computed in 3.9 ms for the same level. The update in the camera position and orientation is made in 5.3 ms. The other algorithms are implemented on CPU. The building of the pyramid is made in 1.6 ms. The load of images I_r and I_{c_i} is made in 3.5 ms. The initial estimate is computed in 31.3 ms.

The system processes 10 fps at the moment, for two iterations of the Levenberg-Mardquart algorithm in the lower level of a four level pyramid. We believe that this performance can be improved by implementing all the algorithms of the localization module on GPU. This is left as future work.

4.2 Convergence

For analysing the performance of the algorithm when it uses a coarse-to-fine scheme, we present the number of iterations required by the algorithm for converging before and after implementing this scheme. In both cases the initial relative transformation is $T_{cr} = \begin{bmatrix} R_{cr} & 0.6t \\ 0 & 1 \end{bmatrix}$, where R_{cr} and t are respectively the ground truth in rotation and translation that relates frame r and c. The images 1 and 10 from the 20 fps dataset was used as I_r and I_c, respectively.

For the case in which a coarse-to-fine scheme is not used, the initial cost is 4386.8. In the iteration 76, the change in the motion vector $d\xi$ is less than $1e-6$ and the algorithm stops, converging to a cost of 481.9.

When a four level pyramid with a scale of 2 is built, the maximum number of iterations in each level is set to: 20, 20, 5 and 2, from higher to lower levels. The initial and final cost for each level is presented in Table 1.

Table 1. Information about convergence using a coarse-to-fine scheme

Level	Size [pix.]	Initial cost	Final cost	Max. iterations
4	60×80	24.27	2.47	20
3	120×160	13.89	12.67	20
2	240×320	72.00	64.20	5
1	480×640	495.44	477.72	2

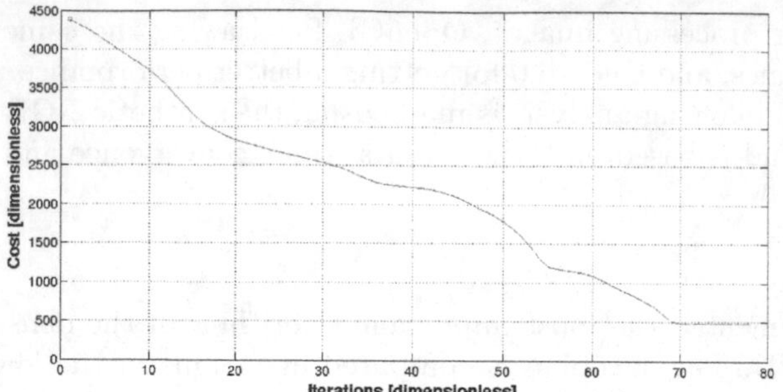

(a) Convergence without using a coarse-to-fine scheme.

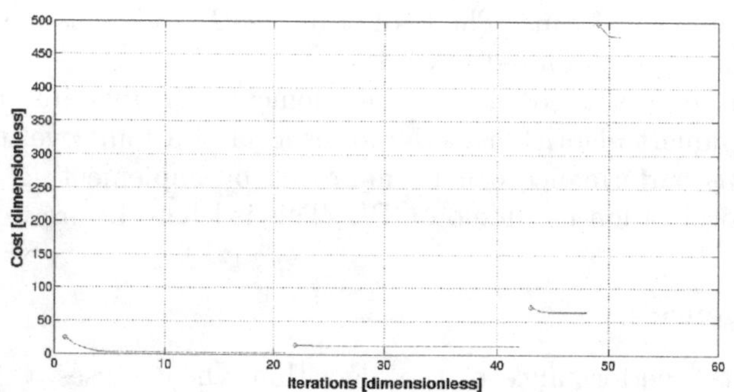

(b) Convergence using a four level pyramid.

Fig. 4. Comparison between using only one image and a four level pyramid. (a) requires 76 iterations with an image of 480×640 pixels for converging. (b) requires 2 iterations in the lower level (image of 480×640 pixels) for converging.

Note that the initial cost in the lowest level (bigger image) is very close to the final cost, unlike in the first case. For this reason only two iterations are required in the lowest level when a coarse-to-fine scheme is implemented (Fig. 4).

4.3 Accuracy

The accuracy in the camera position is quantified using three datasets; two of them consist of perfect ray-traced images taken at 20 fps and 60 fps and the last one consist of images with motion blur and noise, taken at 60 fps (60 fps-blur). We define the error as the Euclidean distance between the ground truth in trajectory and the estimated one. Figure 5(a) shows the estimated trajectory (in orange) and the ground truth (in green) for the 60 fps dataset, obtained in OpenGL and drawn incrementally as the relative transformations are computed and concatenated. The positions of keyframes are drawn as blue points. The analysis in orientation will be made for each axis, comparing the ground truth in orientation with the estimated one along the time. The ground truth in camera rotation around the axis X, Y and Z are shown in Fig. 5(b).

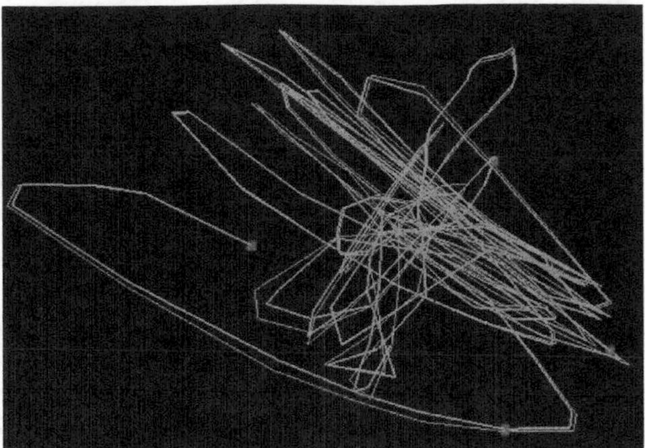

(a) Trajectory estimated (in orange) and ground truth (in green). The position of keyframes are drawn as blue points.

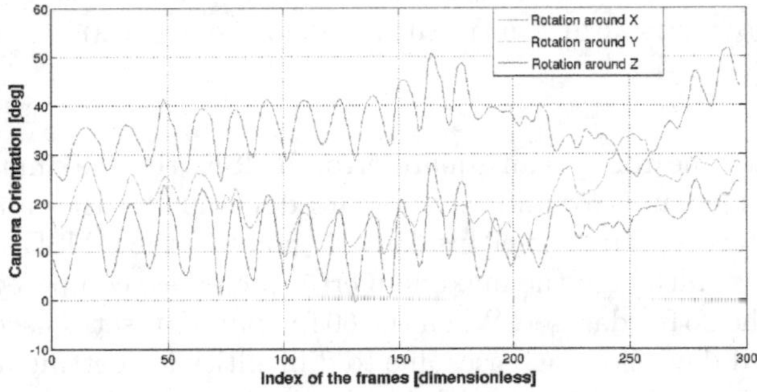

(b) Ground truth in rotations around X, Y, and Z

Fig. 5. Trajectory and orientation of the camera for the 60 fps dataset. The positions of keyframes are drawn as blue points in (a) (Color figure online)

In Fig. 6(a) we draw the error in camera position for the 60 fps dataset (no blur) as a solid red line and the error in camera position for the 60 fps-blur dataset as a dashed blue line, and highlight the errors of keyframes by drawing them as points. Figure 6(b) shows the error in rotations around X, Y, and Z for the 60 fps dataset and Fig. 6(c) shows the corresponding errors for the 60 fps-blur dataset. Note that the errors in both position and orientation for the images with motion blur and noise are greater than the errors for perfect ray-traced images. Errors obtained with the 20 fps dataset are very similar to the solid red line in Fig. 6(a) for camera position and to the errors in Fig. 6(b) for camera orientation so they are not shown.

Table 2. Summary of accuracy in camera position

	20 fps	60 fps	60 fps-blur
RMSE [cm]	*0.74*	*0.52*	*1.12*
Max. error [cm]	4.2	4.82	4.54
Mean error [cm]	0.44	0.33	0.92
Median error [cm]	0.25	0.22	0.87
Standard dev. [cm]	0.59	0.41	0.65

Table 3. Summary of accuracy in camera orientation

	20 fps			60			60-blur		
	X	Y	Z	X	Y	Z	X	Y	Z
RMSE [deg]	0.24	0.21	0.12	0.14	0.16	0.10	0.31	0.34	0.43
Max. error [deg]	0.74	1.52	0.75	0.35	1.63	0.93	1.06	1.34	0.70
Min. error [deg]	−0.64	−0.44	−0.24	−0.57	−0.72	0.40	−1.66	−1.15	−1.90
Mean error [deg]	−0.11	−0.02	−0.04	−4e−3	−0.06	−0.04	−0.02	−0.23	−0.34
Median Error [deg]	−0.13	−0.04	−0.06	−0.02	−0.05	−0.03	−2.9e−3	−0.22	−0.30
Standard dev. [deg]	0.21	0.21	0.11	0.14	0.15	0.10	0.31	0.25	0.27

Table 2 shows the root mean square error RMSE, the maximum error, the mean error, the median error and the standard deviation for the camera position for the three datasets. With the 60 fps dataset (no blur) the RMSE and Standard deviation decrease although the maximum error increases, compared to the ones obtained with the 20 fps dataset. When the 60 fps-blur dataset is used, the RMSE and the standard deviation increases due to difficulties for getting a good initial estimate and for aligning images with blur and noise. Despite these difficulties, the algorithm converges to a good estimate. We can say that the system has high accuracy in position for both frame rates, and for both perfect ray-traced images and images with blur and noise; in the worst case, the error is 0.16 % of the distance navigated by the camera that corresponds to 6.93 m.

Table 3 shows the root mean square error, the maximum error, the minimum error, the mean error, the median error and the standard deviation, for the estimated rotations for the three datasets. The RMSE and the standard deviation decrease with the 60 fps dataset although the maximum error increases for rotations around Y and Z. When the 60 fps-blur dataset is used, the RMSE and the standard deviation increase due to the reasons given for the errors in camera position when the images have blur and noise. Note that the maximum error is 1.63° and the minimum error is −1.90° so we finally can claim that the system has high accuracy in both camera position and camera orientation for different frame rates and even with motion blur and noise.

For comparison purposes, we use the information given in [7] about the RMSE in camera position for several localization systems. Four datasets were

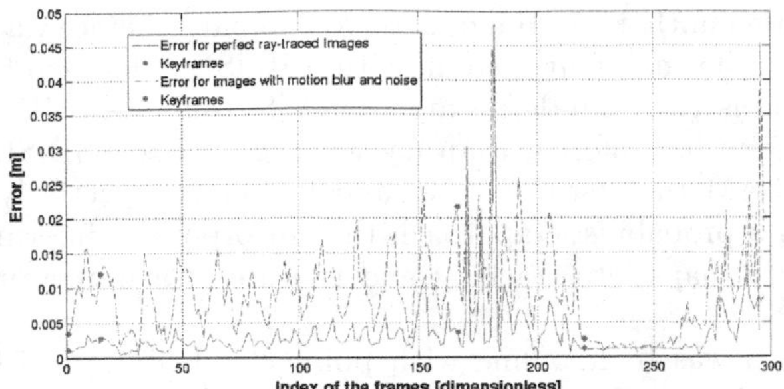

(a) Error in camera position for 60 fps and 60 fps-blur datasets

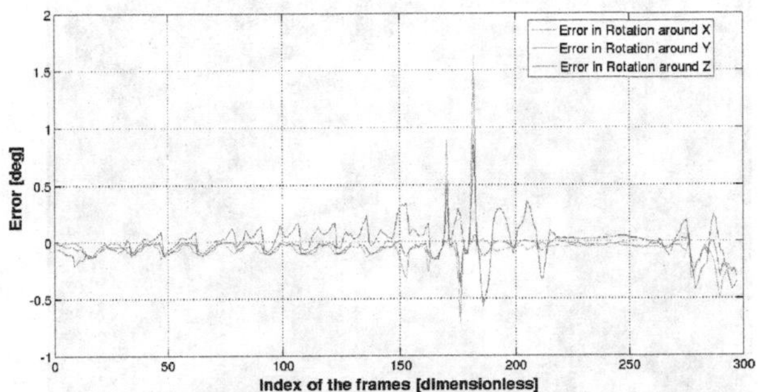

(b) Error in rotations around X, Y, and Z for perfect ray-traced images (60 fps dataset)

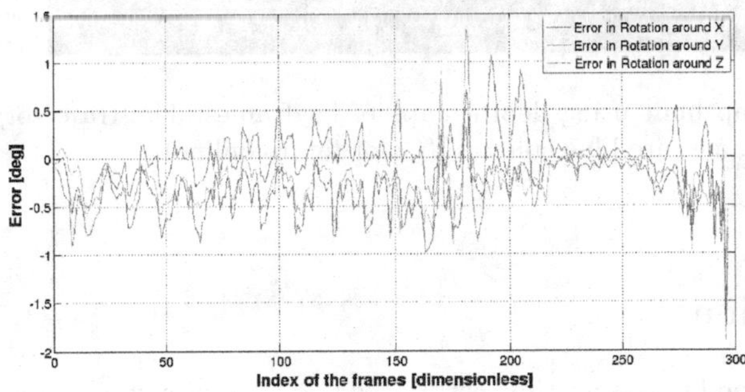

(c) Error in rotations around X, Y, and Z for images with motion blur and noise (60 fps-blur dataset)

Fig. 6. Error in camera pose (position and orientation). The keyframe positions are drawn as blue points in (a) (Color figure online)

used; two real sequences from the TUM RGB-D benchmark [22] and two simulated sequences from [9]. These two synthetic datasets, called sim/desk and sim/slowmo, and the ones that we use in this work, 20 fps, 60 fps and 60 fps-blur, have the same author, A. Handa, so we compare the accuracy using the information of sim/slowmo. Our system, with a RMSE of 0.41 cm for the 60 fps dataset, outperforms the semi-dense mono VO [8] that has a RMSE 2.51 cm. However, our system has inferior accuracy with respect to LSD-SLAM [7] and direct RGB-D SLAM that have 0.35 cm and 0.13 cm, respectively. It may be due to the use of a probabilistic approach that incorporates information about noise on the depth maps into tracking and the use of a framework for graph optimization.

The dense map was built offline with point cloud library 1.7.0 and using depth maps of keyframes. It is shown in Fig. 7. Note that all the local clouds, after being referenced to a global coordinate system, fit tightly.

Fig. 7. Dense map built using depth maps of keyframes. The trajectory is drawn in orange and keyframes are blue spheres. (Color figure online)

5 Conclusion

We have developed a system that estimates the pose of a range camera in 6DOF in a small environment. Our goal is to use this system in augmented reality applications. The proposed algorithm uses a 3D-2D visual odometry technique for computing an initial estimate that is later refined using image alignment. In order to reduce the computational cost, the current camera pose is computed with respect to the closest keyframe stored in the system. Since the localization and mapping algorithms run on different threads, they can be designed and implemented independently and later, they can be joined again to form a more general system. With this goal in mind, in this paper, we have focused on improving the performance of the dense localization algorithm in terms of time and accuracy. The results have been more than satisfactory:

First, using a coarse-to-fine scheme we have improved the convergence of the system, requiring just two iterations of the Levenberg-Marquardt algorithm at the lowest level of our four level pyramid; As a result, we have boosted the performance of the system achieving to process 10 fps implementing most of the steps on a GPU.

Second, using the proposed two step process to localize the camera with respecto to a keyframe –3D to 2D followed by the image alignment- results in high accuracy estimates even when using blurred input images. In the worst case -with blur and noise-, the RMSE is 1.12 cm in translation and 0.31°, 0.34°, and 0.43° in the rotations around X, Y and Z axis respectively.

For future work, we will concentrate on the depth map creation, depth map fusion, and the implementation of the whole system on a GPU.

Acknowledgments. The authors would like to thank Fundanción CEIBA for the financial support that has made the development of this work possible.

References

1. Baker, S., Matthews, I.: Lucas-Kanade 20 years on: a unifying framework: Part 1. Int. J. Comput. Vis. (IJCV) **3**(56), 221–255 (2004)
2. Bylow, E., Sturm, J., Kerl, C., Kahl, F., Cremers, D.: Real-time camera tracking and 3D reconstruction using signed distance functions. In: Robotics: Science and Systems Conference (RSS), June 2013
3. Chambolle, A., Pock, T.: A first-order primal-dual algorithm for convex problems with applications to imaging. J. Math. Imaging Vis. **1**(40), 120–145 (2011)
4. Civera, J., Davison, A.J., Montiel, J.M.M.: Inverse depth parametrization for monocular SLAM. IEEE Trans. Robot. **24**(5), 932–945 (2008)
5. Davison, A.J.: Real-time simultaneous localisation and mapping with a single camera. In: Proceedings of the International Conference on Computer Vision, Nice, October 2003
6. Davison, A.J., Reid, I.D., Molton, N.D., Stasse, O.: MonoSLAM: real-time single camera SLAM. IEEE Trans. Pattern Anal. Mach. Intell. **29**(6), 1052–1067 (2007)
7. Engel, J., Schöps, T., Cremers, D.: LSD-SLAM: large-scale direct monocular SLAM. In: Fleet, D., Pajdla, T., Schiele, B., Tuytelaars, T. (eds.) ECCV 2014. LNCS, vol. 8690, pp. 834–849. Springer, Heidelberg (2014). doi:10.1007/978-3-319-10605-2_54
8. Engel, J., Sturm, J., Cremers, D.: Semi-dense visual odometry for a monocular camera. In: 2013 IEEE International Conference on Computer Vision (ICCV), pp. 1449–1456, December 2013
9. Handa, A., Newcombe, R.A., Angeli, A., Davison, A.J.: Real-time camera tracking: when is high frame-rate best? In: Fitzgibbon, A., Lazebnik, S., Perona, P., Sato, Y., Schmid, C. (eds.) ECCV 2012. LNCS, vol. 7578, pp. 222–235. Springer, Heidelberg (2012). doi:10.1007/978-3-642-33786-4_17
10. Hartley, R.I., Zisserman, A.: Multiple View Geometry in Computer Vision, 2nd edn. Cambridge University Press, Cambridge (2004). ISBN 0521540518
11. Klein, G., Murray, D.W.: Parallel tracking and mapping for small AR workspaces. In: Proceedings of the Sixth IEEE and ACM International Symposium on Mixed and Augmented Reality, November 2007

12. Klein, G., Murray, D.: Improving the agility of keyframe-based SLAM. In: Forsyth, D., Torr, P., Zisserman, A. (eds.) ECCV 2008. LNCS, vol. 5303, pp. 802–815. Springer, Heidelberg (2008). doi:10.1007/978-3-540-88688-4_59
13. Lovegrove, S., Davison, A.J.: Real-time spherical mosaicing using whole image alignment. In: Daniilidis, K., Maragos, P., Paragios, N. (eds.) ECCV 2010. LNCS, vol. 6313, pp. 73–86. Springer, Heidelberg (2010). doi:10.1007/978-3-642-15558-1_6
14. Newcombe, R.A., Davison, A.J.: Live dense reconstruction with a single moving camera. In: 2010 IEEE Computer Society Conference on Computer Vision and Pattern Recognition, pp. 1498–1505, June 2010. http://ieeexplore.ieee.org/lpdocs/epic03/wrapper.htm?arnumber=5539794
15. Newcombe, R.A., Davison, A.J., Izadi, S., Kohli, P., Hilliges, O., Shotton, J., Molyneaux, D., Hodges, S., Kim, D., Fitzgibbon, A.: KinectFusion: real-time dense surface mapping and tracking. In: 2011 10th IEEE International Symposium on Mixed and Augmented Reality, pp. 127–136, October 2011. http://ieeexplore.ieee.org/lpdocs/epic03/wrapper.htm?arnumber=6162880
16. Newcombe, R.A., Lovegrove, S.J., Davison, A.J.: DTAM: dense tracking and mapping in real-time. Department of Computing, Imperial College London, UK (2012)
17. Piniés, P., Paz, L.M., Gálvez-López, D., Tardós, J.D.: CI-graph simultaneous localization and mapping for three-dimensional reconstruction of large and complex environments using a multicamera system. J. Field Robot. **27**(5), 561–586 (2010)
18. Silveira, G., Malis, E., Rives, P.: An efficient direct method for improving visual SLAM, 10–14 April 2007
19. Steinbruker, F., Sturm, J., Cremers, D.: Real-time visual odometry from dense RGB-D images. In: Computer Vision Workshops (ICCV Workshops), 2011 IEEE International Conference on, pp. 719–722 (2011). doi:10.1109/ICCVW.2011.6130321
20. Stewénius, H., Engels, C., Nistér, D.: Recent developments on direct relative orientation. ISPRS J. Photogramm. Remote Sens. **60**, 284–294 (2006)
21. Stühmer, J., Gumhold, S., Cremers, D.: Real-time dense geometry from a handheld camera. In: Goesele, M., Roth, S., Kuijper, A., Schiele, B., Schindler, K. (eds.) DAGM 2010. LNCS, vol. 6376, pp. 11–20. Springer, Heidelberg (2010). doi:10.1007/978-3-642-15986-2_2
22. Sturm, J., Engelhard, N., Endres, F., Burgard, W., Cremers, D.: A benchmark for the evaluation of RGB-D SLAM systems. In: Proceedings of the International Conference on Intelligent Robot Systems (IROS), October 2012
23. Varadarajan, V.S.: Lie Groups, Lie Algebras, and Their Representations. Graduate Text in Mathematics, vol. 102. Prentice-Hall, Englewood Cliffs (1974)

Safe Navigation of Mobile Robots Using a Hybrid Navigation Framework with a Fuzzy Logic Decision Process

Elvis Ruiz[1(✉)] and Raul Acuña[2(✉)]

[1] Mechatronics Research Group, Simón Bolívar University, Caracas, Venezuela
elvis6m@gmail.com

[2] Control Methods and Robotics Lab, Technische Universität Darmstadt, Darmstadt, Germany
racuna@rmr.tu-darmstadt.de

http://mecatronica.labc.usb.ve/, http://www.rmr.tu-darmstadt.de/

Abstract. Autonomous navigation in dynamic environments is one of the most important problems in robotics. The different solutions to achieve this goal may be categorized into two big groups, deliberative methods and reactive methods. Deliberative methods require precise map knowledge, are computationally intensive, but they usually assure a path to the goal, on the other hand reactive methods are fast, dynamic but also subject to local minima among other problems. In this paper we propose a hybrid reactive-deliberative framework for mobile robots navigation which integrates the advantages of a high level deliberative planner with a reactive low-level control. The reactive layer of this new system uses the new map information in an asynchronous way allowing a much more dynamic response of the system to environment changes. For the merging of the reactive and deliberative behaviours a new Fuzzy Logic layer is proposed which defines the contribution of each navigation layer into the final movement of the robotic platform in real-time. The proposed framework was tested in a simulated Amigobot robot with a simulated Kinect sensor using the robotic simulation platform V-REP and the programming of the different layers was implemented in ROS.

Keywords: Navigation · Path-planning · Mobile robots · Control · ROS · Kinect · Fuzzy

1 Introduction

In the past three decades serious efforts have been made to design a strategy for efficient and reliable navigation which can be able to solve problems of obstacle avoidance, reaching objectives, transport, surveillance and exploration [1]. The current navigation algorithms can be categorized into two types: those targeted for cases with full knowledge of the environments and those with incomplete knowledge [2]. In this regard, the initial efforts for finding a plan free of obstacles

F. Santos Osório and R. Sales Gonçalves (Eds.): LARS 2015/SBR 2015, CCIS 619, pp. 39–56, 2016.
DOI: 10.1007/978-3-319-47247-8_3

to reach a goal from a map built by the knowledge of the environment belongs to the first type of navigation algorithms.

However, fully known environments, which are generally solved using deliberative navigation systems, have been widely solved, so the challenge lies in working in dynamic environments [3]. In this sense, reactive navigation can cope better with these scenarios. Nevertheless, in reactive navigation, the prior knowledge or partial knowledge of the world is not used and this becomes a disadvantage for certain cases in real applications.

The concept of hybrid navigation presents the possibility of using both a global knowledge of the environment as well as a reactive behaviour. However, the main problem that arises is how to build an adequate cooperation between the different navigation strategies (deliberative and reactive) in order to combine them into a new one which includes the best characteristics of each of them.

2 Related Works

A hybrid navigation scheme which deals with a partial and imprecise knowledge of the environment can be found in the work of L. Wang 2002 [4], where a global planner based on the distance transformation algorithm and a local control based on potential fields are proposed. For this type of problem other alternatives exist, such as the use of the path planning algorithm A* with fuzzy control, H. Maaref 2002 [5], or using fuzzy neural networks to establish reactive laws, Y. Jiang 2005 [6]. In term of the reactive level, some of the main methods found in the state of the art are: potential fields R. Acuña 2012 [7], histogram of velocity vectors, elastic bands and dynamic windows approaches with the use of a localization based on odometry and the extended Kalman filter, R. Vazquez-Martin 2006 [8].

A global planner for mobile robots (deliberative approach) may be implemented using algorithms based on sampling, graph search, cell decomposition or algorithms based on biological inspirations such as pulse coupled neural networks, H. Qu 2009 [9], or meta-heuristics like genetic algorithms, H. Huang 2011 [10]. An advantage of sample-based planners is that they don't require a complete exploration of every possible action sequence. Additionally, the traditional planners have no direct means of verifying that no solution exists for a particular domain Zickler 2010 [11].

However, the combination of this paradigms brings to the surface three major problems: cooperation mode (synchronous or asynchronous), use of references, and resolving conflicts between deliberative planning and reactive control. In this regard, an independent proposal for asynchronous integration with lists management to place and extract references and give priority to the reactive layer in the presence of a conflict, can be found in Y. Zhu 2012 [2].

3 Methodology

The proposed system is conformed by the following principal parts:

3.1 Reactive Navigation Layer

Given the position of the robot, the objective and the obstacles (known and uncertain), potential fields are generated which will be converted into velocity references for the robot using the negative gradient of the potential field based on techniques described in previous works of our group [7].

3.2 Deliberative Navigation Layer

Given a discrete map, in which each part of the workspace could be unknown or possess an occupation probability value, with a given initial and final position for the robot, this layer will plan a path using sampling based algorithms.

3.3 Hybrid Navigation System

Based on the robot's defined safety variables relative to the environment, for example feasibility of the deliberative planned route or high proximity to obstacle, the results from both methods previously mentioned are integrated into a new hybrid reference for the robot.

The proposed solution is organized as shown in Fig. 1, and it consists of five main blocks. First, the robot has on-board sensors and actuators that allow it to navigate the environment in which it is located, and the sensing system of the robot is used for the construction and update of a map. The representation of the environment for the robot consists in a two-dimensional plane with a arbitrary and finite number of static obstacles. Additionally, it is possible to have an initial knowledge of the map, but this knowledge may not be correct at all. In other words, a certain location on the map can be known or unknown with an uncertain value.

For the implementation of the proposed solution, a 3D model of the robot AmigoBot® was designed using SolidWorks, this model was then used for the simulations. The dynamic model of the robot was implemented in the robot simulation framework V-Rep [12], where it is possible to observe the robot's behaviour and monitor all the variables of interest. The code responsible for carrying out all the necessary computations and sending the appropriate control signals to the actuated elements of the robot was implemented in ROS Groovy [13]. This version is stable in the operating system Ubuntu 12.04.5 LTS (*Precise Pangolin* Long Term Support). Finally a pre-defined Kinect sensor in V-REP was used, with constraints for indoor navigation, implemented as in our previous work [14].

The code implemented in ROS is responsible of generating the velocity reference for the motors. The robot model in V-REP executes the actions ordered from the ROS framework using the V-REP plugin for ROS. Additionally, the

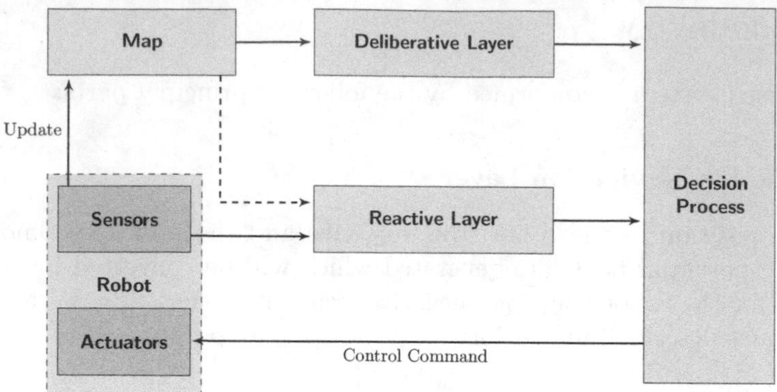

Fig. 1. Hybrid navigation architecture.

simulation in V-REP transmits the information from the simulated sensors onboard of the mobile robot as well as the position of the robot and the desired goal back to the code running in the ROS packages. A connection diagram between both platforms is shown in Fig. 2.

Additionally, the platform Move-it [15] was used, it represents the state of the art software for mobile robotics. Move-it was used for centralizing the planning in the deliberative layer, including the definition of the cinematic model of the robot and the map generation using simulated sensor measurements. Finally, the library fuzzylite [16] was integrated in the system as a fuzzy control framework.

The project consists of nine major nodes using the ROS parallel processing capabilities. The following is a detailed explanation of each node and the interconnection between them.

- **V-REP:** Node that simulates the robot with sensors, actuators and the environment. It is also responsible for passing other nodes the following information: data generated by the Kinect sensor on the robot, the position of the robot, and the position of the target. It receives from other nodes the reference speeds for each of the motors on the robot (left and right).
- **Move-it:** It is the node responsible for creating a map based on the information from the mobile robot sensor data (Kinect) and the path generation by request, based on sampling algorithms, from an initial state to a end state [15]. These states correspond to the actual robot and target positions.
- **Plan Status:** Monitors the validity of the plan. Since the environment is changing during execution, the generated plan may no longer be feasible so it may require replanning.
- **Path following:** This node is responsible for following the path produced by Move-it stage, using a weighted sum of an approach velocity field and a tangential velocity field to the route as explained in the work of Medina-Melendez [17].
- **Potential field:** This node calculates the potential fields associated to the obstacles, according to what is perceived by the sensors and the knowledge

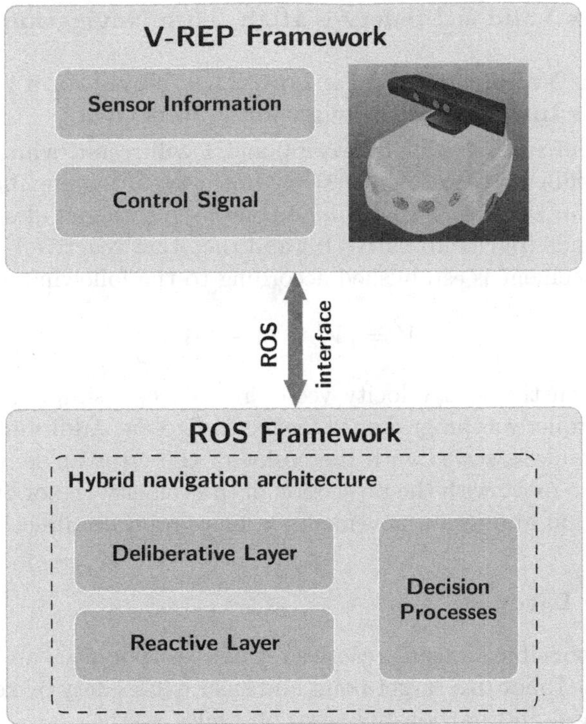

Fig. 2. Connection diagram ROS/V-REP.

of the surrounding map, and sets the attractor field towards the target. The equation to generate a repulsive field should produce a soft landing, with high repulsion in the middle of the obstacle and a limited area of effect in the space around the obstacle defined by some variable. Gaussian repulsive fields in 2D have been previously used for obstacles as they fulfil the conditions, and are generated by the following expression [7,18]:

$$Z(x,y) = Ae^{-((\frac{(x-x_o)^2}{2\sigma_x^2})+(\frac{(y-y_o)^2}{2\sigma_y^2}))} \tag{1}$$

Where A indicates the maximum weight or height of the potential field, σ_x^2, σ_y^2 represent the maximum dispersion of the field in the corresponding axis, which allows the creation of elliptical fields, and x_o, y_o denote the center of the obstacle.

- **Gradient:** This node calculates the gradient of the previously obtained potential field. This will provide a velocity field that will be used to possibly guide the robot safely to a target.
- **Fuzzy logic:** Here the fuzzy rules for implementing the fuzzy control are defined. The following fuzzy rules are proposed based on two input variables: first, the validity of the plan (given by the deliberative layer) and second the security (given by the reactive layer):

- If **Plan** is **Valid** and **Safety** is **High**. Then **Navigation** is **Deliberative.**
- If **Plan** is **Valid** and **Safety** is **Low**. Then **Navigation** is **Reactive.**
- If **Plan** is **Invalid**. Then **Navigation** is **Reactive.**

From this set of rules a value between 0 and 1 will result, which will indicate the level of deliberativeness or reactivity of the resulting command.

- **Fuzzy control:** Using a decision f based on the fuzzy logic rules and two given speed commands (one deliberative V_D and the other reactive V_R), a resulting command agreement is established according to the following expression:

$$V = fV_D + (1 - f)V_R \tag{2}$$

- **Control:** Given the final velocity vector reference, a simple magnitude and phase PI controller was programmed for the Amigobot. Additionally, the velocity vector module is scaled with the cosine of the error angle, this will allow the platform to move with the projection of the velocity vector over its current moving direction and avoid movements in inappropriate directions [18].

3.4 Reactive Layer

This layer is responsible for using reactive techniques (potential and vector fields) for attracting the robot to a target point and ensure it's safety by keeping it away from the perceived and previously known obstacles.

It is composed of the nodes: Move-it, Potential Field and Gradient, it receives as input the sensor information and returns a velocity field, see Fig. 3. In addition, an area of safe operation is set for the robot, by using this area together with the gradient value in that zone the security status variable is determined. As shown in the Fig. 4, for each position of the robot a velocity vector is defined, according to the obstacles that the mobile has sensed and the knowledge that it holds in that moment.

3.5 Deliberative Layer

It is responsible for generating a path from the current position of the robot to a goal in a desired position on the map, it also ensures the validity of the plan generated, if the route is no longer viable a new solution is sought, if any.

This layer is composed by the nodes: Move-it, Plan Status and Path Following. Receives the sensor information as the input (from the simulated Kinect), calculates a plan and returns a valid velocity command for the current position of the robot, see Fig. 5. Due to the fact that the environment may change during execution, it is possible to make online re-planning, leaving the former plan invalid, as shown in Fig. 6. In the event of an update in the map, or if the deliberative layer detects an obstacle in the planned route, a new path plan will be made.

3.6 Decision Process

This block is responsible for observing the validity of the plan and the safety of the robot in order to establish which control command will be sent to the actuators of the robot.

The decision process is composed of the following nodes: Fuzzy Logic and Fuzzy Control, it receives as input two statuses (security and planning) and two velocity vectors associated with a purely deliberative and a purely reactive navigation. It is also responsible for returning a hybrid velocity vector by processing the supplied inputs, see Fig. 7. In this way, the robot's behaviour can be more reactive or more deliberative depending on the surrounding conditions, as shown in Fig. 8.

Along the trajectory the robot will have a dominant reactive or deliberative behaviour depending of the decision made by this block. Based on this decision a hybrid vector (in module-angle representation) is created using Eq. 2, the change of value of this vector along the route obtained from the reactive and deliberative vectors plus the decision can be seen in Fig. 9a and b.

3.7 Test Cases

The proposed solution aims to solve general problems of navigation in environments with uncertainty. The cases shown in the Table 1 were selected based on the knowledge of the environment. Each position on the workspace can be Known Accurately (AK), Unknown (U) or Inaccurately Known (IK), i.e. according to the map an object is in one place but in reality it's not. Environments with variations in these parameters were created to set the initial conditions of navigation. For each one of this four cases the system was simulated 50 number of times with the given initial map conditions for a system completely reactive, another completely deliberative and finally for the proposed hybrid system.

The parameters to consider as a comparison metric are:

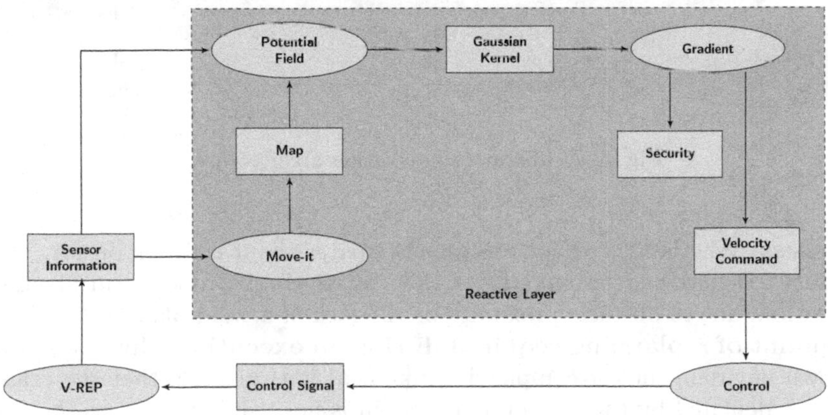

Fig. 3. Reactive navigation architecture.

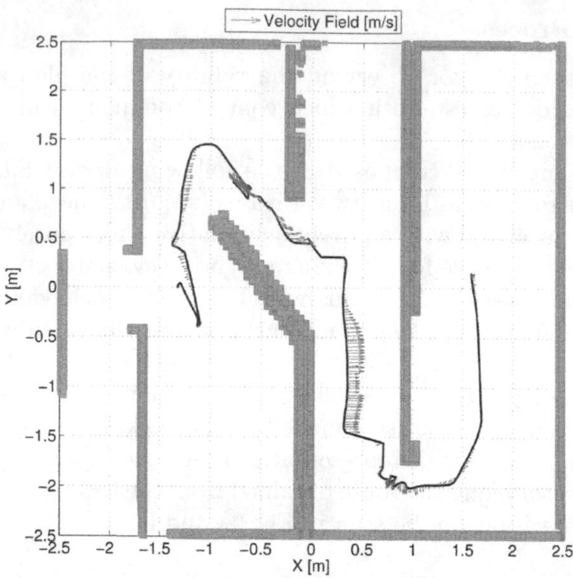

Fig. 4. Reactive vector field for each point in the trajectory executed. Green dots represent obstacles or walls. (Color figure online)

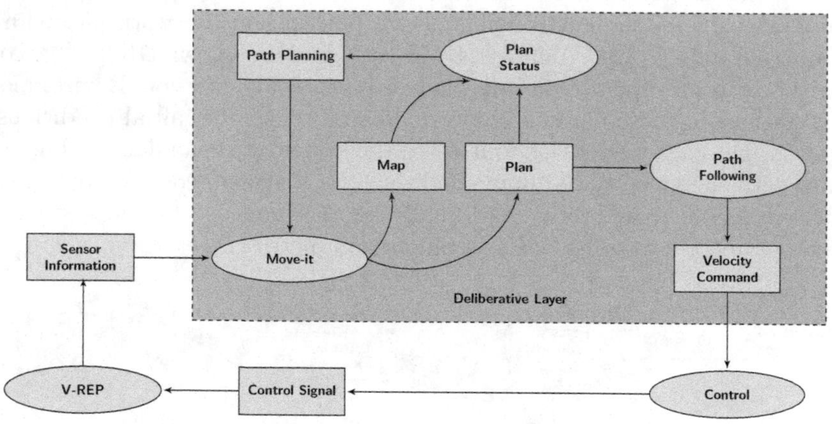

Fig. 5. Deliberative navigation architecture.

- **Success:** Whether the robot reaches the goal without collision or not.
- **Safety:** Defined as the amount of time in which the robot is in an unsafe situation, due to obstacle proximity or the lack of a valid plan.
- **Amount of replanning required during an execution:** Given a partially known environment (or completely unknown) it is possible that the original route calculated by the deliberative system is not valid, as soon as the robot finds a previously unknown obstacle in its path a replanning is required. This

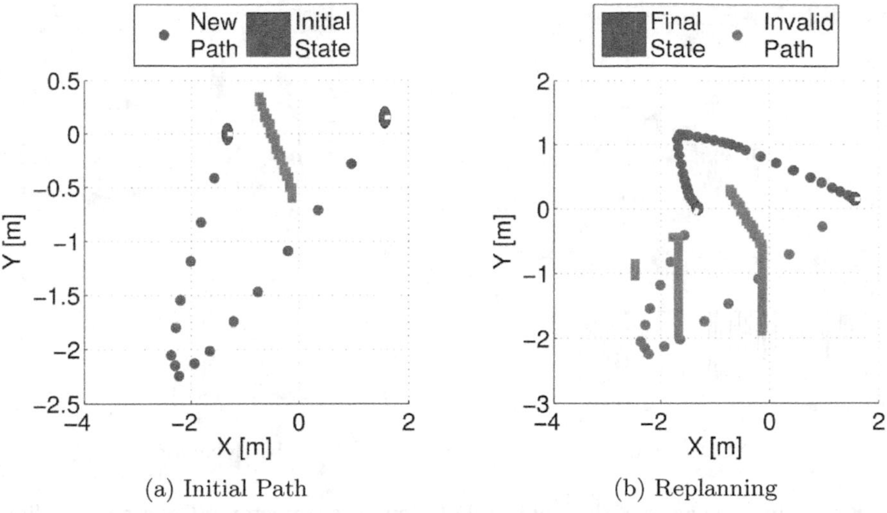

(a) Initial Path (b) Replanning

Fig. 6. Initial path and replanning in cases of an invalid plan.

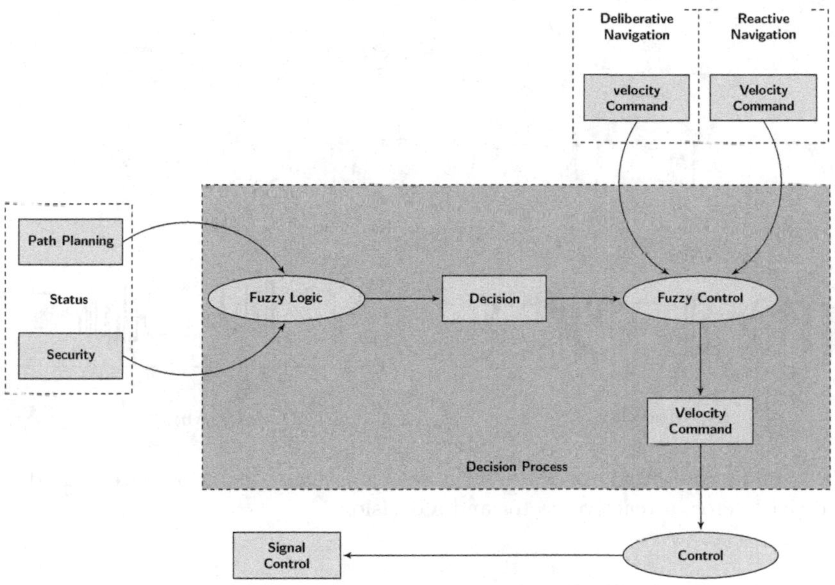

Fig. 7. Decision process architecture.

online replanning is dangerous and can lead to collisions with the obstacles, also during the replanning calculation the reactive system must take control.

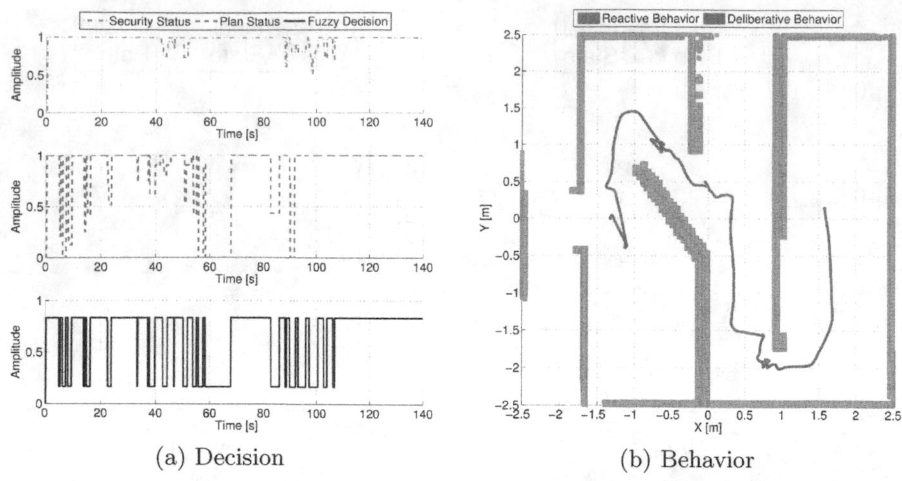

(a) Decision (b) Behavior

Fig. 8. Dominat behavior of the mobile robot along the trajectory. (Color figure online)

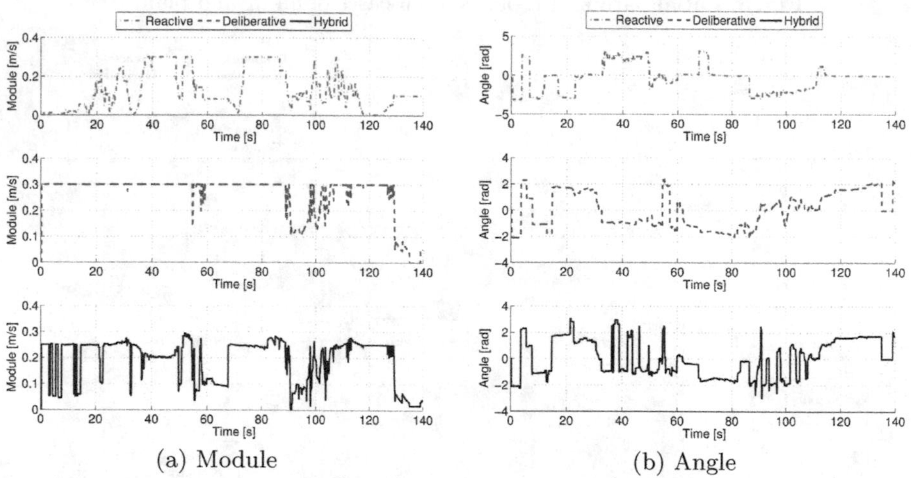

(a) Module (b) Angle

Fig. 9. Module and Angle of the hybrid velocity vector along a route. Based on a deliberative vector, a reactive vector and a decision.

4 Results and Discussion

In this section the results of the defined test cases are presented. As a representative example of the simulations, the initial conditions of test case 1 (Table 1) is shown in Fig. 10. In Fig. 10a the complete V-REP simulation environment is shown and in Fig. 10b the knowledge of the environment held by the robot.

The behaviour of the navigation using only the reactive layer is shown in Fig. 11. A successful navigation is shown in Fig. 3, where the robot is able to reach the goal while avoiding obstacles and going through hallways. In Fig. 11b

Table 1. Environment knowledge

Case	AK[%]	U[%]	IK[%]
1	80	15	5
2	30	35	35
3	10	80	10
4	5	95	0

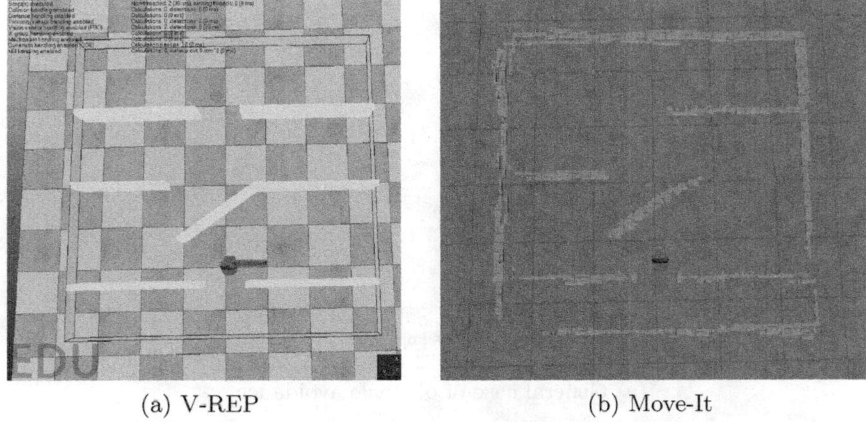

(a) V-REP (b) Move-It

Fig. 10. Environment of work and initial conditions of knowledge.

it is shown an unsuccessful navigation in which the robot is trapped in a local minima, one of the known weakness of the reactive approach.

The results of a pure deliberative navigation are shown in Fig. 12. A succesfull navigation even with the presence of local minima is shown in Fig. 12a. Nevertheless the deliberative layer generates paths in some cases which could lead to collisions and as well the uncertainty during re-planning could also lead to collisions as it is shown in Fig. 12b.

4.1 Average of Planned Paths

This metric is not valid in the case of the pure reactive case since it doesn't have a planning stage, however we can compare the results between the deliberative system and the hybrid system. In the pure deliberative system the average of planned paths required to arrive to the goal varied from 3.6 to 10.9 depending on the amount of correct knowledge of the environment, as it can be observed in Table 2. On the other side, the average of planned paths required using a Hybrid System ranged from 5.2 to 11.8.

Important to note is that the amount of knowledge of the environment has a direct impact in the amount of planned paths required to arrive to the destination. The lack of knowledge of the environment leads to more situations in

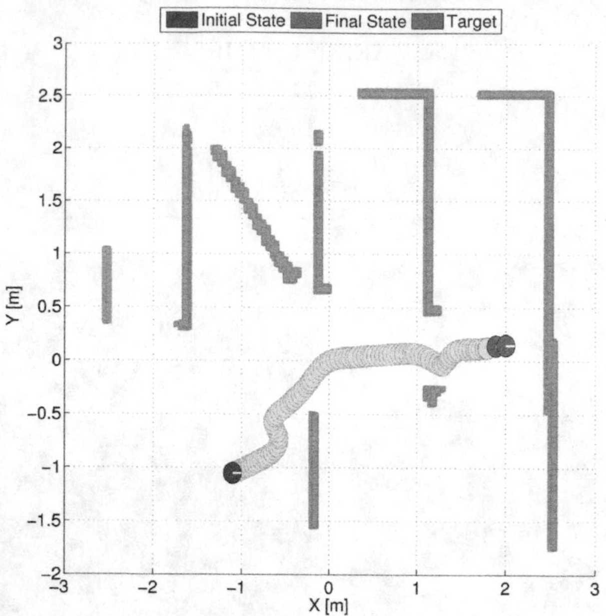

(a) General case of obstacle avoidance

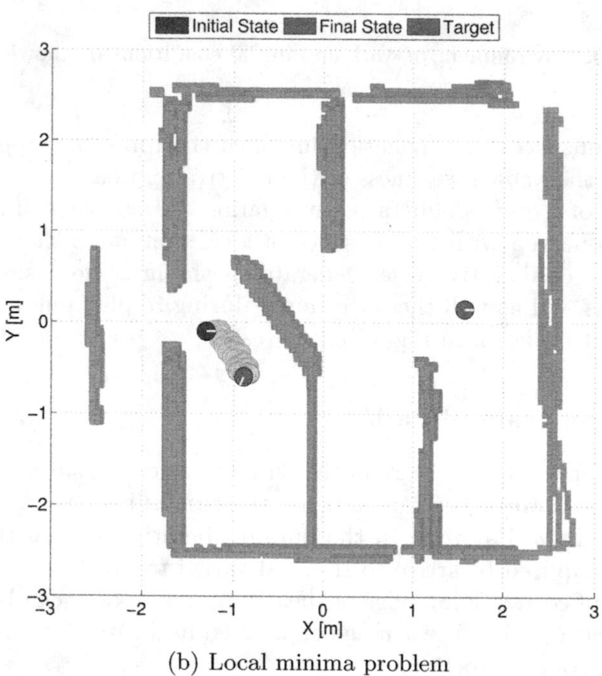

(b) Local minima problem

Fig. 11. A resulting reactive navigation.

which the robot encounters a previously unknown obstacle in its path and needs to find a new route.

Comparing the performance of the Hybrid System versus the pure Deliberative System, it can be observed that in average more planned paths are required for the Hybrid System in each Test Case. This is due to the corrections needed to meet the safety requirements of the route. A plan may be valid in terms of connectivity from start point to goal, but the Decision Process part of the Hybrid System requires a safer option, thus the plan has to be calculated again.

4.2 Time in Situations of Danger

Regarding the amount of time (from the total execution time) in which the robot was in an unsafe situation, either due to obstacle proximity or the lack of a valid plan, in Table 3 can be observed that the pure Deliberative System has a greater percentage of time in situations of danger (44.22[%]), compared to the Hybrid System (14.68[%]) in the Test Case 2, and in general in the other cases the Deliberative System always has more percentage of time in situations of danger.

Table 2. Average of planned paths in each case

Case	Reactive	Deliberative	Hybrid
1	0	3.6	5.2
2	0	6.2	9.6
3	0	10.4	10.2
4	0	10.9	11.8

In certain cases of the Hybrid solution the fuzzy logic decision system may favour the reactive system over the deliberative, not because of the danger of closeness of the robot to obstacles but due to the status of the Deliberative System plan. If a calculated plan becomes invalid during execution (mainly due to previously unknown obstacles) the deliberative system needs to find a new path. Since the search for a path in terms of calculation time is non deterministic due to the characteristics of the path-finding algorithms, the amount of time required to find a new path is not constant and its dependent on the current situation of the robot and the map. During this calculation the robot finds itself in a situation of danger, since the previous path is no longer valid and the new one is not yet available, in this case the reactive system takes control. As it can be seen in Fig. 8a, the switch from deliberative dominant behaviour (blue) into reactive dominant behaviour (red), occurs not only with obstacle proximity but as well during new plan calculation, this behaviour of the hybrid system increases the safety of the robot. This switching is not present in the pure deliberative navigation scheme and thats the main reason why in the hybrid system the time in situations of danger is smaller. This mean that the reactive actions of

Table 3. Time in situations of danger

Case	Reactive[%]	Deliberative[%]	Hybrid[%]
1	0	25.21	10.89
2	0	26.09	10.69
3	0	44.22	14.68
4	0	34.51	17.81

the system were executed timely helping to reduce the exposure of the robot to unsafe situations.

Additionally, in Fig. 13a, in the space between coordinates $[0, 1] \in X$ and $[1.5, 0] \in Y$, it can be noticed an interesting behaviour of the reactive layer of the Hybrid System. In this segment of the route the red vectors of the reactive layer are trying to bring the robot to the middle of the hallway, however, since the plan obtained from the deliberative layer is correct and there is no danger of collision, the Hybrid System favours the deliberative layer instead of the reactive. This shows that the Fuzzy decision process achieves the proper integration of the reactive and deliberative system, giving the control to the appropriate layer during the execution of the task.

4.3 Percentage of Success

From the proposed cases in terms of percentage of success (robot reaching the goal safely), the results presented in Table 4 were obtained. For a purely reactive navigation the robot is unable to reach the target due to the presence of local minima in the potential fields associated to the environment.

The deliberative navigation with replanning presented high levels of success, however, in terms of safety, it fails in cases where the robot strikes or rubs the walls. This was caused by a lack of available time for replanning or because the planned route was too adjusted to the dimensions of the mobile robot and the imprecisions in the movement control caused a collision. It is worth noting that the behaviour of the deliberative path-planner may be improved by increasing the safety radius around the obstacles for the planning (expanding the borders), or by using a greater restriction in the maximum movement velocity of the robot, but that would require the system to specialize for a unique set of cases.

Finally, the proposed hybrid navigation system achieved a hundred percent success in the test cases. In times when deliberative planning was very risky, the reactive layer took control and assured the safety of the mobile robot. In the case when there was no valid plan, the reactive layer allowed a reduction of the risk as it can be seen in Fig. 13b.

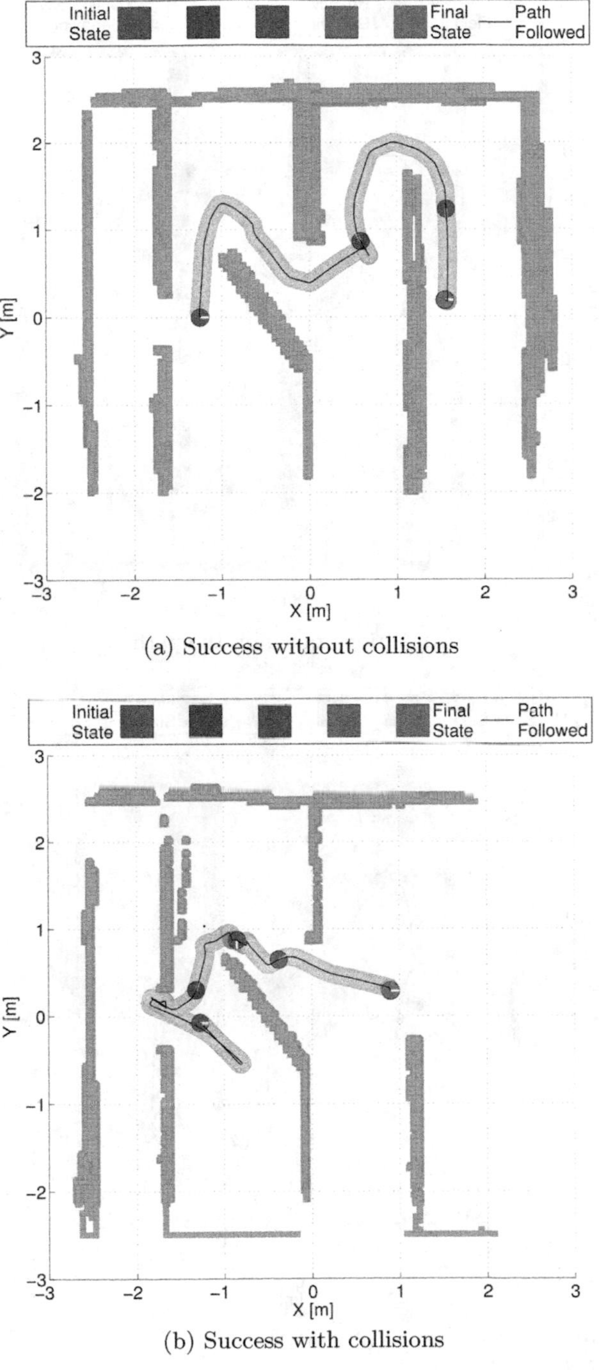

(a) Success without collisions

(b) Success with collisions

Fig. 12. A resulting deliberative navigation to the case 1 of knowledge.

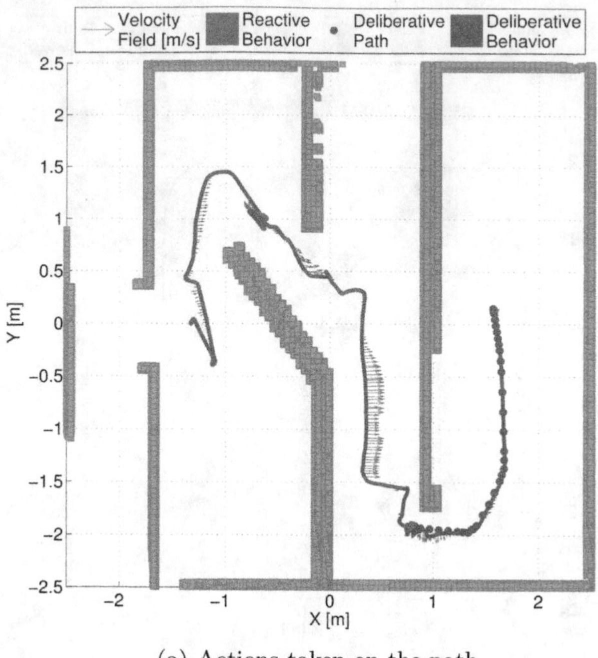

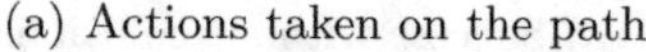
(a) Actions taken on the path

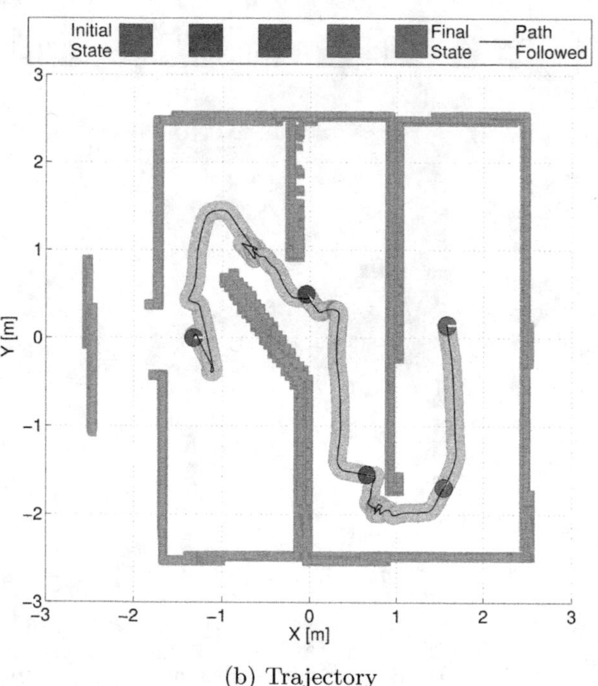

(b) Trajectory

Fig. 13. A resulting hybrid navigation in Test Case 4. (Color figure online)

Table 4. Percentage of success

Case	Reactive[%]	Deliberative[%]	Hybrid[%]
1	0	80	100
2	0	90	100
3	0	60	100
4	0	40	100

5 Conclusion

A hybrid navigation architecture that could benefit from the advantages of reactive and deliberative navigation, mitigating their individual disadvantages like local minima and planning time, was presented. Also a strategy based on Fuzzy Logic to merge the commands computed by each layer and prioritize them in case of conflicts was defined.

In contrast to the works of Y. Zhu [1,2] and H. Maaref [5], the reactive layer used the map information in an asynchronous way and in an area nearby the mobile robot, in order to exploit the newly acquired environment knowledge. The deliberative navigation used a sampling-based algorithm in which the concept of probabilistic completeness in dynamics environments is taken into account [11], taking advantage of the time spent in re-planning.

Finally, the proposed architecture was tested successfully in environments with different levels of uncertainty using ROS, Move-It and V-REP as framework. Furthermore, thanks to the modularity of the proposed navigation system, an implementation in a physical robot will be straight forward and it will be the next step to execute, the only thing that will change is the robot simulation module implemented in V-REP, which will be replaced by the actual robot. Consequently, a low cost approach could be made using the Kinect Sensor and the AmigoBot, both widely used in universities and research laboratories. Nevertheless, the architecture could be applied to other mobile robots and sensors.

References

1. Zhu, Y., Zhang, T., Song, J., Li, X.: A hybrid navigation strategy for multiple mobile robots. Robot. Comput.-Integr. Manuf. **29**(4), 129 141 (2013). http://linkinghub.elsevier.com/retrieve/pii/S0736584512001408
2. Zhu, Y., Zhang, T., Song, J., Li, X.: A new hybrid navigation algorithm for mobile robots in environments with incomplete knowledge. Knowl.-Based Syst. **27**, 302–313 (2012). http://dx.doi.org/10.1016/j.knosys.2011.11.009
3. Laugier, C., Vasquez, D., Yguel, M., Fraichard, T., Aycard, O.: Geometric and bayesian models for safe navigation in dynamic environments. Intell. Serv. Robot. **1**(1), 51–72 (2008). http://dx.doi.org/10.1007/s11370-007-0004-1
4. Ang, M.: Hybrid of global path planning and local navigation implemented on a mobile robot in indoor environment. In: Proceedings of the IEEE Internatinal Symposium on Intelligent Control, pp. 821–826. IEEE (2002). http://ieeexplore.ieee.org/xpls/abs_all.jsp?arnumber=1157868, http://ieeexplore.ieee.org/lpdocs/epic03/wrapper.htm?arnumber=1157868

5. Maaref, H., Barret, C.: Sensor-based navigation of a mobile robot in an indoor environment. Robot. Auton. Syst. **38**(1), 1–18 (2002). http://www.sciencedirect.com/science/article/pii/S0921889001001658, http://linkinghub.elsevier.com/retrieve/pii/S0921889001001658
6. Zhao, M.-Y., Wang, H.-G., Fang, L.-J.: Hybrid navigation for a climbing robot by fuzzy neural network and trajectory planning. In: 2005 International Conference on Machine Learning and Cybernetics, vol. 2, pp. 1069–1075. IEEE (2005). http://ieeexplore.ieee.org/lpdocs/epic03/wrapper.htm?arnumber=1527102
7. Acuña, R., Terrones, A., Certad-H, N., Fermín-León, L.: Dynamic potential field generation using movement prediction. In: Proceedings of the 15th International Conference on Climbing and Walking Robots and the Support Technologies for Mobile Machines, vol. 23, p. 26. World Scientific (2012)
8. Vazquez-Martin, R., Perez, E., Urdiales, C., del, Toro, J., Sandoval, F.: Hybrid navigation guidance for intelligent mobiles. In: 2006 IEEE 63rd Vehicular Technology Conference, vol. 6, pp. 2992–2996. IEEE (2006). http://webpersonal.uma.es/EPEREZ/files/VTC06.pdf, http://ieeexplore.ieee.org/lpdocs/epic03/wrapper.htm?arnumber=1683417
9. Qu, H., Yang, S.X., Willms, A.R., Yi, Z.: Real-time robot path planning based on a modified pulse-coupled neural network model. IEEE Trans. Neural Netw. Publ. IEEE Neural Netw. Counc. **20**(11), 1724–39 (2009). http://ieeexplore.ieee.org/xpls/abs_all.jsp?arnumber=5256181, http://www.ncbi.nlm.nih.gov/pubmed/19775961
10. Huang, H., Tsai, C.: Global path planning for autonomous robot navigation using hybrid metaheuristic GA-PSO algorithm. 2011 SICE Annual Conference (SICE), pp. 1338–1343 (2011). http://ieeexplore.ieee.org/xpls/abs_all.jsp?arnumber=6060543
11. Zickler, S.: Physics-based robot motion planning in dynamic multi-body environments, Ph.D. dissertation (2010)
12. Rohmer, M.F.E., Singh, S.P.N.: V-REP: a versatile and scalable robot simulation framework. In: Proceedings of The International Conference on Intelligent Robots and Systems (IROS) (2013)
13. Quigley, M., Conley, K., Gerkey, B.P., Faust, J., Foote, T., Leibs, J., Wheeler, R., Ng, A.Y.: ROS: an open-source robot operating system. In: ICRA Workshop on Open Source Software (2009)
14. Ruiz, E., Acuna, R., Certad, N., Terrones, A., Cabrera, M.E.: Development of a control platform for the mobile robot roomba using ROS and a kinect sensor. In: 2013 Latin American Robotics Symposium and Competition, pp. 55–60. IEEE, October 2013. http://ieeexplore.ieee.org/lpdocs/epic03/wrapper.htm?arnumber=6693270
15. Sucan, I.A., Chitta, S.: Moveit (2014). http://moveit.ros.org
16. Rada-Vilela, J.: fuzzylite: a fuzzy logic control library (2014). http://www.fuzzylite.com
17. Medina-Meléndez, W., Fermín, L., Cappelletto, J., Murrugarra, C., Fernández-López, G., Grieco, J.C.: Vision-based dynamic velocity field generation for mobile robots. In: Kozłowski, K. (ed.) Robot Motion and Control 2007. LNCIS, vol. 360, pp. 69–79. Springer, Heidelberg (2007)
18. Estevez, P., Capelletto, J., Alvarez, F., Acuña, R., Fernandez-Lopez, G.: Coordinated navigation using dynamically varying velocity fields. In: 9th International Conference on Artificial Intelligence and Soft Computing (2008)

A Harris Corner Detector Implementation in SoC-FPGA for Visual SLAM

Victor Hugo Schulz, Felipe Gustavo Bombardelli(✉), and Eduardo Todt

VRI, Department of Informatics, Federal University of Parana, Curitiba, Brazil
{victor.h.schulz,f.g.bombardelli,todt}@ieee.org
http://web.inf.ufpr.br/vri

Abstract. The present paper discusses the implementation of the Harris and Stephen corner detector algorithm optimized for an embedded system-on-a-chip (SOC) platform that integrates a multicore ARM processor and FPGA fabric in a single chip, the Xilinx Zynq-7000. The algorithm is implemented as a hardware co-processor on the FPGA portion of the SoC. As a whole, the SoC is used as a stereo vision preprocessing module to retrieve depth information from the features in order to compose 3D landmark points for Visual SLAM, speeding up feature extraction and relieving this highly parallelizable process from the main embedded processor. The optimizations of the algorithm's hardware implementation take into account the particularities of the SoC, such as compliance with its I/O requirements and FPGA's constraints on the amount of logical elements available for hardware synthesis. Also, optimizations done in order to reduce the time of execution of the algorithm in hardware, such as parallelization and introduction of a pipeline, are also presented in the article. A speedup of 1.77 was achieved when comparing the time of execution of the algorithm in the hardware coprocessor with the algorithm running in software in the dual-core ARM processor.

Keywords: Coprocessors · Feature detection · Field programmable gate arrays · Robotics and automation · Stereo vision · Computer vision

1 Introduction

The use of cameras as the main sensors in Simultaneous Localization and Mapping, what is called Visual SLAM, has increased recently due to the drop in camera prices. While images bring richer information than other typical SLAM sensors, such as lasers and sonars, there is significant extra processing cost when they are used. During the process of obtaining unique landmarks for SLAM, the use of a camera projects the 3D world into a 2D image plane. This process reduces information, and one of its consequences is depth perception loss. One way to recover it is capturing different images simultaneously with two cameras, what is called stereo vision. As long as the relative position between the cameras

F. Santos Osório and R. Sales Gonçalves (Eds.): LARS 2015/SBR 2015, CCIS 619, pp. 57–71, 2016.
DOI: 10.1007/978-3-319-47247-8_4

is known, the perceived horizontal disparity between the same projected points in the two images can be used to recover their depth information [26].

The problem of determining which point in one image corresponds to its respective pair in the other image from a stereo pair is called the Stereo Correspondence problem, and its solutions are distinguished in two main categories, area-based and feature-based. In the former, a small window of the image is looked for in the other image in order to find the most similar one. In the latter, unique features are detected within both images and then matched according to local descriptors. When comparing the two methods, in general, feature-based algorithms are faster and more robust than area-based ones [26].

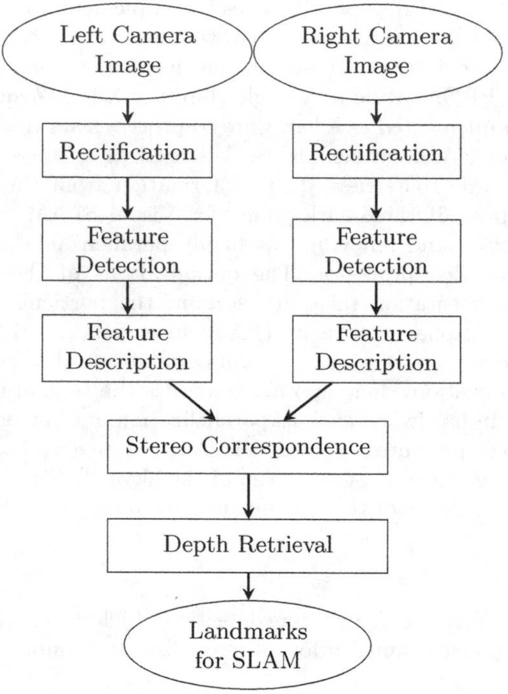

Fig. 1. Landmark acquisition system for Binocular Visual SLAM with feature-based correspondence. Features are detected on both images, rectified according to epipolar geometry and described uniquely. The descriptors are then matched, and the correspondences used to determine horizontal disparity, which is finally converted into depth information.

In one hand, the pre-processing steps involved in depth retrieval from stereo images add extra computational load to a complete Visual SLAM solution when compared to other solutions that employ main sensors which intrinsically provide depth, such as lasers and sonars. On the other hand, using cameras as the main sensors add richer information valuable for landmark matching and loop-closure situations. They are also cheaper, smaller, weight less, and have lower power consumption, making them easily embedded in mobile robots [18,21].

The main goal of this paper is to develop an embedded hardware system that performs the pre-processing routines included in depth retrieval for stereo Visual SLAM using feature-based correspondence, in order to obtain relative landmark positions with depth. The typical processing steps included in the system are feature detection in the image pair, feature description, rectification of the points, finding correspondence (matching) the features and ultimately recovering the depth from horizontal disparity. A block structure of the system is shown in Fig. 1. Within the components of a feature-based stereo correspondence system, the feature detection is usually the most costly in terms of computational resources, what is later measured in Sect. 2, thus becoming the main concern for optimization.

There are some implementations of stereo vision systems on SoC. Goldberg and Matthies [15] developed a stereo and IMU system, including corner extraction, on an OMAP3530 ARM Cortex A8 running GNU/Linux. Hybrid implementations followed using a dual-core Intel Core2Duo processor with a Xilinx Spartan 6 LX75 FPGA connected through the PCI interface [24,25], where an auxiliary OMAP3530 ARM processor was also used for visual processing. Later, an implementation of a stereo vision system applied to autonomous mobile robot navigation [20] included a Xilinx Spartan 6 FPGA to perform stereo correspondence and a quad-core Exynos 4412 ARM Cortex-A9 processor responsible for the navigation system. While the system presented in our paper is also based on the ARM processor and FPGA combination, a modern embedded SoC system is used, the Zynq-7000 XC7Z020, which integrates a main hard processor with FPGA fabric inside a single chip. The Zynq SoC was used within the ZedBoard development kit as the target for the landmark acquisition system embedded design.

The Zynq's ARM Cortex A9 processor runs a complete GNU/Linux system with the OpenCV computer vision library, which are used as the base for the custom software that processes all the needed steps shown in Fig. 1, with the exception of the feature extraction process, which is implemented on hardware on the FPGA using the Harris and Stephens corner detector [16]. The image acquisition is made using two Logitech C525 consumer grade webcams, which are supported on Linux, using the USB-OTG interface on ZedBoard.

There are many different hardware implementations of the Harris algorithm on FPGA [4–7,10,23]. The architecture used in this work is a sliding window approach, similar to the one presented by Amaricai et al. [4], but differs from it as to the nature of the target hardware platform, that integrates a processor on which the main tasks and coordination are done and that only delegates the feature detection step to the FPGA. Particularities of the architecture are also different. Here, the Gaussian filter was replaced by a simpler all-ones matrix filter (block filter) inspired by the OpenCV implementation of the Harris algorithm. Also, an adaptive threshold was included, in order to address the problem of the number of detected interest points dependent on the scene illumination, brought up on [6]. The adaptive threshold uses the previous frame to calculate the current threshold, a technique which has already been used on the Harris algorithm on [10] for space applications.

Table 1. Feature-based stereo correspondence profile measured on target SoC processor.

Task	Time (ms)
Image acquisition and rectification	53.32
Feature detection	295.33
Feature description	40.01
Correspondence (k-NN)	17.79

2 System Profiling

A simple system for landmark data acquisition, based on the steps described in Fig. 1 was implemented, initially only in software, using the OpenCV computer vision library in order to determine which portion of a feature-based stereo correspondence system would benefit more from optimization.

To compose the system, the Harris and Stephens corner detection algorithm was chosen for being one of the most widely used interest point detectors [7], which was used in numerous works [1–3,9,22,27] for the same purpose due to its high repeatability rate and speed. With the embedded platform in mind, the BRIEF descriptor [8] was chosen to provide a unique identifier to each feature, based on the fact that it has in most cases an accuracy close to SIFT [19], currently the best performing descriptor in accuracy, while the extraction time is improved by a factor of seven [17]. Here, accuracy is considered in terms of position error for visual odometry and SLAM. The descriptors are matched using the k nearest neighbor (k-NN) algorithm [11].

The images were acquired using the two Logitech C525 cameras with a resolution of 640×360. The measurements were made using GNU/Linux kernel clock sampling functionality on the ZedBoard's ARM Cortex A9 processor running at 866 MHz. The clock time was sampled before and after each portion of the code, and the time difference is shown in Table 1. The results indicated that the feature detection portion of the system was the most computationally expensive, and thus it was chosen to be optimized with the construction of a hardware co-processor on the FPGA.

3 Architecture of the Hardware Co-processor

The FPGA part of Zynq-7000, available on the chosen board, has 36 Kb of Block and operates at 100 MHz, with the clock period being 10 ns. Since the addressing considers a byte to be composed as 9 bits, 8 bits plus an extra bit for parity, 4 KB of memory are available. A gray-scale pixel is typically represented as a 0-255 decimal value representing its intensity, which is represented as a single byte. This limits the pixels that can be stored in the board to 4,000, leaving a square 64×64 image as the upper limit resolution to be processed if the whole image is stored into the FPGA block memory.

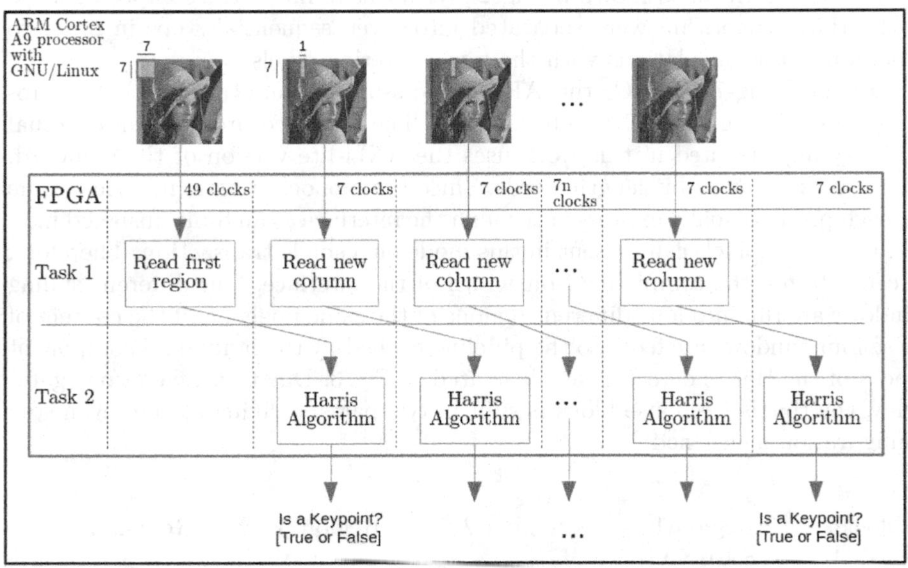

Fig. 2. A global vision of the system. The image read from one of the USB cameras on the ARM processor is transferred to the FPGA through the 32-bit wide AXI interface [28]. For each image line, a 7×7 window is firstly transferred, with a cost of 49 clocks. However, in the following steps, just the next column of seven pixels is transferred. Inside the FPGA, there are two tasks running in parallel: the input converter (Task 1) and the execution of the Harris Algorithm (Task 2). The ARM processor then reads the result of the algorithm, if a certain pixel is considered a corner (keypoint) or not.

Since the image sizes are very large compared to what can be stored locally for processing in the FPGA, the proposed architecture was designed to avoid using Block RAM and to rely in the minimum data dependency required for calculating if a certain pixel is considered a corner or not.

The Harris and Stephens corner detector [16] is comprised of these sequential steps and shown in Fig. 4: the Sobel filter, calculation of the M matrix coefficients, the Gaussian Smooth (replaced by the Block Filter), Harris response calculation, threshold, and non-maximum suppression. Working backwards, from the end to the beginning, with a non-maximum suppression within a 3×3 region, the data dependency from all the stages when using 3×3 filter masks implies in a sliding window with minimum size of 7×7. This was also independently determined by Amaricai et al. [4]. Therefore give a 7×7 matrix, the Harris Algorithm can calculate, if the center of this matrix is a keypoint.

The image read from one of the USB cameras by GNU/Linux on the ARM processor is transferred to the FPGA. For each image line, a 7×7 window is firstly transferred, with a cost of 49 clocks. However, in the following steps, this window is shifted to the right and filled with the next column of 7 pixel. So, while the co-processor is receiving the next column in 7 clocks, it also executes the Harris Algorithm as shown in Fig. 2. With this in mind, the composing steps of the Harris algorithm were separated into seven sequential steps in order to synchronize the calculations with the seven entering pixels.

On the Zynq-7000 SoC, the ARM processor communicates with the programmable through the AXI4 interface [28]. The Harris co-processor intellectual property implemented in this work uses the AXI4-lite version of the standard, which works with 32-bit length words. Since the monochromatic intensity value for each pixel is usually expressed through the interface as an 8-bit unsigned integer, up to four pixels can be sent in this mode on a single transaction. Therefore, in order to use the complete 32-bit width of the interface, four different sliding windows are run into four different regions of the same image, and the content of these four windows are fed into the pipeline formed by the individual component blocks of the Harris detector, as illustrated in Fig. 3. Due to propagation (gate) delay, the Harris Response block is split in two parts, arriving at the seven step configuration mentioned.

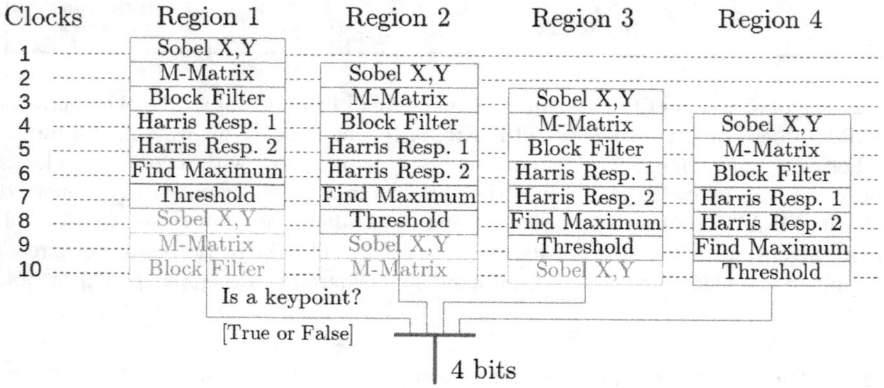

Clocks	Region 1	Region 2	Region 3	Region 4
1	Sobel X,Y			
2	M-Matrix	Sobel X,Y		
3	Block Filter	M-Matrix	Sobel X,Y	
4	Harris Resp. 1	Block Filter	M-Matrix	Sobel X,Y
5	Harris Resp. 2	Harris Resp. 1	Block Filter	M-Matrix
6	Find Maximum	Harris Resp. 2	Harris Resp. 1	Block Filter
7	Threshold	Find Maximum	Harris Resp. 2	Harris Resp. 1
8	Sobel X,Y	Threshold	Find Maximum	Harris Resp. 2
9	M-Matrix	Sobel X,Y	Threshold	Find Maximum
10	Block Filter	M-Matrix	Sobel X,Y	Threshold

Is a keypoint?
[True or False]
4 bits

Fig. 3. Pipeline of the Harris algorithm implemented to FPGA. Each individual block that composes the pipeline (from a total of 7) stays active for 4 clock periods, processing the information that comes from the 4 different image region, in series, and idle for the subsequent periods. Since 3 extra clocks are needed for the completion of the processing of the 4 regions, an output is ready every $7n + 3$ clock periods, where $n = 1, 2, 3, \ldots$

In the following subsections, each composing block of the proposed architecture is briefly explained.

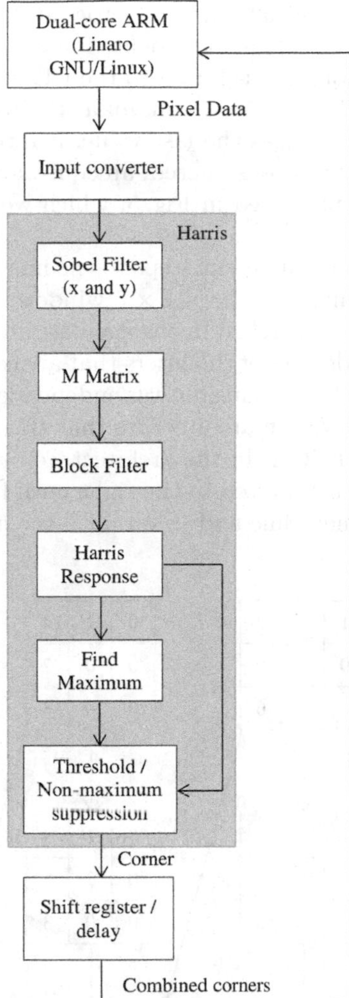

Fig. 4. Simplified block diagram for the proposed hardware architecture for the Harris Algorithm on FPGA. A GNU/Linux distribution runs on the dual-core processor (first block), which connects to the remaining blocks (implemented on FPGA logic) through the AXI4 interface. The blocks inside the gray area belong to the Harris algorithm. The corners that are determined by the algorithm are combined in the shift register, and then read again through the AXI interface by the main processor.

3.1 Multiplexed Input Converter

Internally, the multiplexed input converter is composed of four identical blocks (Input converters). Each one receives a single pixel at a time serially and builds the 7×7 window required to determine if a pixel is or is not a corner.

The initial 7×7 matrix is initialized with zeroes at reset. A counter that goes from 1 to the number of rows (7) is incremented after each received pixel. The row corresponding to the counter is shifted to the left, while simultaneously the received pixel is added to the rightmost column at the row corresponding to the counter. When the counter reaches the last value, it returns to 1. To exemplify the order in which the matrix is constructed, an equivalent 3×3 matrix showing the entering pixel order is presented in Fig. 5, which works similarly to what a 7×7 matrix would.

Due to the filters applied in different steps to the image, the minimum square region required was determined to be a 7×7 window. This assumes that the non-maximum suppression is searched in the smallest area possible of 3×3, and considers all de data dependencies of the filters that are components of the Harris and Stephens algorithm. This conclusion was independently determined in the article of [4], that also explores an architecture that tries to minimize the use of Block RAM on a different FPGA. In the architecture used, sliding window, the 7×7 window is processed then moved to the right until it reaches the end of the line, when it starts at the next line and so on until the end of the image.

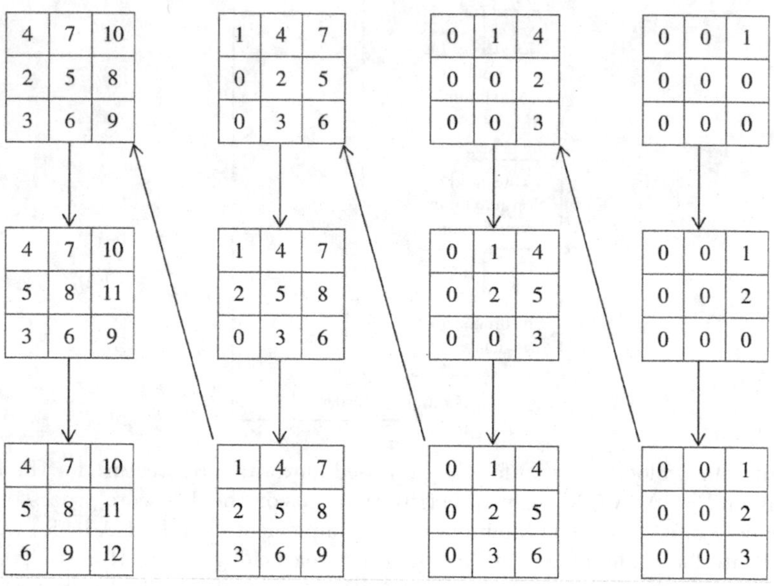

Fig. 5. Order of entering pixels for a 3×3 window.

The AXI4 lite interface works with 32-bit input and output registers. In the proposed platform, a gray scale pixel uses only 8 bits, representing the intensity in decimal value between 0 and 255. By concatenating 4 pixels from different regions of the same image together, 4 different 7×7 windows can be constructed simultaneously. The 4 different window matrices are processed serially in the remaining blocks that work as a pipeline. A multiplexer selects each matrix when the matrix is needed by the next step (Fig. 6). The output of the multiplexer is connected to the input of the next stage, the Sobel x and y blocks.

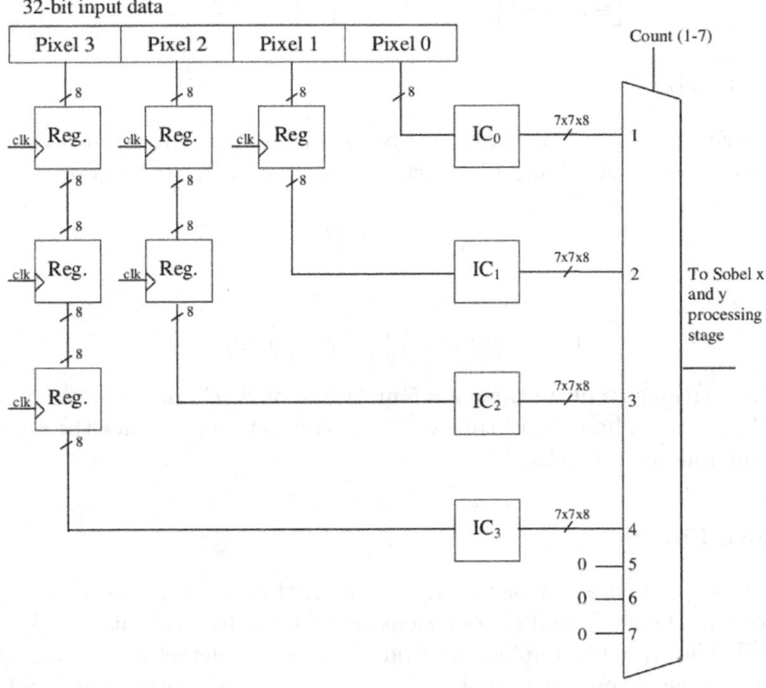

Fig. 6. Multiplexing four 7×7 window input converters (IC 0 to IC 3).

Using this structure, it is possible to increase the input data being processed by a factor of four without the need to change the blocks that come after the multiplexed input converter, except for doing a similar concatenation of the output.

By delaying the input being fed to the individual input converters, the matrices are correctly filled exactly when they are ready to be loaded into the pipeline by the multiplexer. In contrast to delaying with registers the whole output matrix of the converters, this saves logic elements in the FPGA.

3.2 Sobel Filters

The Sobel filter approximates the gradient function of the image intensity function, by convoluting the Sobel operator (mask) in the x and y directions with the image intensity data coming from the input converters, as it can be seen in (1).

The output is two 5×5 11-bit signed matrices, one for each direction, which themselves are calculated concurrently.

$$I_x = \begin{bmatrix} -1 & 0 & +1 \\ -2 & 0 & +2 \\ -1 & 0 & +1 \end{bmatrix} * I \; ; \quad I_y = \begin{bmatrix} [r] - 1 & -2 & -1 \\ 0 & 0 & 0 \\ +1 & +2 & +1 \end{bmatrix} * I \tag{1}$$

3.3 M Matrix

The M Matrix was defined by Harris and Stephens as shown in (2). It is composed by three coefficients, A, B and C, which are calculated from the gradient values I_x and I_y, as shown in (3).

$$M = \begin{bmatrix} A & C \\ C & B \end{bmatrix} \tag{2}$$

where:

$$A = I_x^2 \; ; \quad B = I_y^2 \; ; \quad C = I_x \cdot I_y \tag{3}$$

The calculations result in three 5×5 matrices, of A, B and C values, with 21-bit signed integers. These were truncated to 16-bit values to reduce the necessary logic for the following blocks.

3.4 Block Filter

The classic implementation of the Harris algorithm uses a Gaussian filter to reduce noise in the A, B and C coefficients calculated for the autocorrelation M Matrix [16]. The OpenCV implementation of the corner detector uses a simplified mask that averages the 3×3 neighborhood around the selected pixel, which is called a Block Filter (Bf), and shown in (4). The implementation proposed in this article uses the same approach.

$$Bf = \begin{bmatrix} 1 & 1 & 1 \\ 1 & 1 & 1 \\ 1 & 1 & 1 \end{bmatrix} \tag{4}$$

After the convolution (in parallel) of the three 5×5 matrices internally with the Bf kernel, without considering the boundary values, the result is three filtered 3×3 matrices with 20-bit signed values. The output values are once again truncated to 16-bit to save resources.

3.5 Harris Response

The Harris Response (R) is calculated using the determinant (Det) and trace (Tr) of the M matrix, where k has a typical value between 0.04 and 0.06 [5], as seen in the Eqs. (5), (6) and (7). The reference implementation on OpenCV uses a default value of 0.04, which was approximated by the sum

$k = 1/2^5 + 1/2^7 = 0.0391$ in this implementation, realized internally in the FPGA with two bit shifts followed by an adder.

$$R = Det - k \cdot Tr^2 \tag{5}$$
$$Det = \bar{A} \cdot \bar{B} - \bar{C}^2 \tag{6}$$
$$Tr = \bar{A} + \bar{B} \tag{7}$$

By making the substitutions of (6) and (7) in (5), the resulting equation can be seen in (8).

$$R = \bar{A} \cdot \bar{B} - \bar{C}^2 - k \cdot (\bar{A} + \bar{B})^2 \tag{8}$$

Due to the cascading of multipliers with adders in the resulting synthesized hardware, the entire calculation of the Harris response cannot be realized in a single clock period on the target FPGA, resulting in timing constraint errors when synthesizing. For such reasons this step was further divided into two stages. The resulting calculations can be expressed as a single 3×3 32-bit signed integer matrix.

3.6 Find Maximum

This stage keeps in memory the maximum value from the Harris Response found within the whole image, so that it is available to be used in the next frame as a reference for the adaptive threshold stage. A threshold step is needed in order to select what values of the Harris responses are high enough to be considered as corners. The use of an adaptive threshold reduces the impact of variations of scene illumination on the number of pixels detected as corners [6].

3.7 Adaptive Threshold

Although most Harris implementations in hardware use a fixed threshold value, this approach doesn't give efficient results when illumination changes in a large range [6].

The approach followed in OpenCV uses by default an adaptive threshold of 0.01 times the maximum value of the response found within the current image. This would require the entire response being calculated within the whole image before the threshold could be applied to the image, so an approach like the proposed here where just a 7×7 image window is sent to the FPGA cant benefit from an adaptive threshold like the one used in OpenCV.

To overcome this limitation, given that the application of the Harris Algorithm is a sequence of frames in which the difference in illumination is small between consecutive frames, instead of using the actual maximum value for calculating the threshold value, the maximum value from the previous frame is used. This allows the adaptive threshold to be calculated with only the 7×7 region being available from the second frame onwards. The use of the previous image information to compute the threshold for the next image was already applied for space applications with success by [10].

Again, avoiding using a floating point multiplier, the constant used for calculating 79 the adaptive threshold is approximated as $value = 1/2^7 + 1/2^9 \approx 0.00977$, using two bit shifts followed by an adder. All values below the threshold multiplied by the maximum from the previous frame are changed to zero. Adjusting this constant effectively changes how high a response needs for an image pixel to be to be considered a corner.

3.8 Non-maximum Suppression

From the output of the previous adaptive threshold stage, the center response value is considered a corner if it is the maximum value when compared to its 8 neighbors. The output from this stage is a single bit signaling if a corner was found. This step is taken in the same clock period as the adaptive threshold step, thus represented as a single block in Fig. 4.

3.9 Shift Register and Delay

This last step concatenates the 4 bits that signal if a corner was detected in one of the 4 windows processed by the pipeline. Due to the timing requirements, the output will be ready after 12 clock steps. To simplify the design of the software that controls the I/O in the main processor, a delay is induced so the data will be ready within 14 clock steps. This allows the corner status to be read after 7 write operations are done in the hardware. Thus, when the 7 pixels are written to apply the algorithm for each of the 4 windows that are introduced simultaneously, the corner status of the center pixel of these windows will be ready after two more windows are written to the hardware and so on.

4 Results

Two tests were performed with the Harris co-processor hardware. The latter was compared to the reference implementation of OpenCV to ensure that it was performing correctly in terms of quality. Also, the time of execution was measured in order to compare with the software counterpart, so it could to measure the speedup of the hardware implementation related to the ARM processor present on the Zynq-7000 SoC.

To perform such comparisons, the KITTI Vision Benchmark Suite dataset [12–14] was selected for providing real world stereo sequential images from a typical SLAM problem. The dataset was comprised of 22 image sequences, labeled from 00 to 21. The first sequence, 00, was used in this test, which is a real world video of a car driven outdoors on the street of a city, with a total of 4,541 stereo frames with each frame having the resolution of 1241×376 pixels. Due to the architecture details, the first and last frames are not computed on the calculations.

The time of the execution measurements was made considering only the feature extraction step of the Harris and Stephens corner detector, using

Table 2. Comparison with software: time of execution.

OpenCV Time	Hardware time	Speedup
588 ms	332 ms	1.77

Table 3. Comparison based on the OpenCV ground truth.

Data	Frames	2*4,539
	Processed pixels	4,235,940,048
Results	True positives	10,437,007
	True negatives	4,224,750,170
	False positives	372,423
	False negatives	380,448
Statistics	Recall	0.965
	Specificity	0.999
	Precision	0.966

the OpenCV implementation on the ARM processor and the synthesized co-processor on the FPGA. In the latter, the transfer of the images from the main processor's DDR3 memory to the co-processor, the data processing and the transferring of the results back from the co-processor to the main processor were included in the measurements. The results are shown in Table 2.

The quality comparison with OpenCV summed the number of true positives, true negatives, false positives and false negatives, and the recall, specificity and precision metrics were also calculated from the previously mentioned values. The results can be seen in Table 3.

5 Discussion and Conclusion

This work explored a practical analysis of a SoC architecture that incorporates a FPGA with an embedded ARM processor for Visual SLAM applications. Within the tasks of acquiring landmarks for SLAM with a visual system based on stereo cameras, the slowest performing task for feature-based correspondence in software was shown to be the feature detection.

The re-implementation in hardware of this task, being solved using the Harris and Stephens corner detection algorithm, showed a significant execution time improvement over software, around half the time needed for the software reference. The slower clock of the FPGA is compensated by the parallelism introduced internally on the stages of the pipeline, and the pipeline structure itself, which essentially allows for processing four regions of the image simultaneously in search for corners, which only impacts the complexity of the control structure used on the input and output converters. Therefore, the usage of logical elements remains similar to a construction that would process only a single region of the image at once.

Thus, the use of SoC platforms that integrate both a high performance embedded processor and an FPGA fabric is promising for speeding up highly specific tasks in the field of Mobile Robotics, such as the case investigated in this paper, of image preprocessing for the landmark acquisition task in a Visual SLAM application.

References

1. Ahn, S., Choi, J., Doh, N.L., Chung, W.K.: A practical approach for EKF-SLAM in an indoor environment: fusing ultrasonic sensors and stereo camera. Auton. Robot. **24**(3), 315–335 (2008)
2. Ahn, S., Lee, K., Chung, W.K., Oh, S.R.: SLAM with visual plane: extracting vertical plane by fusing stereo vision and ultrasonic sensor for indoor environment. In: 2007 IEEE International Conference on Robotics and Automation, pp. 4787–4794, April 2007
3. Alcantarilla, P., Bergasa, L., Dellaert, F.: Visual odometry priors for robust EKF-SLAM. In: 2010 IEEE International Conference on Robotics and Automation (ICRA), pp. 3501–3506, May 2010
4. Amaricai, A., Gavriliu, C.E., Boncalo, O.: An FPGA sliding window-based architecture Harris corner detector. In: 24th International Conference on Field Programmable Logic and Applications (FPL), pp. 1–4, September 2014
5. Aydogdu, M., Demirci, M., Kasnakoglu, C.: Pipelining Harris corner detection with a tiny FPGA for a mobile robot. In: 2013 IEEE International Conference on Robotics and Biomimetics (ROBIO), pp. 2177–2184, December 2013
6. Birem, M., Berry, F.: FPGA-based real time extraction of visual features. In: 2012 IEEE International Symposium on Circuits and Systems (ISCAS), pp. 3053–3056, May 2012
7. Birem, M., Berry, F.: DreamCam: a modular FPGA-based smart camera architecture. J. Syst. Archit. **60**(6), 519–527 (2014). http://www.sciencedirect.com/science/article/pii/S1383762114000228
8. Calonder, M., Lepetit, V., Strecha, C., Fua, P.: BRIEF: binary Robust independent elementary features. In: Daniilidis, K., Maragos, P., Paragios, N. (eds.) ECCV 2010, Part IV. LNCS, vol. 6314, pp. 778–792. Springer, Heidelberg (2010)
9. Choi, J., Lee, K., Ahn, S., Choi, M., Chung, W.K.: A practical solution to SLAM and navigation in home environment. In: 2006 International Joint Conference on SICE-ICASE, pp. 2015–2021, October 2006
10. Di Carlo, S., Gambardella, G., Prinetto, P., Rolfo, D., Trotta, P., Lanza, P.: FEMIP: a high performance FPGA-based features extractor amp; matcher for space applications. In: 23rd International Conference on Field Programmable Logic and Applications (FPL), pp. 1–4, September 2013
11. Fix, E., Hodges Jr., J.L.: Discriminatory analysis-nonparametric discrimination: consistency properties. Technical report, DTIC Document (1951)
12. Fritsch, J., Kuehnl, T., Geiger, A.: A new performance measure and evaluation benchmark for road detection algorithms. In: International Conference on Intelligent Transportation Systems (ITSC) (2013)
13. Geiger, A., Lenz, P., Stiller, C., Urtasun, R.: Vision meets robotics: the KITTI dataset. Int. J. Robot. Res. (IJRR) **32**(11), 1231–1237 (2013)

14. Geiger, A., Lenz, P., Urtasun, R.: Are we ready for autonomous driving? The KITTI vision benchmark suite. In: Conference on Computer Vision and Pattern Recognition (CVPR) (2012)
15. Goldberg, S., Matthies, L.: Stereo and IMU assisted visual odometry on an OMAP3530 for small robots. In: 2011 IEEE Computer Society Conference on Computer Vision and Pattern Recognition Workshops (CVPRW), pp. 169–176, June 2011
16. Harris, C., Stephens, M.: A combined corner and edge detector. In: Proceedings of Fourth Alvey Vision Conference, pp. 147–151 (1988)
17. Hartmann, J., Klussendorff, J., Maehle, E.: A comparison of feature descriptors for visual SLAM. In: European Conference on Mobile Robots (ECMR), pp. 56–61, September 2013
18. Lee, S., Lee, S.: Embedded visual SLAM: applications for low-cost consumer robots. IEEE Robot. Autom. Mag. **20**(4), 83–95 (2013)
19. Lowe, D.: Distinctive image features from scale-invariant keypoints. Int. J. Comput. Vis. **60**(2), 91–110 (2004)
20. Mattoccia, S., Macri, P., Parmigiani, G., Rizza, G.: A compact, lightweight and energy efficient system for autonomous navigation based on 3D vision. In: IEEE/ASME 10th International Conference on Mechatronic and Embedded Systems and Applications (MESA), pp. 1–6, September 2014
21. Munguia, R., Castillo-Toledo, B., Grau, A.: A robust approach for a filter-based monocular simultaneous localization and mapping (SLAM) system. Sensors **13**(7), 8501–8522 (2013)
22. Paz, L.M., Pinies, P., Tardos, J., Neira, J.: Large-scale 6-DOF SLAM with stereo-in-hand. IEEE Trans. Robot. **24**(5), 946–957 (2008)
23. Possa, P., Mahmoudi, S., Harb, N., Valderrama, C., Manneback, P.: A multi-resolution FPGA-based architecture for real-time edge and corner detection. IEEE Trans. Comput. **63**(10), 2376–2388 (2014)
24. Schmid, K., Hirschmuller, H.: Stereo vision and IMU based real-time ego-motion and depth image computation on a handheld device. In: 2013 IEEE International Conference on Robotics and Automation (ICRA), pp. 4671–4678, May 2013
25. Schmid, K., Tomic, T., Ruess, F., Hirschmuller, H., Suppa, M.: Stereo vision based indoor/outdoor navigation for flying robots. In: 2013 IEEE/RSJ International Conference on Intelligent Robots and Systems (IROS), pp. 3955–3962, November 2013
26. Siegwart, R., Nourbakhsh, I.R., Scaramuzza, D.: Introduction to Autonomous Mobile Robots, 2nd edn. The MIT Press, Cambridge (2011)
27. Spampinato, G., Lidholm, J., Ahlberg, C., Ekstrand, F., Ekstrom, M., Asplund, L.: An embedded stereo vision module for industrial vehicles automation. In: 2013 IEEE International Conference on Industrial Technology (ICIT), pp. 52–57, February 2013
28. Xilinx: UG761 AXI reference guide v. 13.1. http://www.xilinx.com/support/documentation/ip_documentation/ug761_axi_reference_guide.pdf

Comparison Among Experimental PID Auto Tuning Methods for a Self-balancing Robot

Marcus Romano Salles Bernardes de Souza, Rodrigo Hiroshi Murofushi, José Jean-Paul Zanlucchi de Souza Tavares(✉), and José Francisco Ribeiro

Manufacturing Automated Planning Laboratory (MAPL), Universidade Federal de Uberlândia, Uberlândia, Brazil
mrsbs.mecatronica@gmail.com, hiroshirhm@gmail.com, {jean.tavares,jribeiro}@ufu.br

Abstract. A self-balancing robot, also known as two-wheeled vehicle is an unstable system and it can be approximated to inverted pendulum, so there is a need of a suitable controller so that it can be stabilized. This paper compares five PID design techniques without mathematical model of the system in order to remain it stand. The PID tuning methods discussed are Manual, Ziegler-Nichols, Relay, Augmented Ziegler-Nichols and Augmented Relay. The augmented method modifies the PID constants online depending on the error value and use a Ziegler-Nichols or Relay PID tuned controller as initial one. Some experimental results presented suggest that the Ziegler-Nichols tuning method is slightly better than the other techniques. All the electronic gadgets and algorithms are embedded in the prototype.

Keywords: Two-wheeled vehicle · PID tuning · Relay method · Augmented Ziegler-Nichols PID, Augmented Relay PID

1 Introduction

In recent years, the performance and quality specifications have become more difficult to reach in industries. This increase is due to the largest global competitiveness, the frequent fluctuation in economic conditions and the more rigorous environmental and safety regulations. Modern plants, considering its high complexity and high integration of subsystems, have become more and more challenging to operate [1]. In this context, the control systems is gaining more importance.

Given the desired operating conditions, the main objective of control systems is to keep the process within acceptable performance, safely and efficiently, coordinating the interactions in the subsystems that comprise it.

So far, the great advances in process control techniques have not diminished the popularity of industrial PID controller due to simplicity and ease of on-line re-tuning provided by this controller. Aström and Hägglund [2] have suggested the use of an ideal (on-off) relay to generate a sustained oscillation of the controlled variable and to get the ultimate gain (K_u) and the ultimate frequency

F. Santos Osório and R. Sales Gonçalves (Eds.): LARS 2015/SBR 2015, CCIS 619, pp. 72–86, 2016.
DOI: 10.1007/978-3-319-47247-8_5

(ωu) directly from the relay experiment. This method is an alternative to the conventional method of Ziegler-Nichols for closed loop systems and the success of this method is due to the simplicity of the mechanisms of identification and calibration, and also its applicability in slow or highly nonlinear systems [3].

Some advantages of the relay method are that it requires little mathematical processing, it identifies the system characteristics around its critical frequency, its application does not require knowledge of the system mathematical model and it avoids the trial and error procedure in determining the critical gain.

The two-wheeled vehicle behaves like an inverted pendulum control. The inverted pendulum is an inherently a nonlinear unstable system, then a control action is necessary to keep the wheeled pendulum in a vertical position. For this reason, it is used as an object of study in several studies [4–7].

In [4], the authors propose a control strategy for implementing a mobile inverted pendulum with two wheels, the plant model is unknown and the system is influenced by external disturbances. In [5] it is shown a two-wheeled vehicle movement control as well a sits stability analysis. The authors proposed a self-tuning PID control strategy, based on a model obtained by the vehicle dynamics analysis. The developed implementation enables to maintain the vehicle in its vertical position and make it respond to motion commands. In [6], the control strategy presented consists on a neural network incorporated with PID controllers and in [6–8] the authors propose a fuzzy logic coupled to the classical PD controller.

The authors in [9] proposed an auto tuning algorithm for PI and PID controllers based on relay experiments to minimize the load disturbance integral error by maximizing the integral gain and a minimum required gain margin constraint. In addition, it says that this approach is applicable to any linear model structure, including dead time and non-minimum phase systems. The algorithm has been tested on a real experimental thermal process, and it worked well with real measurement noise and disturbances.

In [10] it is proposed an improved relay auto tuning of PID controllers for critically damped first order plus time delay model and in [11] for a critically damped second-order plus time delay model. The both proposed methods give good results even when noise is present in the output variable and a load separately enters the system along with the input variable.

This paper introduces the design of a PID for a two-wheeled vehicle using three different tuning methods having no mathematical model of it. The first method is the manual tuning and the other one is through the relay method.

This paper is organized as follows. After defining the wheeled inverted pendulum problem in Sect. 2, the methodology used is described in Sect. 3. The experimental results and discussion are presented in Sect. 4. Then the conclusion and future works are presented in Sect. 5.

2 Problem Statement

The two-wheeled vehicle is an inherently unstable system so it must have a controller so that it becomes stable.

The pitch angle (θ) provides the angular position of the vehicle relative to its static equilibrium position. Hence, θ is the parameter to be controlled and it can be corrected through the action of the motors coupled to each wheel so that the vehicle can stand.

The pitch angle is measured by MPU 6050 Arduino sensor module, which its value is given by a combination of the measurements from an accelerometer and a gyroscope present in the sensor module.

All peripheral devices are linked to an Arduino Mega 2560 R3 which is responsible for the data processing and system control. The PID controller algorithm is implemented and embedded in the Arduino.

A bluetooth module (HC 05 module) is used for the communication between the vehicle and a computer in order to visualize and store the experiments data. Still, there is a driver (Motor Shield L293D) coupled to Arduino to operate the two DC Motors (3–6 V) of the vehicle.

The vehicle chassis is assembled using a LEGO bricks kit as it can be seen at Fig. 1 and the vehicle schematic is presented in Fig. 2.

Fig. 1. Two-wheeled vehicle.

3 Methods

In this paper it is used three different methods tuning a PID controller. The first one is the manual method, the second is the relay method proposed by [2,3],

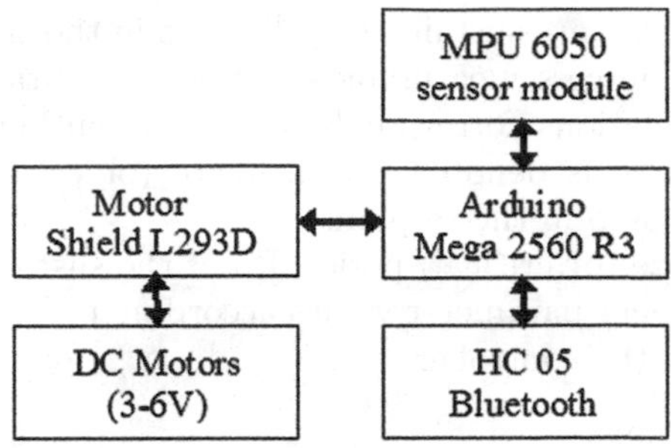

Fig. 2. Vehicle schematic.

and the last one is the Augmented Ziegler and Nichols tuned PID controllers (AZNPIDs). All experiments were initiated with the vehicle at rest (zero angular velocity) in the vertical position ($\theta = 0$) and the only perturbation considered in the system response was the acceleration of gravity.

3.1 Manual Tuning

This method consists on estimating the values K_P, K_I and K_D of the PID controller through successive attempts to find a satisfactory system performance.

The manual tuning is a trial and error method consisting of the following steps: First set K_I and K_D values to zero and gradually increase the K_P until the output of the process oscillates. Given the fact that an inverted pendulum is a nonlinear system, the proportional coefficient must provide a fast response. Then divide the K_P by two and gradually increase the value of K_D until the system respond quick enough to reach the setpoint. Notice that small values of K_D provide a smooth oscillation, despite the fact that the amplitude rises and the process become slow. Then increase gradually the value of K_I until the setpoint be corrected with a small displacement. The parameter K_I has to be large enough to not to end up falling; although, high K_I values tend to let the system unstable.

3.2 Ziegler-Nichols PID (ZNPID) Tuning Procedure

The Ziegler-Nichols' closed loop method is based on experiments executed on an established control loop (a real system or a simulated system).

1. The tuning procedure is as follows:
2. Bring the process to (or as close to as possible) the specified operating point of the control system to ensure that the controller during the tuning is "feeling" representative process dynamic and to minimize the chance that variables during the tuning reach limits [12].
3. Turn the PID controller into a P controller by setting set $T_i = \infty$ and $T_d = 0$. Initially set gain $K_P = 0$. Close the control loop by setting the controller in automatic mode.

4. Increase KP until there are sustained oscillations in the signals in the control system, e.g. in the process measurement, after an excitation of the system. (The sustained oscillations corresponds to the system being on the stability limit.) This K_P value is denoted the ultimate (or critical) gain, K_u. The excitation can be a step in the setpoint.
5. Measure the ultimate (or critical) period Pu of the sustained oscillations.
6. Calculate the controller parameter values according to Table 1, and use these parameter values in the controller.

Table 1. ZNPID tuning constants as a function of ultimate gain and period [12].

Specification	ZNPID constants		
	K_p	T_i	T_d
P controller	$0.5K_u$	∞	0
PI controller	$0.45K_u$	$P_u/1.2$	0
PID controller	$0.6K_u$	$P_u/2$	$P_u/8$

3.3 Relay PID Tuning (RPID)

The RPID method makes use of an ideal (onoff) relay to generate a sustained oscillation of the controlled variable and to get the ultimate gain (K_u) and the ultimate period (P_u) directly from the experiment.

The success of this method is due to the simplicity of the mechanisms of identification and calibration and its applicability in slow or highly nonlinear systems.

1. Substitute a relay with amplitude d for the PID controller as shown in 0.
2. Kick into action, and record the plant output amplitude a and period P.
3. The ultimate period is the observed period, $P_u = P$, while the ultimate gain is inversely proportional to the observed amplitude.
4. Compute the K_u parameter according to (1).

$$K_u = \frac{4d}{\pi a}. \tag{1}$$

where a is the amplitude, d is the control signal amplitude and P_u is the period of oscillations in the system output as it can be seen in 0 (Figs. 3 and 4).

Having established the ultimate gain and period with a single experiment, it can use the ZN tuning rules (or equivalent) to establish the PID tuning constants. Incidentally, the modified values given in Table 2 are improved versions of the original constants given in most textbooks that have been found to be excessively oscillatory. It was used the original specification to design the controllers in this paper.

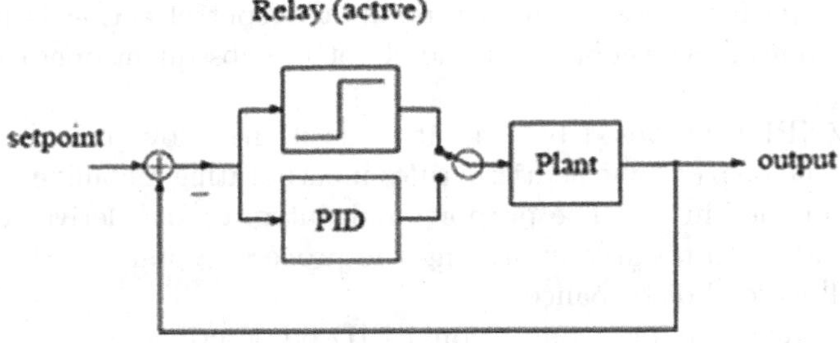

Fig. 3. Block diagram of the relay method [12].

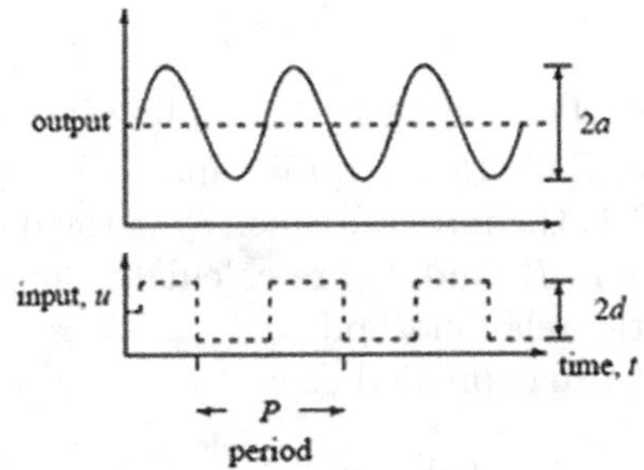

Fig. 4. Parameters of relay method [12].

Table 2. Modified ZNPID tuning constants as a function of ultimate gain and period [13].

Specification	ZNPID constants		
	K_p	T_i	T_d
P controller	$0.6K_u$	$P_u/2$	$P_u/8$
PI controller	$0.33K_u$	$P_u/2$	$P_u/3$
PID controller	$0.2K_u$	$P_u/2$	$P_u/3$

3.4 Augmented Ziegler and Nichols Tuned PID Controllers (AZNPID)

The AZNPID proposes a real time update of the PID parameters K_P, K_I and K_D and procedure implies a low computing cost, and the performance of the resulting control loop is somehow near the optimum that could be obtained if the full information of the process model were available [9].

To tune a PID in manual mode, an operator generally adjusts the controller gains according to the current process trend to attain the desired response. The basic idea behind such gain manipulation strategy is that, when the process variable is moving away from the set point, the controller takes aggressive action to bring it back to the desired value as soon as possible. Moreover, when the process is moving fast towards the set point, the control action is reduced to restrict the potential overshoot and undershoot in subsequent operating phases [14].

In the AZNPID, proposed by [14], it is used the above gain modification strategy with the help of some heuristic rules incorporating an online gain updating factor α, defined in (5). The proportional, integral, and derivative gains of AZNPID are adjusted towards improving the process response during set point change as well as load disturbance.

Let the discrete form of a conventional PID be described as

$$u^c_{(k)} = K_P e\left(k\right) + K_I \sum_{i=0}^{k} e\left(i\right) + K_D \Delta e\left(k\right). \tag{2}$$

In (2), $u^c_{(k)}$ is the control action at k-th sampling instant, K_P is the proportional gain, $K_I = K_P\left(\Delta t/T_i\right)$ is the integral gain, and $K_D = K_P\left(T_d/\Delta t\right)$ is the derivative gain. Where T_i is the integral time, T_d is the derivative time, and Δt is the sampling interval. K_P, T_i, and T_d are calculated according to ZN ultimate cycle tuning rules using the relay method.

Here, $e\left(k\right)$ and $\Delta e\left(k\right)$ are expressed as

$$e\left(k\right) = r - y\left(k\right) \tag{3}$$

$$\Delta e\left(k\right) = e\left(k\right) - e\left(k-1\right) \tag{4}$$

Where r is the set point, and $y(k)$ is the process output. The update factor α is defined by

$$\alpha = e_N\left(k\right) \times \Delta e_N\left(k\right) \tag{5}$$

Here

$$e_N\left(k\right) = \frac{e\left(k\right)}{\left|r\right|} \tag{6}$$

And

$$\Delta e_N\left(k\right) = e_N\left(k\right) - e_N\left(k-1\right) \tag{7}$$

are the normalized values of $e\left(k\right)$ and $\Delta e\left(k\right)$ respectively.

In the proposed AZNPID, KP, KI, and KD will be continuously modified by the gain updating factor α with the following simple heuristic relations:

$$K_P^m\left(k\right) = K_P\left(1 + k_1\left|\alpha\left(k\right)\right|\right) \tag{8}$$

$$K_I^m\left(k\right) = K_I\left(0.3 + k_2\left|\alpha\left(k\right)\right|\right) \tag{9}$$

$$K_D^m\left(k\right) = K_D\left(1 + k_3\left|\alpha\left(k\right)\right|\right) \tag{10}$$

Thus, from (2) and (8)–(10), AZNPID can be expressed as

$$u^a\left(k\right) = K_P^m\left(k\right) + K_I^m\left(k\right) \sum_{i=0}^{k} e\left(i\right) + K_D^m\left(k\right) \Delta e\left(k\right) \tag{11}$$

Where, $K_P^m(k)$, $K_I^m(k)$ and $K_D^m(k)$ are the modified proportional, integral, and derivative gains respectively at k-th instant, and $u^a(k)$ is the corresponding control action.

In (8)–(10), k_1, k_2, and k_3 are three positive constants, which will make the required variations in K_P^m, K_I^m and K_D^m around their respective initial values.

The Fig. 5 shows the simplified block diagram of the PID controller proposed by [14].

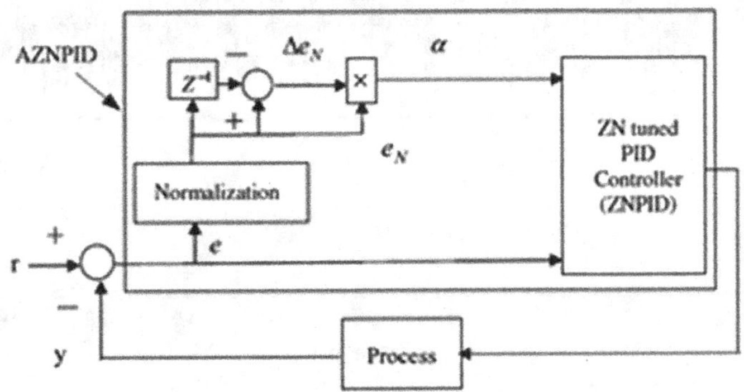

Fig. 5. AZNPID block diagram [14].

3.5 Augmented Relay Tuned PID Controllers (ARPID)

The ARPID method is the combination of RPID and AZNPID methods. This is the application of the techniques used in AZNPID method but using the control parameters calculated by the RPID method.

4 Experimental Results and Analysis

4.1 Manual Tuning

With the manual tuning method, the PID coefficients adopted for a satisfactory system response are $K_P = 0.10$, $K_D = 0.30$, $K_I = 0.01$. The system response plot is shown in Fig. 6.

The PID controller designed using the manual tuning stabilized the system and the vehicle could remain stand.

4.2 ZNPID

The ultimate gain obtained is $P_u = 0.336$ s and the PID coefficients adopted for a satisfactory system response are $K_P = 0.060$, $K_D = 0.126$, $K_I = 0.007$. The system response plot is shown in Fig. 7.

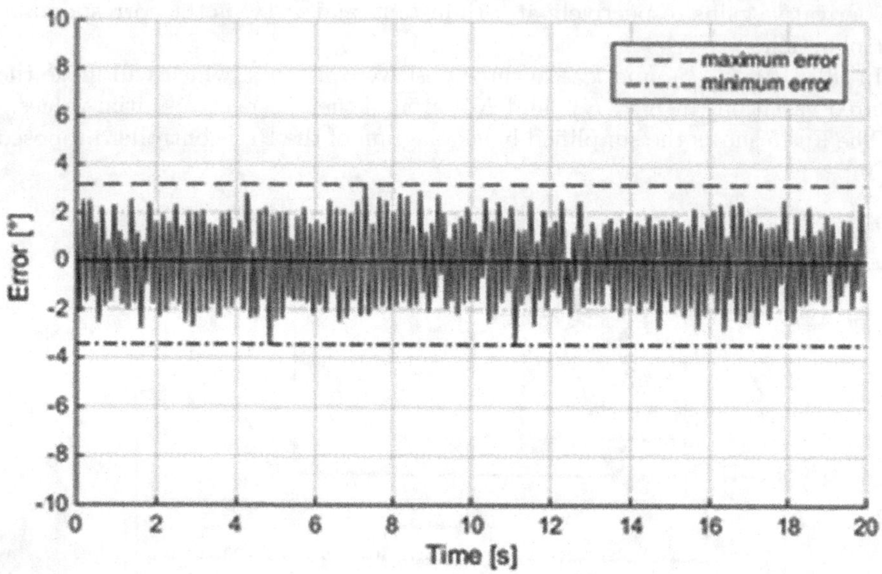

Fig. 6. Pitch response of the vehicle using the PID manual tuning.

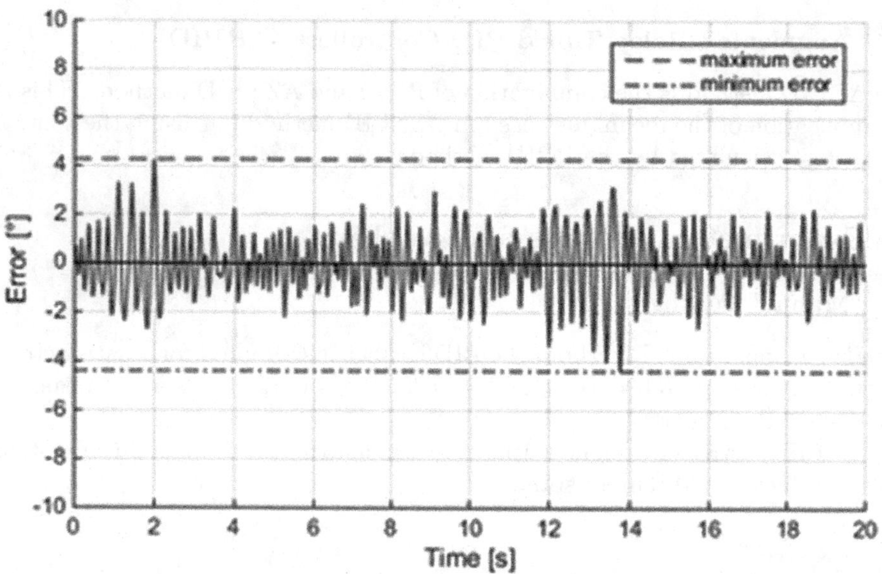

Fig. 7. Pitch response of the vehicle using the ZNPID.

4.3 RPID

Due the fact this system is unstable, it is difficult to control with a single relay type input. Small values of d results in a divergent process response and large system input values make the system oscillates with large amplitude variation. Therefore, the optimal value of d is the one that gives the process an oscillation with a constant amplitude for considerable time. For $d = 0.3$ on a range of 0 to 1, the system had the best acceptable oscillatory response. The average amplitude value is $a = 3.75°$ and the period is $P_u = 0.359\,\text{s}$.

The system output obtained from the relay test is shown in Fig. 8. Using (1), $K_u = 0.07$ and according to Table 2, using the originals specifications for PID coefficients results in $K_P = 0.042$, $K_D = 0.094$, $K_I = 0.005$. The vehicle response obtained experimentally is shown in the graphic of the Fig. 9.

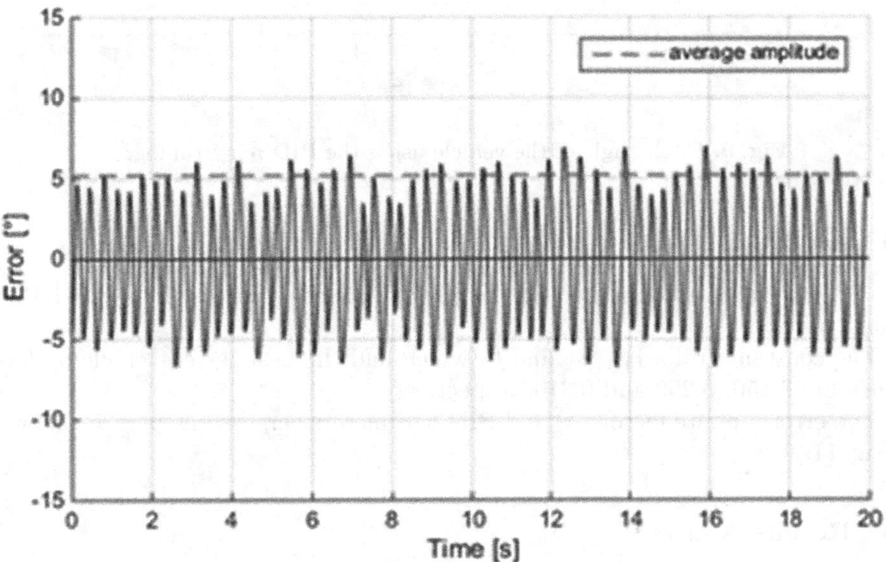

Fig. 8. Pitch response of the vehicle in relay test.

4.4 AZNPID

The PID coefficients considered for the AZNPID are the same found on the ZNPID tuning method. The constants values k_1, k_2 and k_3 which had the best tested vehicle performance are 0.100, 0.265 and 0.300, respectively.

The error, update factor (α) and PID parameters versus time plot are shown in Fig. 10.

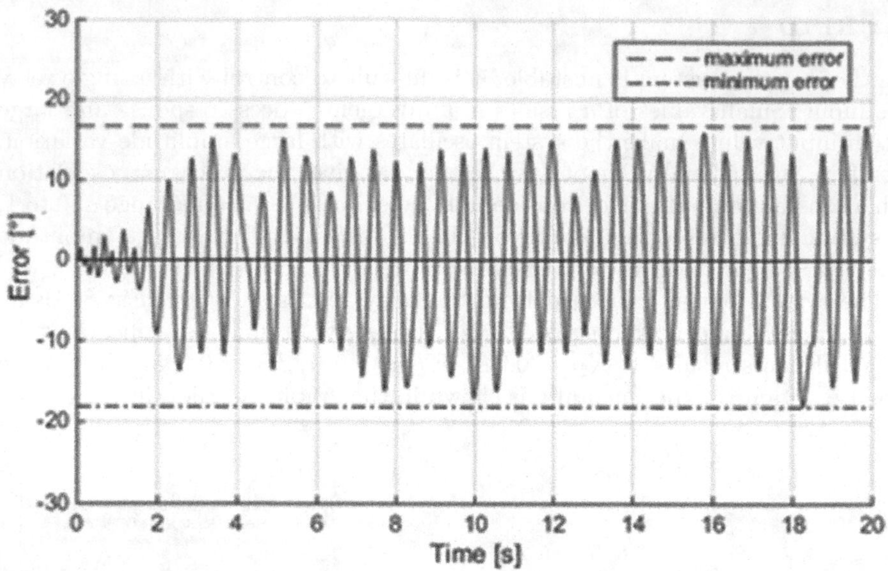

Fig. 9. Pitch angle of the vehicle using the PID relay tuning.

4.5 ARPID

The PID coefficients considered for the ARPID are the same found on the RPID tuning method.

The constants values k_1, k_2 and k_3 which had the best tested vehicle performance are 0.150, 0.250 and 0.310, respectively.

The error, update factor (α) and PID parameters versus time plot are shown in Fig. 11.

4.6 Results Analysis

In order to compare the results of the three methods discussed in this paper it is computed the root mean square (RMS) of the pitch angle measurements according to the Eq. (12) and the data standard deviation for ten tests and the mean of the data test is disposed at Table 3.

$$\Theta_{RMS} = \sqrt{\frac{1}{N}\sum_{i=1}^{N}\Theta_i^2} \tag{12}$$

Where N is the amount of points considered in the RMS calculation.

According to Table 3, the ZNPID method has the lowest Θ_{RMS} mean value, followed by the manual tuning and, after the ARPID, and then the AZNPID and the last one is the RPID.

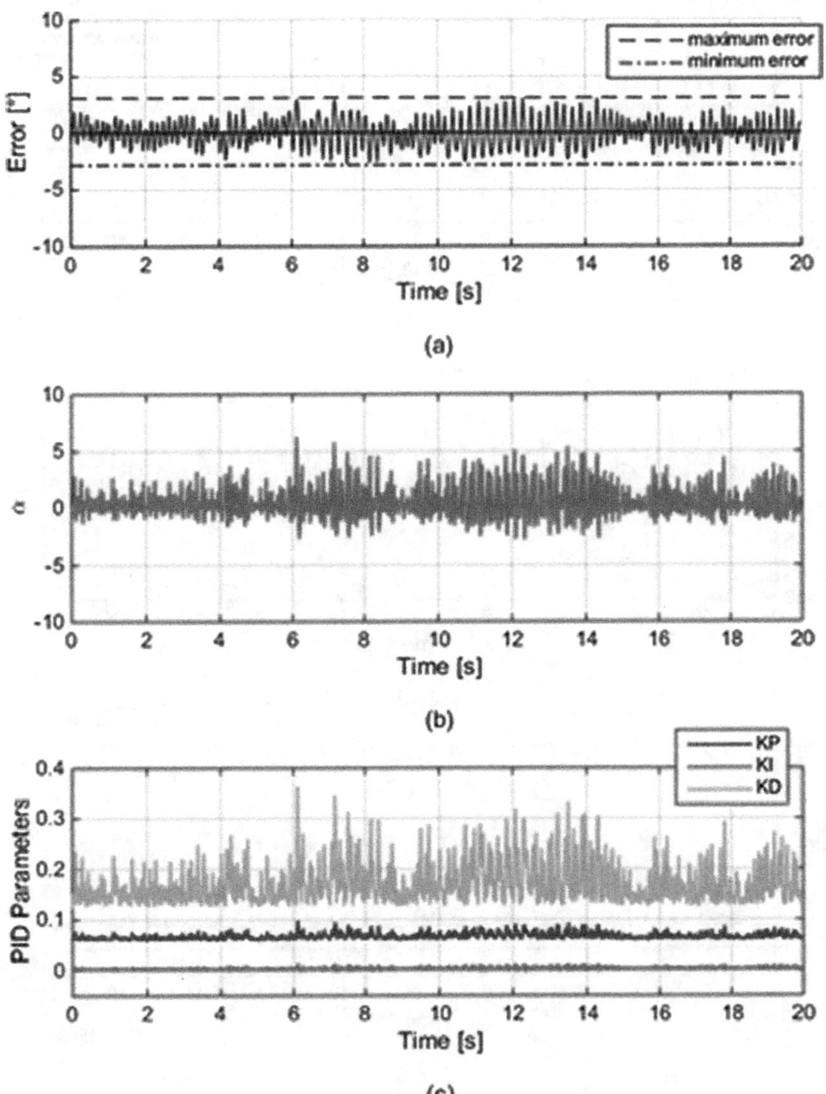

Fig. 10. AZNPID (a) Error × Time; (b) a × Time; (c) PID parameters × Time.

Table 3. Results: RMS and percentage difference.

	Manual tuning	ZNPID	RPID	AZNPID	ARPID
Θ_{RMS} mean	1.35	1.25	7.40	1.59	1.52
Θ_{RMS} standard deviation	0.05	0.14	1.05	0.15	0.16

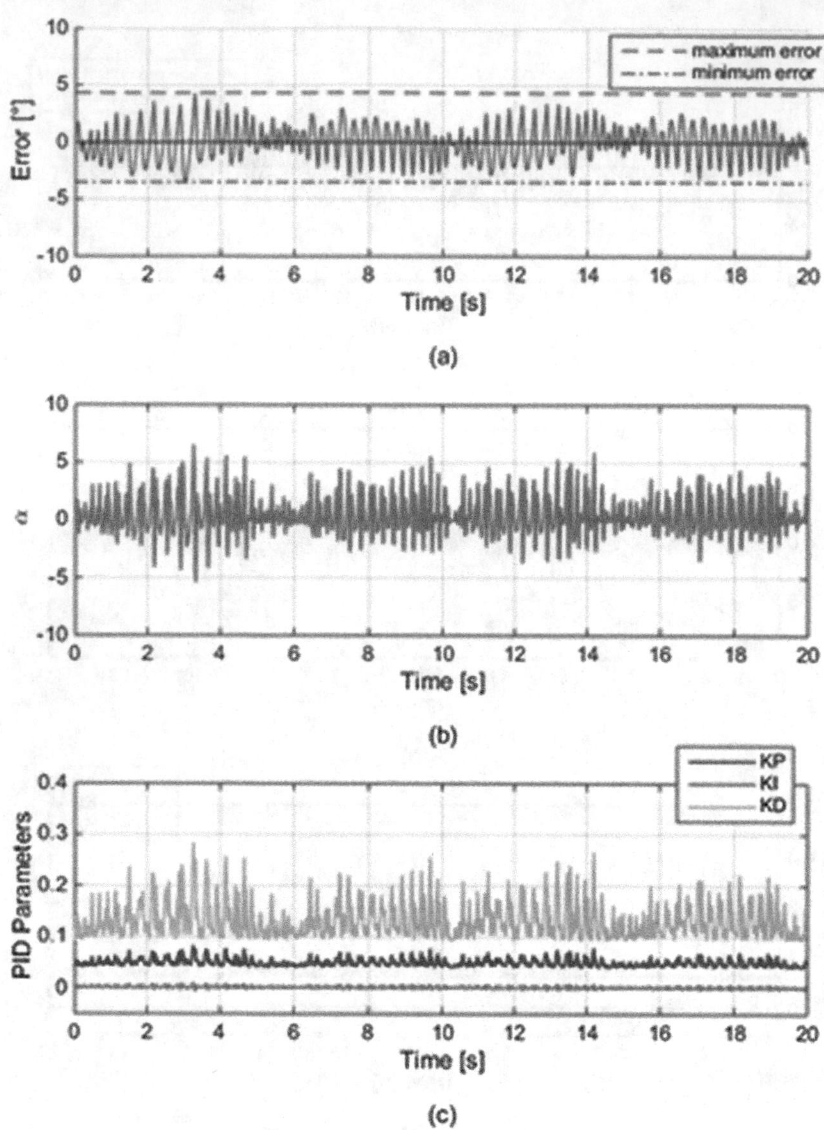

Fig. 11. ARPID (a) Error × Time; (b) a × Time; (c) PID parameters × Time.

The manual tuning PID design demonstrated to be a good option to control the vehicle because this technique demonstrated a good capability on controlling the vehicle and it had a better performance than the PID designed through the relay method. Although it is needed a good background of the system and the PID controller, especially in cases of nonlinear unstable systems like the two-wheeled vehicle used in the experiments of this paper.

The ZNPID is the best technique among the ones presented in this paper because it presented the lowest Θ_{RMS} value. Its good performance allows comparison among the methods studied and demonstrate that the augmented methods are possible to improvement since optimal values for k_1, k_2 and k_3 constants are utilized.

On the other hand, the relay method tuning demonstrated not being as efficient as the ZNPID or even the AZNPID, ARPID and manual method because it had the highest pitch angle error around the vehicle equilibrium angle. However, the relay method still demonstrated to be able to control the vehicle so that it can remain stand.

For last, the ARPID showed substantial improvement when used with the augmented technique from a mean pitch angle error of 7.40° to 1.52° using the same initial PID parameters. This behavior proves that the use of the heuristic technique is a powerful tool aiming a performance improvement.

5 Conclusion and Further Works

Thus, the ZNPID method is the best technique among the ones presented in this paper since it is a benchmark method for estimating the PID parameters. Although, like in manual tuning, the attempts on finding the appropriate constants may become costly.

For future works, more advanced techniques can be design to control the two-wheeled vehicle, e.g., fuzzy logic couple to PID, neural networks, and some adaptive methods.

Also using the AZNPID and the ARPID, optimization tools as genetic algorithm and ant colony may be applied to find their best possible values of k_1, k_2 and k_3, with a view to achieving the desired performance with some pre-specified performance indices.

In addition, it is possible to add the horizontal displacement control of the vehicle while it tries to remain stand, although it becomes a multiple input and multiple output system (MIMO) and a different control technique is needed.

Acknowledgments. We gratefully acknowledge the support of the FEMEC, CAPES, CNPQ, FAPEMIG, Prof. Dr. R. Fortes and Prof. Dr. M.V. Duarte.

References

1. Seborg, D., Edgar, T., Mellichamp, D.: Process Dynamics and Control. Wiley, New York (2004)
2. Aström, K.J., Hägglund, T.H.: The future of PID control. Control Eng. Pract. **9**(11), 1163–1175 (2001)
3. Aström, K.J., Hägglund, T.H.: New tuning methods for PID controllers. In: Proceedings of 3rd European Control Conference, pp. 2456–2462 (1995)
4. Chiu, C., Peng, Y., Lin, Y.: Intelligent backstepping control for wheeled inverted pendulum. Expert Syst. Appl. **38**, 3364–3371 (2011)

5. Ren, T., Chen, T., Chen, C.: Motion control for a two-wheeled vehicle using a self-tuning PID controller. Control Eng. Pract. **16**, 365–375 (2008)
6. Jung, S., Kim, S.S.: Control experiment of a wheel-driver mobile inverted pendulum using neural network. IEEE Trans. Control Syst. Technol. **16**(2), 297–303 (2008)
7. Su, K., Chen, Y., Su, S.: Design of neural-fuzzy-based controller for two autonomously driven wheeled robot. Neurocomputing **73**, 2478–2488 (2010)
8. Oltean, S.E., Duka, A.V.: Balance control system using microcontrollers for a rotational inverted pendulum. In: The 7th International Conference Interdisciplinarity in Engineering, Tirgu Mures-Romania. Procedia Technology, vol. 7, pp. 11–19 (2013)
9. Romero, J.A., Sanchis, R., Balaguer, P.: PI and PID auto-tuning procedure based on simplified single parameter optimization. J. Process Control **21**, 840–851 (2011)
10. Vivek, S., Chidambaram, M.: An improved relay auto tuning of PID controllers for unstable FOPTD systems. Comput. Chem. Eng. **29**, 2060–2068 (2005)
11. Vivek, S., Chidambaram, M.: An improved relay auto tuning of PID controllers for unstable SOPTD systems. Chem. Eng. Commun. **199**, 1437–1462 (2012)
12. Ziegler, J.G., Nichols, N.B.: Optimum settings for automatic controllers. Trans. ASME **64**, 759–768 (1942)
13. Wilson, D.I.: Relay-based PID tuning. Automation & Control (2005)
14. Dey, C., Mudi, R.K.: An improved auto-tuning scheme for PID controllers. ISA Trans. **48**, 396–409 (2009)

RoSoS - A Free and Open-Source Robot Soccer Simulator for Educational Robotics

Felipe N. Martins[1(✉)], Ivan S. Gomes[2], and Carmen R.F. Santos[3]

[1] IFES - Federal Institute of Education, Science and Technology of Espírito Santo, Serra Campus, Serra, ES, Brazil
felipe.n.martins@gmail.com

[2] UFABC - Federal University of ABC, Santo André, SP, Brazil

[3] UFES - Federal University of Espírito Santo, Vitória, ES, Brazil

Abstract. The use of robots as educational tools provides a stimulating environment for students. Some robotics competitions focus on primary and secondary school aged children, and serve as motivation for students to get involved in educational robotics activities. Although very appealing, many students cannot participate on robotics competitions because they cannot afford robotics kits. Hence, several students have no access to educational robotics, especially on developing countries. To minimize this problem and contribute to education equality, we have created RoSoS Robot Soccer Simulator, in which students program virtual robots in a similar way that they would program their real ones. In this chapter we explain some technical details of RoSoS and discuss the implementation of a new league for the robotics competitions: Junior Soccer Simulation league (JSS). Because soccer is the most popular sport in the world, we believe JSS will be a strong motivator for students to get involved with robotics.

Keywords: Educational robotics · Robotics competitions · Soccer Simulation · Simulation software

1 Introduction

Educational Robotics is a valuable tool for interdisciplinary teaching. Several researchers have shown that the use of robots as educational tools provide a stimulating environment for the students, with very positive results [10,11,13,20]. The implementation of Educational Robotics on technical courses (at high-school level) was discussed in [19], where some results were presented with emphasis on the participation of teams in robotics competitions. Some researchers claim that the social impact of robotics competitions is also important because it motivates integration between schools, university and the community [12]. Some examples of robotics competitions are RoboCup, FIRST Lego League, FIRA Cup and the Latin-American Robotics Competition (LARC), all being used as a motivation to get students involved in robotics. One of the biggest and most important one is the RoboCupJunior (RCJ), that focuses on primary and secondary school aged

F. Santos Osório and R. Sales Gonçalves (Eds.): LARS 2015/SBR 2015, CCIS 619, pp. 87–102, 2016.
DOI: 10.1007/978-3-319-47247-8_6

children. RCJ is part of the RoboCup, an annual international robotics competition that aims to promote robotics and Artificial Intelligence research [1]. About 45 countries select teams to participate in the international RoboCup. In Brazil the selection of teams for the RoboCup international competition is done by the Brazilian Robotics Olympiad (OBR) and the Brazilian Robotics Competition (CBR). While CBR targets undergraduate and graduate students, OBR is an annual competition for primary and secondary students in Brazil. OBR has two main categories: practical and theoretical. In 2015, around 8,000 students from about 1,500 schools have participated on the OBR practical competition [4]. In the same year, OBR theoretical competition had 90,000 students enrolled.

RoboCup has several different categories, such as Rescue, @Home and Soccer, among others. It also has some simulation leagues, in which virtual robots compete in a simulated environment. With no hardware, simulation league's focus is on artificial intelligence and team strategy [9]. This makes it possible for teams to fully concentrate on the development of higher level behavior for its robots, and to improve performance of their software by running thousands of tests on several simulations. For example, [23] used simulated soccer to study and compare several multiagent reinforcement learning algorithms, and concluded that in some scenarios direct search in policy space can offer advantages over approaches based on evaluation functions learning. By its turn, [15] presents a comparison between several machine learning techniques to identify and classify robot formations of the opponent soccer team in order to improve playing strategy. Other works show the application of reinforcement learning to improve the defense behavior [16] and to select the best action in setplays [14] of simulated soccer teams.

Like in major competition, RoboCupJunior also has soccer category. But, in all Junior categories (soccer, dance and rescue) students have to deal robot hardware, and there is no pure simulation league. Important Brazilian competitions like OBR and CBR also lack a simulation league dedicated to primary and secondary students. In such a context, we propose a Junior Soccer Simulation (JSS) league for primary and secondary students that could be adopted by OBR, CBR, RoboCupJunior and any other robotics competition. Soccer is the most popular sport in the world [3], therefore we believe this will be a strong motivation for students to get involved in robotics. Because CBR and RoboCup already have a Soccer Simulation league for students over 19 years old, the ones that choose to participate in the JSS would have the opportunity to move to a major league after they finish secondary school. JSS can also be used in classrooms as a tool to motivate students to develop computational thinking, which is arguably a fundamental skill to virtually any future professional [24].

Perhaps the most important reason for the adoption of a simulation league by OBR and RCJ is the fact that several schools and students, especially on developing countries, do not have financial means to maintain robotics projects. As we mentioned, in Brazil OBR had around 8,000 students participating on the practical competition in 2015, while more than 90,000 were enrolled in the theoretical competition. It is possible that many of those 90,000 students did not register for the practical competition only because they did not have access to robotics kits. A simulation league would allow those schools and students to get involved with robotics without the need of buying hardware. Because our simulator is free and can run on free operating systems, the only cost for the school is the computer to run the simulation.

On the following Sections we present the general concept of a Junior Soccer Simulation League, give an explanation of the RoSoS software and show some details of how it is organized. Finally, we discuss some ideas and future work. Parts of the work presented here were also presented in two conference papers [21,22]. In this Chapter, however, we give a more detailed explanation of RoSoS structure and present new examples of code to better explain RoSoS implementation and usage.

2 Junior Soccer Simulation League

In this section we give a brief description of the proposed Junior Soccer Simulation League. The main idea is similar to the 2D Simulation League that already exists in RoboCup [9], but adapted to primary and secondary student level. We propose a competition in which two teams of N autonomous software programs (virtual robots) play soccer in a two-dimensional virtual soccer field. The dimensions of the field, ball and robots are adjustable, and can be made proportional to the dimensions of their correspondent real counterparts in the actual RoboCupJunior competition, for example. Game rules for the simulation league can be similar to the rules for the real leagues, but this can be adapted. The size of the field and number of robots per team can also be changed. In fact, different categories could have different field sizes, ball sizes, mass and speed of robots etc.

During a match, the simulator knows the status of everything on the game, such as position, orientation and speed of all robots, position and speed of the ball, the laws of physics etc. Students have to program their independent agents individually (each virtual robot) to communicate with the simulator. Each virtual robot receives some information from the simulator, like noisy input of its virtual sensors, and have a limited amount of time to send some commands in order to move. If no command is sent, the simulator assumes that all desired speeds are zero.

3 Robot Soccer Simulator - RoSoS

The simulator platform was programmed using Processing, which consists of a programming language and a development environment [5]. Processing language is built on Java, but it uses a simplified syntax, in a way similar to the Arduino language [2]. This software was created on a project initiated in a group from the MIT Media Lab and, since 2012, is maintained by the Processing Foundation, whose primary goal is "to empower people of all interests and backgrounds to learn how to program..." [6]. Today, Processing is being used for learning, prototyping, and production by tens of thousands of students, artists, designers, researchers, and hobbyists. It is free, open-source, and runs under GNU/Linux, Mac OS X, and Windows [5]. Its popularity, flexibility and similarity to the Arduino language were the main reasons for choosing Processing as the simulator language.

While developing RoSoS, the main guidelines were: (1) the program for a virtual robot should be similar to the program of a real robot; and (2) users should have the possibility to change the simulator operation (physics and game rules), and to personalize their robots. We built the simulator considering the above guidelines and the resulting software has the following features:

- Processes physics and collisions in the engine, and allows users to alter physics parameters;
- Allows changes on the game rules;
- Allows customized shapes for Robots and environments;
- Allows the use and positioning of different sensors for each one of the virtual robots;
- Each robot of a team can have its own set of sensors and its own program.

Figure 1 illustrates the graphical appearance of the simulator in action. It shows a screen-shot of RoSoS running a match between two teams of three virtual robots, each.

In the following subsections we present a brief explanation about how to use the simulator to build a soccer team and give some technical information about how the simulator works.

3.1 Using RoSoS to Build a Team

First of all, it is important to state that students can program its virtual robots without a deep understanding about how the simulator was programmed. The structure of the simulated team was built in such way that, for each virtual robot, students use a *setup()* function to implement instructions that should be executed once (at the beginning), and a *loop()* function to implement the main code (that is repeated indefinitely). This is to mimic the structure of the Arduino language, which is very popular among students. To implement specific behaviour, students can use RoSoS's main base classes and methods to build and program its virtual robots. RoSoS classes were designed to simulate the behaviour of real robots so that students can focus on how an actual robot would work.

RoSoS is distributed with an empty team class that contains a pre-built structure with comments explaining what students need to fill in to start programming their virtual robots. The package also contains some examples of teams to give students a better idea of how they can use the provided functions. The following code shows one of the provided examples. The public class *CustomTeamA* contains two types of robots, an *Attacker* and a *Goalie*. Both *Attacker* and *Goalie* have a sensor to detect the ball. The *Goalie* also has ultrasonic distance sensors pointing forward, backwards, to the left and to the right. The *Attacker* and the *Goalie* have different programs in their respective loop functions. Students can use this example as a base structure, modifying only the contents of the *setup()* and *loop()* functions of each robot. A more detailed explanation of the program and its functions will be given in the following subsections.

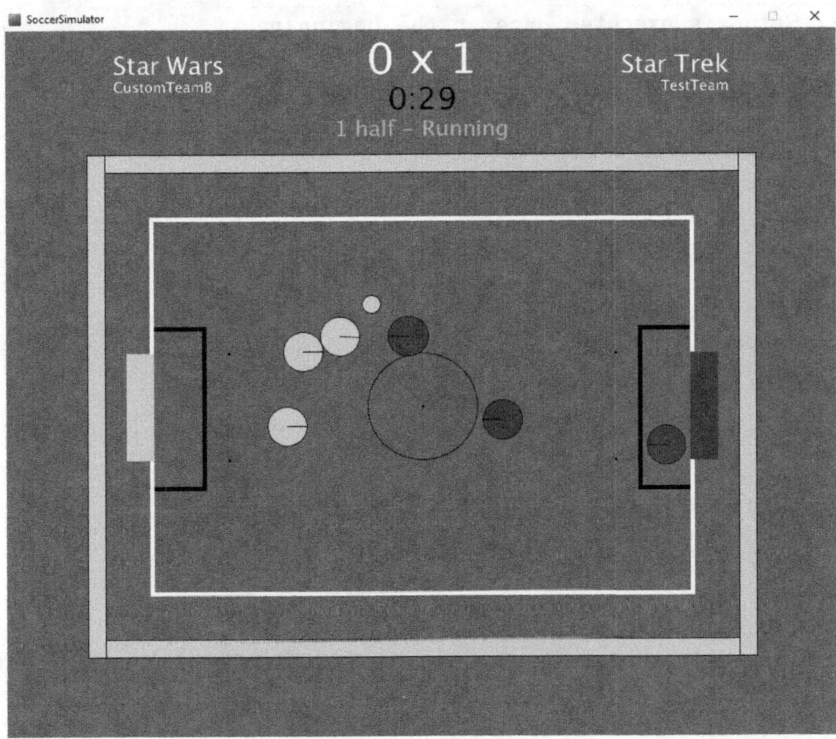

Fig. 1. Screen-shot of RoSoS running a match between teams Star Wars and Star Trek, in which Star Trek is winning (obviously).

Example team CustomTeamA with two robots.

```
public class CustomTeamA implements Team{

  public String getTeamName(){
    return "Emerotecos"; }

  public Robot buildRobot(GameSimulator s, int index){
    if(index == 1)
      return new Goalier(s);
    // By default, return a new attacker
```

```
    return new Attacker(s);
  }

  class Attacker extends RobotBasic{
    Attacker(GameSimulator s){
      super(s); }

    // Setup is executed once at the beginning.
    public void setup(){
      System.out.println("Attacker is built!");
    }

    // Loop is called continuously
    public void loop(){
      float angle = getSensor("BALL").readValue(0);

      setRotation(angle * 7f);
      setSpeed(0.5f, 0);
      delay(100);
    }
  }

  class Goalier extends RobotBasic{
    Goalier(GameSimulator s){
      super(s); }

    public void setup(){
      float usDistance = getSensor("ULTRASONIC_FRONT").readValue(0);

      System.out.print("Goalier is built!");
      System.out.println("My US_FRONT dist is:"+usDistance);
    }

    public void loop(){
      System.out.println("Running!");
      float angle = getSensor("BALL").readValue(0);

      if(Math.abs(angle) < 90)
        setSpeed(0f, angle / 150f);
      else
        setSpeed(0f, 0f);
      delay(100);
    }
  }
}
```

(Example code of a team. The class *CustomTeamA* contains two types of robots, an Attacker and a Goalie. Both Attacker and Goalie have a sensor to detect the ball, but the Goalie also has ultrasonic distance sensors. The Attacker and the Goalie have different programs in their respective loop functions.)

3.2 Robot Class

Because Arduino is very popular in robotics competitions, we believe that implementing a similar structure will make it easier for students to learn how to program in RoSoS. Therefore, the following three basic methods are used on the programming of the virtual robots:

setup()
Code inside setup() block runs only once when the robot is started;
loop()
Code inside loop() runs continuously until the robot stops;
run()
Overrides the Arduino-like behavior of setup-loop. This is the method called from the Robot Thread.

The *Robot* class provides methods to control the virtual robots, such as:

setSpeed(xSpeed, ySpeed)
Sets the target linear speeds $\dot{x}$ and $\dot{y}$ in X and Y directions of the robot's reference frame, respectively (see Fig. 2);
setRotation(angularSpeed)
Sets the target angular speed ω around the robot center;
stopMotors()
Stops the robot.

It should be noticed that the X and Y directions in the method *setSpeed* are related to the reference frame that is fixed on the robot, with X axis pointing forwards from the robot center (see Fig. 2). RoSoS does not deal with individual motor speeds. Instead, it considers only linear and angular speeds, specified with respect to the robot's reference frame. Therefore, to make the robot move forward the user only needs to provide a positive xSpeed value when calling setSpeed.

The kinematic model implemented by RoSoS for the robots is the omnidirectional model. This model was chosen based on the wide use of omnidirectional platforms by students on the RoboCupJunior Soccer competition. Such platform allows the robot to move in any direction regardless its orientation.

With the omnidirectional model, the user does not need to change the robots' orientation to implement linear move in Y direction. In other words, the virtual robot is able to move sideways with no need to rotate around its own axis to change its own orientation. On the other hand, it is also possible for the user to simulate a differential steered robot (unicycle model). For that, it is just a mater of keeping the ySpeed value as zero. By doing so, the robot will behave almost like a differential-drive, but it will move sideways if it is pushed by other robots.

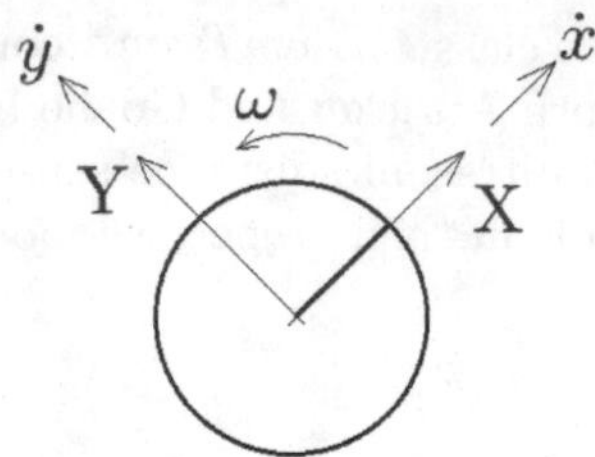

Fig. 2. Position and orientation of the reference frame fixed on the robot.

In the real world, robots do not react instantaneously to a given command. To simulate this behavior, RoSoS imposes limits on speed and acceleration for the robots. As a result, the target speed values are not reached instantaneously and the virtual robots' behavior is similar to the real ones.

Virtual sensors like Ball sensor, Compass sensors and Distance sensors are provided by the *RobotBasic* class and can be used by the virtual robots. The method *getSensor(String id)* returns the value provided the specified sensor.

The following example shows the implementation of a simple robot that rotates around its own axis and moves towards the ball if it is far away. Notice that the only sensor used by this robot is the ball sensor. The code inside the *loop()* function first reads the ball sensor to get the orientation of the ball with respect to the robot, which is stored in variable *angle*. Then, the robot is commanded to turn to orient itself in the direction of the ball. After a delay of 1000 ms, the rotation speed is set to zero and the robot is commanded to go forward if the distance to the ball is bigger than 0.5 units.

```
class RobotFollower extends RobotBasic {
  RobotFollower(GameSimulator s) { super(s); }

  Sensor locator;
  public void setup() {
    // Save sensor for further use
    locator = getSensor("BALL");
  }

  public void loop() {
    // Read the sensor angle
    float angle = locator.readValue(0);

    // Rotate ANGLE for a second and aim at the ball
    setRotation(angle);
    delay(1000);
    setRotation(0);
```

```
    // Read ball distance
    float dist = locator.readValue(1);

    // Move forward if ball is far away (> 0.5m)
    if (dist > 0.5f)
      setSpeed(0.2f);

    // Wait and stop motors
    delay(200);
    stopMotors();
  }
}
```

(Code that simulates a simple robot that rotates around its own axis to follow the ball, and moves towards the ball if it is far away.)

3.3 Technical Details

For students to implement their teams, it is not necessary that they know the technical details presented in this section. But, the interested student can dive into the simulator code if he/she wants to understand how it works.

A program for a typical real robot used on junior soccer competitions might consist of a single thread running on a microcontroller. Such a program might implement a state-machine, infinite loops, delays, as well as access to hardware devices. We have implemented RoSoS in such a way that the student is able to program its virtual robots as if he/she was programming a real one. For that, we had to guarantee that the simulator code runs independently of the robots' code.

To avoid hanging problems, every Robot extends a *Runnable* class, meaning that the program written to the Robot is going to run inside a thread, parallel to the Simulator code, but controlled by the Simulator. Running robots in threads is essential for the simulator to maintain control of the program flow. It also means that users can write infinite loops and long delays with no effect on the rest of the simulation.

GameSimulator is a global object responsible for creating physical elements such as walls, goals, ball and all field-related parts. It is also responsible for handling simulation time. By physical elements we mean anything that physically interacts with the environment.

Simulatable contains basic methods for checking collisions and for running simulation with discrete time steps. This class is extended by all "simulatable" objects such as *Ball*, *Block*, *GoalWall* and every *Robot*. Most of the "simulatable" objects can also be *"Drawable"*, which means that they can be drawn on the screen. *Drawable* is a simple interface with the method *draw()*. Notice that some "simulatable" objects don't need to be shown on the screen, like as the *GoalWall* that is used only to check collision with the ball. *Simulatable* objects can be modified to collide only with specific types of objects, or to be physically static (such as walls). It is also possible to control the type of collision for each object, from non-elastic to fully elastic collisions. The physics engine simulates the interaction of objects in a two-dimensional space according to the values of force, acceleration, speed and position of objects in every simulation step.

The following code shows the part of the program where the physics simulation takes place inside the *Simulatable* object.

Physics simulation.

```
public void simulate(float dt){
  // Saves last Speed and Position to calculate real Accel and Speed
  PVector lastSpeed = speed.get();
  PVector lastPosition = position.get();

  // Simulate Acceleration
  accel = getForce(dt);
  accel.div(getMass());
  force.set(0,0);

  // Simulate Speed
  PVector dSpeed = accel.get();
  dSpeed.mult(dt);
  speed.add(dSpeed);

  // Simulate Position
  PVector dPos = speed.get();
  dPos.mult(dt);
  position.add(dPos);

  // Calculate real Accel and Speed
  realAccel = speed.get();
  realAccel.sub(lastSpeed);
  realAccel.div(dt);

  realSpeed = position.get();
  realSpeed.sub(lastPosition);
  realSpeed.div(dt);
}
```

(Code inside *Simulatable* showing how physics rules are applied.)

During the simulation, the following set of instructions is executed for every time step:

1. Calculate the time dt since the previous iteration;
2. Obtain the new position of the simulatable object;
3. Resolve collision by separating objects placed over each other.

The part of the code responsible for handling the simulation flow is presented as follows.

Simulation flow.

```
public void simulate(float t){
  float dt = t - lastT;
  lastT = t;
  if(dt > 0.5 || dt <= 0f){ return; }

  // Simulate Physics
  for(Simulatable s:simulatables){
    // Simulate object
    s.simulate(dt);

    // Resolve Collisions
    for(Simulatable b:simulatables){
      // Check if is NOT colliding
      if(b == s || !s.canCollide(b))
        continue;

      // Resolve collision if so
      if(s.colliding(b))
        s.resolveCollision(b);
    }
  }
}
```

(Part of the *GameSimulator* where simulation flow is executed.)

An important part of the simulation concerns the Robot. For a matter of simplicity, *Robot* inherits from *Simulatable*. It has properties like force, acceleration, speed, position and orientation. It also extends *Drawable*, which allows users to even change their robot's appearance.

3.4 Sensors

Each virtual robot can use several different sensors by using the method *getSensor(String ID)*. Sensors are built as a basic unit that can do measurements and return vectors of values. A class called *RobotSensor* was implemented from *GameSimulator* and *Robot*. The public method *readValue(index)* returns the index'th value from the vector of values corresponding to the instantaneous sensor reading.

Sensors most required by real soccer robots are:

- Infrared sensors: placed around the robot to detect the direction and even the distance to the ball (considering that the ball emits infrared light, which is the case in junior competitions like RoboCupJunior and Latin American Robotics Competition);

- Compass sensors: detect absolute robot orientation and are used, for example, to identify the orientation of the target goal;
- Distance sensors: Can be used for positioning the robot inside the field and to avoid going outside the field limits by measuring distances to the field walls.

We present the following code to illustrate the implementation of a sensor. It is part of *BallLocator* class and is responsible for implementing a sensor that detects the distance and the relative orientation from the robot to the ball. Notice that some limits to the readings are implemented.

Implementation of ball sensor.

```
class SensorBall extends Sensor{

  float[] values = new float[2];
  float sensorLimit = 1f;

  SensorBall(GameSimulator g, Robot r){ super(g, r); }

  float lastRead = 0;
  public float[] readValues(){
    // Avoid multiple readings within 100ms
    if(game.getTime() >= lastRead + 0.1f)
      doReading();

    return values;
  }

  private void doReading(){
    Robot thisRobot = getRobot();
    Ball ball = getGameSimulator().ball;

    // Set values to 0's if ball is off
    if(!ball.isOn()){
      values[0] = 0;
      values[1] = 0;
      return;
    }

    // Find relative distance from Ball to Robot
    PVector dist = PVector.sub(ball.position, thisRobot.position);
    dist.rotate(-thisRobot.getOrientation());

    // index 0 contains the Angle of the ball
    values[0] = (float)Math.toDegrees(dist.heading());
    // index 1 contains the distance to the ball
    values[1] = (float)Math.min(dist.mag(), sensorLimit);
```

```
  }
}
```

(Code responsible for implementing a sensor that detects the distance and orientation from the robot to the ball.)

To sum up, to participate on a competition using RoSoS every group of students shall program a class that implements a *Team*. This class is responsible for distributing robots given an index. Those robots can be implemented from *BasicRobot* or even *Robot* classes. Each *Team* object will be notified of actions taken by the simulator, like change in side, robot removal etc. The *Team* class must include all classes required by itself to run, and that set will compose a single file with the name of the Team.

4 Discussion

RoSoS was successfully tested using Processing version 2.2.1 on Windows 7, Windows 8.1, Windows 10, Ubuntu 14.04 (64 bits), Linux Mint 17.1 and Mac OS X computers. To test the performance of the software, and with the support of the Federal University of ABC (UFABC) branch of IEEE RAS, we have organized a small competition with some students at the UFABC campus, Brazil, in July 2015. For this competition, university students between 18 and 20 years old were divided into 8 groups, with 4 students each, to program teams of 3 robots to play games of 5 min (2.5 min before switching sides). Besides the short notice (only two weeks between the announcement and the actual competition), students were able to come up with very interesting ideas on their code, like proportional control, communication between robots, central control class for a team etc. The simulations were ran in a MAC OS X computer. In January 2016, we have organized a workshop at Campus Party Brasil 2016 [18] where people learned about RoSoS and were able to do some basic programming. Finally, in June 2016 RoSoS was used in the Robotics course of the Control and Automation Engineering degree of the Federal Institute of Espirito Santo (IFES) - Serra Campus, Brazil. Students had to program their robots to play soccer, but using some of the mobile robot controllers studied during the course. At the end of the semester, a small competition was held. The above mentioned activities were a success and proved that RoSoS is a valuable tool for educational robotics.

A junior simulation league brings some interesting possibilities to enrich robotics competitions. For example, game rules could be changed during competition days, and teams could have a limited number of hours to adapt their programs. Such changes might even include adding more robots to the team. Superteam competitions could also be held with a big virtual field and several robots from different teams to form one virtual superteam. Finally, students that participate on real Robot Soccer competitions (like RoboCupJunior Soccer) can use RoSoS as a test platform to improve its game strategy.

It is worthy to make a brief comparison between RoSoS and the RoboCup Soccer Simulator 2D (RoboCup Sim) to state the main similarities and differences between the two simulators. In general, RoSoS and RoboCup Sim are very similar. Both simulate a soccer match in which two teams of autonomous virtual robots play soccer in a two-dimensional virtual soccer field. In both simulators, the server knows the status of everything and controls the game. Each virtual robot has its own program and uses information from its own virtual sensors to decide how to move. But, there are some differences on the game rules and on the way the simulator was built that make RoSoS a more appropriate choice for younger students. First, RoSoS was built to mimic the real RoboCupJunior Soccer competition rules, so that students participating on the real competition can use it as a test platform for their team strategy. Second, RoSoS is much simpler than RoboCup Sim regarding installation, use and programming. While in RoboCup Sim one has to program and compile its own team to communicate with the server using UDP/IP socket, in RoSoS students create their teams on a single file and use simple pre-built functions to read sensor values and send speed commands. This makes a RoSoS program more similar to a real robot program (when considering the robots used on the RCJ Soccer competition). RoboCup Sim programs are more complicated not only because of communication between the agents and the simulator, but also because of game rules. In RoboCup Sim robots have to deal with lots of situations that are not present in RoSoS, like corner kicks, lateral backs, the amount of energy a player has (stamina), when to hold the ball and how to interpret coach instructions. In RoSoS there are no coaches and game rules are simpler, which makes it ideal for students that are beginners on programming.

For now, RoSoS implements an omnidirectional robot model, which is used by many RCJ Soccer teams. But, as we mentioned, the simulator program can be adapted to simulate a differential drive robot, if desired. Also, if a more realistic simulation is desired, the simulator can be adapted to include more complex robot dynamics and to use more complex physics modelling. The software is open-source and is available on Github, so motivated students can collaborate with the development of the simulator itself [17]. The Github page has links for the documentation page [7] and for the RoSoS YouTube channel [8], that contains videos with explanation and short code examples.

5 Conclusion and Future Work

We have presented RoSoS, a Robot Soccer Simulator designed for education robotics and robotics competitions. We also presented a proposal for a new league for the robotics competitions, named Junior Soccer Simulation (JSS). This proposal was also presented to the organization committee of the Brazilian Robotics Olympiad (OBR), which agreed to try implementing a JSS league using RoSoS during the next edition of OBR. First, we are going to start with a demo competition, with some selected teams. Then, we expect to have a national competition using RoSoS the following year. We have produced some documentation for using RoSoS, including a series of YouTube videos. Besides that, we also intend to prepare training material for teachers, so that they can motivate their students to use RoSoS and participate on the competitions. RoSoS can also be used in undergraduate courses. For instance, in the first semester of 2016 RoSoS was be used for the first time on the Robotics course of the Control and Automation Engineering career at IFES, where students had to implement several concepts learnt during classes and a small competition was organized at the end of the semester. This increased students' motivation to work harder while keeping a nice environment during classes.

It is important to point out that a text-based programming language might be difficult for primary students, especially the youngest ones. Because of that, the actual implementation of RoSoS is more suitable for secondary students. But, it is possible to make it easier to program by creating macros for high-level commands. We also intend to implement a graphic-based interface for creating the virtual robots' programs. By doing so, we believe that primary students over 10+ years-old would be able to program their robots in RoSoS.

An important aspect of a simulation league is that it requires much less financial investment from schools and students. Therefore, it facilitates the inclusion of robotics in their curriculum, especially on developing countries. In robotics competitions students develop teamwork and gain experience, while exchanging knowledge and information with other students from different backgrounds and culture. We believe that JSS league and RoSoS will contribute to education equality by giving all primary and secondary students the possibility to have the above mentioned experiences.

Acknowledgments. We would like to express our gratitude to everybody that is giving support to this project. Special thanks to: OBR, Campus Party Brazil, IEEE RAS (branch UFABC), UFABC, IFES - Serra Campus, Renato Ferreira, João Pedro Vilas, Otacílio Neto, Shander Lyrio, and Marek Šuppa.

References

1. About RoboCup. http://www.robocup.org/about-robocup/objective/. Accessed 6 Jan 2016
2. Arduino software. https://www.arduino.cc/en/Main/Software. Accessed 6 Jan 2016
3. Football - soccer. http://www.britannica.com/sports/football-soccer. Accessed 6 Jan 2016
4. Olimpíada brasileira de robótica. http://www.obr.org.br/. Accessed 6 Jan 2016
5. Processing. https://processing.org/. Accessed 6 Jan 2016
6. Processing foundation. https://processingfoundation.org/. Accessed 6 Jan 2016
7. Robot Soccer Simulator - RoSoS - Documentation. http://ivanseidel.github.io/Robot-Soccer-Simulator/. Accessed 30 Jan 2016
8. Robot Soccer Simulator - RoSoS - Videos. https://www.youtube.com/channel/UCZekRTPIwhe56lbicQpO-vg. Accessed 30 Jan 2016
9. Soccer simulation league. http://wiki.robocup.org/wiki/Soccer_Simulation_League. Accessed 6 Jan 2016
10. Ahlgren, D.J.: Meeting educational objectives and outcomes through robotics education. In: Automation Congress, 2002 Proceedings of the 5th Biannual World, vol. 14, pp. 395–404. IEEE (2002)

11. Alves, S.F.R., Ferasoli Filho, H., Pegoraro, R., Caldeira, M.A.C., Rosário, J.M., Yonezawa, W.M.: Educational environment for robotic applications in engineering. In: Obdržálek, D., Gottscheber, A. (eds.) EUROBOT 2011. CCIS, vol. 161, pp. 17–28. Springer, Heidelberg (2011)
12. Aroca, R.V., Aguiar, F.G., Aihara, C., Tonidandel, F., Montanari, R., Fraccaroli, E., Silva, M., Romero, R.A.F.: Olimpíada Brasileira de Robótica: relatos da primeira regional em São Carlos-SP. In: V Workshop de Robótica Educacional, p. 35 (2014)
13. Aroca, R.V., Gomes, R.B., Tavares, D.M., Souza, A., Burlamaqui, A.M., Caurin, G., Goncalves, L.M., et al.: Increasing students' interest with low-cost cellbots. IEEE Trans. Educ. **56**(1), 3–8 (2013)
14. Fabro, J., Reis, L.P., Lau, N., et al.: Using reinforcement learning techniques to select the best action in setplays with multiple possibilities in robocup soccer simulation teams. In: 2014 Joint Conference on Robotics: SBR-LARS Robotics Symposium and Robocontrol (SBR LARS Robocontrol), pp. 85–90. IEEE (2014)
15. Faria, B.M., Reis, L.P., Lau, N., Castillo, G.: Machine learning algorithms applied to the classification of robotic soccer formations and opponent teams. In: 2010 IEEE Conference on Cybernetics and Intelligent Systems (CIS), pp. 344–349. IEEE (2010)
16. Gabel, T., Riedmiller, M., Trost, F.: A case study on improving defense behavior in soccer simulation 2D: the NeuroHassle approach. In: Iocchi, L., Matsubara, H., Weitzenfeld, A., Zhou, C. (eds.) RoboCup 2008. LNCS, vol. 5399, pp. 61–72. Springer, Heidelberg (2009)
17. Gomes, I.S., Martins, F.N., Ferreira, R., Vilas, J.P.: Robot Soccer Simulator - RoSoS. https://github.com/ivanseidel/Robot-Soccer-Simulator. Accessed 6 Jan 2016
18. Gomes, I.S., Silva, J.P.: Workshop: Programe Seu Próprio Time de Futebol de Robôs Simulado. http://campuse.ro/events/campus-party-brasil-2016/workshop/programe-seu-proprio-time-de-futebol-de-robos-simulado-cpbr9/. Accessed 28 Jan 2016
19. Martins, F.N., Oliveira, H.C., Oliveira, G.F.: Robótica como meio de promoção da interdisciplinaridade no ensino profissionalizante. In: Anais do Workshop de Robótica Educacional (2012)
20. Martins, F.N., Oliveira, H.C.G., Amaral, E.: NERA - a center for research on educational robotics and automation. In: WEROB 2012 - Workshop on Educational Robotics, 16th RoboCup International Symposium, Mexico City (2012)
21. Martins, F.N., Gomes, I.S.: Soccer simulation league - a proposal for the RoboCupJunior competition. In: WEROB - Workshop on Educational Robotics at the RoboCup Symposium (2015)
22. Martins, F.N., Gomes, I.S., Santos, C.R.F.: Junior soccer simulation - providing all primary and secondary students access to educational robotics. In: XII LARS Latin American Robotics Symposium (2015)
23. Sałustowicz, R.P., Wiering, M.A., Schmidhuber, J.: Learning team strategies: soccer case studies. Mach. Learn. **33**(2–3), 263–282 (1998)
24. Wing, J.M.: Computational thinking. Commun. ACM **49**(3), 33–35 (2006)

Path Planning with Collision Avoidance for Free-Floating Manipulators: A RRT-Based Approach

João R.S. Benevides and Valdir Grassi Jr.(✉)

Department of Electrical and Computer Engineering,
São Carlos School of Engineering (EESC), University of São Paulo (USP),
São Carlos, Brazil
{jrsbenevides,vgrassi}@usp.br

Abstract. The difficulty of creating a path planner with collision avoidance for Space Manipulators (SMs) is well known due to the presence of dynamic singularities and because of its non-holonomic behaviour. Furthermore, the main contributions in the field of motion planning of SMs are often concentrated in the point-to-point strategy, with special interest in the complex dynamics of such systems. In fact, planners for space manipulators generally count on a previously computed path in order to modify it to avoid collisions. Nonetheless, the computing of the previous path still lacks robust formulations, specially in the case of free-floating manipulators. Our goal consists in creating a path planner with collision avoidance for a free-floating planar manipulator. The dynamic model is based on the Dynamically Equivalent Manipulator and the concept of Rapidly-Exploring Random Trees serves as a framework for the developed algorithm. A combination of a method that reduces the metric sensitivity with a bidirectional approach is proposed in order to achieve a solution convergence. Details of the collision checking algorithm are provided. The system is validated by simulating the path planning task for a three-link planar free-floating manipulator, while considering the presence of an obstacle. The results are then discussed and promising directions for future works are presented.

1 Introduction

Space missions are often related to hostile environments, which are also connected to extreme temperatures, radiation and lack of gravity. These factors endanger and complicate human mobility in Extra-Vehicular Activities (EVAs). In order to assist in assembly services, space manipulators (SMs) are playing a key role in this matter. Activities like substitution of components, satellite repair and refueling are fundamental in the sense of making more flexible on-orbit operations and increasing the overall mission lifespan [1].

Path planning is known to be a major challenge in the field of general robotics. In the case of space robots, this difficulty is magnified by the dynamic coupling and the non-holonomic behavior of such systems, due to the nonintegrability of

F. Santos Osório and R. Sales Gonçalves (Eds.): LARS 2015/SBR 2015, CCIS 619, pp. 103–119, 2016.
DOI: 10.1007/978-3-319-47247-8_7

the angular momentum [2]. Another challenging task is the handling of dynamic singularities (DS), which had their existence proven in [3]. These differ from singularities of fixed-based manipulator because their location cannot be simply predicted from the kinematic manipulator structure. In fact, dynamic singularities are product of the dynamic properties of space robots and depend on the path taken.

Space manipulators are normally classified into two major categories. First, free-flying manipulators count on an active position and attitude control. This compensates the displacement generated by the joint motions in order to maintain a stable basis. Therefore, most of the control laws for fixed-base manipulators also apply. However, excessive fuel consumption compromises the duration of on-orbit missions. On the second group are the free-floating manipulators, which allow the satellite to freely move in response to the arm's motion. In that case, no reaction wheels or propulsion jets are used. Thus, the system can save fuel and energy. Nonetheless, this advantage comes with an extra challenge regarding the description of its dynamics and behavior.

In spite of so many difficulties, researchers gave valuable contributions in the matter of motion planning of space robots, specially in the sense of avoiding DSs and dealing with their special nature. The work presented in [4] adopts unit quartenions in order to represent dynamic singularities and avoid them. Using this representation, inverse kinematics algorithms are formulated based on geometric variables. An analytical path planning method for free-floating manipulators is presented in [5]. The cartesian control of the end-effector is achieved along with the system's attitude control. Nevertheless, trajectory points are supposed known and collision avoidance is not considered in this planning. [6] proposes a path planning technique that yields the appropriate initial system configurations to avoid DSs. However, the approach is based on a reference path prior to the application of the proposed technique, specifically, a straight line is considered in this evaluation. As one can notice, path-planning of free-floating robots in the presence of obstacles reveals a vast scenario to be exploited.

With the goal of evaluating space systems on earthly environments, a platform for assessing different control approaches for a free-floating planar manipulators was built in [7]. The UnderActuated Robot Arm-E (UARM-E) consists of a mechanical-electronical system that floats over an Ealing-like table. This manipulator is remotely connected to a simulation and control environment, also developed in this work. Moreover, this platform can be configured with up to six active joints with one or two arms. A typical configuration of the UARM-E is shown in Fig. 1.

The path planning of free-floating manipulators does not count on solid formulations when random obstacles are considered. This encouraged us to pursue a methodology of a planner that autonomously computes a path between two configurations of a free-floating manipulator. The Rapidly-Exploring Random Tree (RRT) algorithm, widely known as a powerful path-planning tool, acts as a framework for the proposed architecture. Two other major RRT-based approaches are considered in order to solve convergency problems and

Fig. 1. UARM-E configured with a single arm and two active links.

core details about the implementation are given. These concepts were blended and modified to finally form the structure of the final planner.

This paper is organized as follows. Section 2 covers the key concepts about the dynamically equivalent manipulator. Interesting properties, which are later exploited, are derived here. Section 3 introduces the basic RRT algorithm along with two modifications. This section also proposes and discusses some of the adaptation to the problem. Section 4 describes how all the elements are integrated, provides details of the collision detection method we used and presents the architecture of the proposed planner. Section 5 shows the results of planning tasks in a simulated environment. Finally, Sect. 6 gives the proposed method a general overview and discusses promising directions for future works.

2 Dynamically Equivalent Manipulator

The concept of DEM was introduced in [8] as an alternative to the complex kinematic-dynamic approaches when modelling space robots. This method maps a free-floating manipulator into a conventional fixed-base manipulator, preserving both its kinematic and dynamic properties. This equivalence not only allows the modelling of free-floating arms through traditional methods, but also enables the experimental study of space platforms in more feasible environments, without the need for complex structures that emulate space conditions.

The DEM is originally based on the concept of Virtual Manipulator (VM), presented in [9], in which a kinematically equivalent manipulator is proposed. However, the VM equivalent model is considered to be an ideal kinematic chain with null mass. Therefore, practical experiments are unfeasible for this approach. The DEM exploits that fact and proposes an equivalent model that can be physically built and adopted in experimental studies regarding the dynamic behavior of space manipulators. Besides, it may be used as a tool for developing the dynamic model itself. The DEM is applicable for both free-floating and free-flying manipulators.

Consider a n-links rigid manipulator mounted over a free-floating base. Let C_i be the center of mass of link i and the base from the space manipulator named as link 1. Furthermore, the following links are named from 2 until $n+1$. Assuming that forces and external torques are non-existent, the center of mass

C_o remains fixed in inertial space and is also chosen as main frame's origin (depicted as frame 0, in Fig. 2).

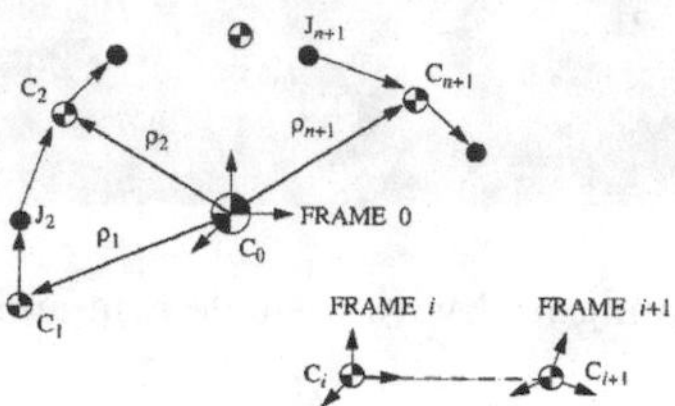

Fig. 2. Frame fixed to SM links.

Briefly, the model derivation uses Lagrange equations in the space manipulator to obtain the dynamic model of a fixed-based manipulator. In this case, the fixed base is replaced by a passive spherical joint. Considering that the DEM works in the absence of gravity and that its base is located at the center of mass of the space robot, the conditions under which both models are equivalent satisfy the following algebraic equations:

$$\begin{aligned} m_i^{'} &= \frac{M_t^2 m_i}{\sum_{k=1}^{i-1} m_k \sum_{k=1}^{i} m_k}, && i = 2, \ldots, n+1 \\ I_i^{'} &= I_i, && i = 1, \ldots, n+1 \\ W_1 &= r_1, \\ W_i &= r_i + l_i, && i = 2, \ldots, n+1 \\ l_{c1} &= 0, \\ l_{ci} &= \frac{\sum_{k=1}^{i-1} m_k}{M_t} L_i, && i = 2, \ldots, n+1 \end{aligned} \tag{1}$$

In (1), W vectors represent the DEM link lengths and their inertial orientations with respect to the space manipulator (SM) inertial frame; $m_i^{'}$ corresponds to the mass of DEM's i-th link; $I_i^{'} \in \mathbb{R}^{3\times3}$ denotes the inertial tensor corresponding to DEM's i-th link; l_{ci} represents the vector from DEM's i-th joint to the center of mass of i-th link. Additionally, [8] demonstrates that the value of $m_1^{'}$ does not influence in dynamic equivalency. Thus the mass of the DEM's first link might be arbitrarily assigned as a positive non-null value.

Let the generalized coordinates vector be $q = \left[\phi \; \theta \; \psi \; \theta_2 \cdots \theta_{n+1}\right]^T \in \mathbb{R}^{(n+3)}$ decomposed as $q = \left[q_b^T \; q_m^T\right]^T$, where b and m represent the base and manipulator components, respectively. Similarly to classic Euler-Lagrange equations, the system model assume the special form:

$$M(q_m)\ddot{q} + C(q_m, \dot{q})\dot{q} = \tau \tag{2}$$

As the gravity effects are neglected in a spatial environment, so is the gravity vector. In (2), $M(q_m) \in \mathbb{R}^{(n+3)\times(n+3)}$ denotes the symmetric and positive definite inertia matrix, which is dependent exclusively of manipulator coordinates; $C(q_m, \dot{q}) \in \mathbb{R}^{(n+3)\times(n+3)}$ represents the matrix of Coriolis and centrifugal forces. Finally, $\tau = \begin{bmatrix} 0 \; 0 \; 0 \; \tau_2 \cdots \tau_{n+1} \end{bmatrix}^T \in \mathbb{R}^{(n+3)}$ denotes the vector of applied torques over DEM joints.

Because the DEM coordinate frames are parallel to the SM corresponding frames, and its base is located at the center of mass of the SM, the DEM is identical in geometry to the VM. Therefore, it inherits the fundamental properties of the VM, which are:

- The DEM end-effector coincides with the SM's end-effector.
- The axis of DEM's i-th joint is parallel to the axis of the i-th SM joint.
- During motion, the displacement of each of the DEM's joints is identical to the displacement of the corresponding SM joint.

These properties have their importance later demonstrated in the task of path-planning of free-floating manipulators.

3 Rapidly-Exploring Random Trees

Aiming to contribute in the field of sampling-based algorithms, the basic RRT algorithm, introduced in [10], stands out for its natural support to non-holonomic systems and several DOFs. The algorithm provides simple, but powerful search concepts that are explained as follows:

Let T be a tree rooted at its initial state, x_{init}. In each iteration, a random state x_{rand} is uniformly sampled in the free workspace, X_{free}. The algorithm then applies a search method in order to find the state in T that is closest to x_{rand}, based in a certain metric ρ. This state is now called x_{near}. Then, a valid random input vector u is sampled among all set of possible inputs U. Each of the inputs in u is applied in x_{near} and integrated over a certain time interval. After evaluating all the expansions from x_{near} that are collision-free, the one with lowest cost to x_{rand} is chosen as x_{new}. This new state is then added to the nodes of tree T. Likewise, the edge connecting the node x_{near} to x_{new} is added to the edges of T. This process is repeated until the state x_{new} is close enough to x_{goal}, meaning that the algorithm successfully found a solution for the path planning problem. In that case, the sequence of branches that reaches x_{goal} with minimum total cost according to the metric ρ is returned by the algorithm together with the sequence of inputs used to create those branch sequence. However, a solution may not exist or it is extremely hard to find. In order to avoid endless searches for a solution, a limit on the size of T is imposed to restrain the search time. So the algorithm stops when T achieves a user-defined maximum number of branches N. If that limit is achieved, the algorithm does not return a valid solution for the path planning problem. The pseudo-algorithm is shown as Algorithm 1.

Algorithm 1. Basic RRT Planner

BuildRRT(N, ΔT,x_{init})

Add_State_To_Tree(x_{init},$Tree$);
for $n = 1$ to N **do**
    $x_{rand} \leftarrow$ Sample_State();
    $x_{near} \leftarrow$ Find_Closest_Neighbor($Tree$,x_{rand});
    $u \leftarrow$ Sample_Input(x_{near},x_{rand},U) ;
    $x_{new} \leftarrow$ New_State(x_{near},ΔT,u) ;
    Add_State_To_Tree(x_{new},$Tree$);
    Add_Edge_To_Tree(x_{near},x_{new},$Tree$);
end for
return $Tree$

In Algorithm 1, the function *New_State* is responsible for finding the next state of the system x_{new} by integrating the dynamic model of the robot, represented by a state-pace transition equation $\dot{x} = f(x, u)$, for fixed time ΔT applying input u from state x_{near}. Techniques with higher degree of integration, such as Runge-Kutta or the Euler method, are preferred for solving that integration [10].

The sequence of inputs applied to the dynamic model used for generating the path from x_{ini} to x_{goal} is also returned. In an ideal case, if the dynamic model used for path planning would be a perfect representation of the real system, the application of these sequence of inputs would make the robot to describe the desired path. However, due to modelling uncertainties and unpredictable disturbance, a path-following feedback controller must be used to follow the planned path. Nonetheless, the planned path is feasible because the dynamic model of the system was used for path planning.

Some of the main aspects and advantages of using the RRT algorithm, according to [10,11], are: (1) Strong bias to not yet explored regions; (2) Probabilistic completeness, this means that the probability of finding a solution tends to 1 as time tends to infinity; (3) States always remain connected; (4) Independence of explicit description of X_{free}; (5) Few heuristics or random parameters are required.

The RRT planner, seems to present better convergence ratio when a subtle bias p is applied to the final configuration. It means that the algorithm has probability p to choose final configuration over a random state. However, if p is set too large, it is likely that the search gets trapped in local minima [12].

3.1 Reduction of Metric Sensitivity

As a criterion, the chosen metric ρ is based on a cost function that should translate the cost of bringing one state to another. The ideal metric is represented by the optimal displacement cost between two states. It is known that performance degrades substantially when the chosen metric does not reflect the real cost of motion between two configurations. In fact, computing the ideal metric was proven to be as hard as solving the trajectory problem itself [11].

In order to remedy this situation, a modification in Algorithm 1 is proposed in [13]. The main goal of this method is to refine the exploration strategy, even in the absence of a proper metric. The contribution of this approach does not lie in developing a specific metric, but collecting information along the exploration and growth of the tree.

Basically, a record of all set of inputs is kept for each node. This allows the discard of inputs that were already evaluated. Also, the constraint violation frequency (CVF) is collected for each state. The CVF estimates the probability of a node expansion to result in a collision. Initially, every node has CVF equals to zero. Once a state is selected as x_{near}, its CVF is increased a quantity c/N, where c stands for the number of inputs that result in collision or movement restriction and N the number of inputs. Furthermore, when a $CVF > 0$ is computed, all father-nodes that lead to that state have their CVF incremented accordingly. That is, its m-th father will be added a CVF of c/N^{m+1}. The CVF is then used as probability of not choosing some state when selecting the closest neighbor. By punishing regions that lead to nodes that are likely to collide, the exploration information helps the planner to pick a better x_{near}.

In order to implement this method and consequently assign a CVF to a node, it is necessary to keep a record of all the possible inputs that can be applied to that node for its expansion. This ca only be done if instead of a continuous input we restrict the choices of inputs to a discrete.

Following we present a method of organizing a set of discrete inputs for the problem of a generic free-floating manipulator, when torques are the only input to be considered. Let $x \in X$ a node, whose father is x_{father}. Consider the set of inputs necessary to take x_{father} to x, over a certain time interval Δt, represented by a vector $\tau = \left[\tau_1\ \tau_2\ \cdots\ \tau_n\right]^T$, where n denotes the number of links of the space manipulator. Let us impose $+\Delta_i$ and $-\Delta_i$ as a threshold in the increase and decrease of τ_i, respectively. Also, as the torque τ_1 is null due to DEM equivalence conditions, the resulting set of inputs is organized as shown in Table 1, where each row comprises the set of possible torques to be applied.

Table 1. Organization of the discrete inputs

0	$\cdots$	0	$\cdots$	0
$\tau_2 - \Delta_2$	$\cdots$	τ_2	$\cdots$	$\tau_2 + \Delta_2$
$\tau_3 - \Delta_3$	$\cdots$	τ_3	$\cdots$	$\tau_3 + \Delta_3$
$\vdots$	$\vdots$	$\vdots$	$\vdots$	$\vdots$
$\tau_{n-1} - \Delta_{n-1}$	$\cdots$	τ_{n-1}	$\cdots$	$\tau_{n-1} + \Delta_{n-1}$
$\tau_n - \Delta_n$	$\cdots$	τ_n	$\cdots$	$\tau_n + \Delta_n$

k columns

As k stands for the degree of discretization of each input in τ, the amount of possible combinations equals k^{n-1}. The algorithm will then need at least k^{n-1} *bits* to keep track of the expansion of each set of inputs. A naive approach would also consider to keep record of the set of inputs that lead to that expansion. However, for several degrees of freedom and many iterations this may represent an unnecessary allocation of memory, degrading the overall computational performance.

As a workaround, we created for each state a single vector V, with size k^{n-1}, built as follows: The indices $\left(i_2\ i_3\ \cdots\ i_n\right)$, relative to the columns in the torques matrix presented in Table 1, are stored. This set of indices (that correspond to inputs) is associated to element L from vector V as computed in (3).

$$L = 1 + \sum_{j=2}^{n} (i_j - 1)k^{n-j} \tag{3}$$

Another adaptation was made to the method for reduction of metric sensitivity. An heuristic was adopted in order to eliminate, from the closest neighbour search, nodes with high probability of producing child nodes with collision. Consider a set of all possible inputs U for a given state x, and a set of inputs that was already sampled and verified, $U_s \in U$. If U_s reaches a considerable size with respect to the size of the set U, and all input in U_s resulted in collision, then the node x is not selected for further expansion. This is done because is highly probable that other inputs in U that were still not tested would also result in collision, meaning that this node is not able to produce valid children.

The reduction of metric sensibility, combined with the modifications above described, collaborated to find a better expansion during the initial experiments. However, the task of reaching a goal position while avoiding obstacles could not find a single solution, even though 50000 iterations were run in each trial. The unidirectional RRT seems to have special sensibility to local minimas. Depending on a single tree to proper explore the environment and converge to a solution has been verified to be a hard task, specially in the presence of obstacles. This motivated us to also incorporate the bidirectional approach.

3.2 Bidirectional RRTs

Using a single RRT from x_{init} to connect with x_{goal} works well for state spaces of low dimension. The bidirectional RRT, proposed originally by [11], grows two RRTs independently. This approach improves the efficiency for state spaces of high dimension at the expense of having to connect a pair of nodes between two trees. The main aspects are described as follows:

Like the basic algorithm, the first tree has its root at the initial state x_{init}. A second tree is then grown from final state x_{goal}. At each iteration, a random state x_{rand} is sampled and an expansion attempt is made from the first tree. After that, the second tree takes the same x_{rand} as parameter and tries to expand in that direction. The process is repeated until two states x_1 and x_2 are

Algorithm 2. Dual RRT Planner

```
BuildBiRRT()
Add_State_To_Tree(x_init,Tree_1);
Add_State_To_Tree(x_goal,Tree_2);
while not Connected do
    x_rand ← Sample_State();
    x_i ← Create_State(Tree_1,x_rand,Forward);
    if x_i ∈ X_free and close enough from Tree_2 then
        connect Tree_1 to Tree_2 through x_i;
    end if
    x_g ← Create_State(Tree_2,x_rand,Backward);
    if x_g ∈ X_free and close enough from Tree_1 then
        connect Tree_2 to Tree_1 through x_g;
    end if
end while
```

close enough. This proximity is verified if $\rho(x_1, x_2) < \delta$, for a small $\delta > 0$. The pseudo-algorithm can be seen in Algorithm 2.

Notice that, because the second tree is created from final configuration towards the initial one, its integration must be computed backwards in time. Considering Δt the step size used to grow the main tree, this issue can be solved by using another $\Delta t^{'} = -\Delta t$, everytime a backward integration is necessary. The reason for this lies in the fundamental equality $\int_a^b f(x)dx = -\int_b^a f(x)dx$.

Another important aspect to highlight is that the bidirectional RRT also require the solution for the inverse kinematic of the manipulator at the desired final pose of the end-effector in order to integrate and compute new states for the second tree. However, it is difficult to find an analytic solution for the space manipulator. For this reason as both the free-floating manipulator and the DEM have the same end-effector location, the inverse kinematic was handled by iteratively computing the joint positions of the DEM manipulator from the end-effector. After finding the DEM joint positions, the real manipulator is constructed from the DEM through direct kinematics and the collision detection algorithm performs the validation of such configuration.

The knowledge of the joint positions provides a thorough description of every configuration. This is particularly interesting because it allows the program to consider more parameters, thus enabling a more accurate estimation of the cost-to-go.

4 Creating the Path Planner

In order to use consistent data for the simulated environment, the test platform UARM-E had its parameters, shown in Table 2, incorporated into the algorithm.

The DEM parameters are then computed with (1) and shown in Table 3:

In order to detect collision of an obstacle with the space manipulator at a given configuration, we decided to use the Separating Axis Theorem (SAT)

Table 2. SM parameters of UARM-E robot

	$m_i(kg)$	$I_i(kgm^2)$	$L_i(m)$	$R_i(m)$
Base	4.780	0.0404	0	0.150
Link 2	1.380	0.0182	0.144	0.111
Link 3	1.011	0.0115	0.103	0.112

Table 3. DEM parameters of UARM-E robot

	$m_i(kg)$	$I_i(kgm^2)$	$W_i(m)$	$l_{ci}(m)$
Base	4.780	0.0404	0.100	0
Link 2	2.410	0.0182	0.191	0.096
Link 3	1.177	0.0115	0.201	0.089

Algorithm [14]. The main goal was to perform a rapid collision check among the manipulator and the obstacles. Additionally, a self-collision check is performed using the same approach. As the SAT applicability is restricted to convex forms, we chose to treat each robot link as a convex polygon and to represent the space manipulators as an open kinematic chain. An illustration of the SM and its DEM is depicted in Fig. 3.

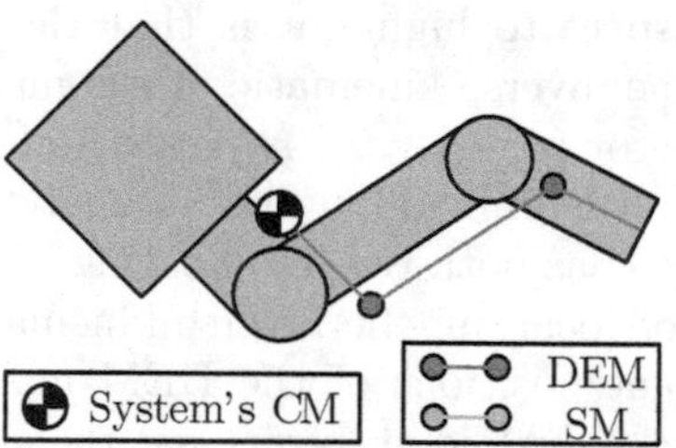

Fig. 3. SM constructed from the DEM.

The SAT is a special case of Minkowski's separating hyperplan theorem applied to solve a collision detection problem. Essentially, the theorem states that two convex objects do not collide if there is at least one line (called here as separating axis), upon which the objects projections do not overlap (see Fig. 4). Algorithm 3 presents the pseudocode of the Separating Axis Theorem regarding two bidimensional objects O_1 and O_2. On this algorithm, a sweep is done around the β angle of the separating axis over a step size ξ between each collision check.

Regarding the execution of SAT algorithm described in Algorithm 3, some particularities were observed and then modifications were introduced in the sense of reducing computational cost. First, the axis is no longer sampled. Instead, it is always perpendicular to a line containing an object's side. The reason for

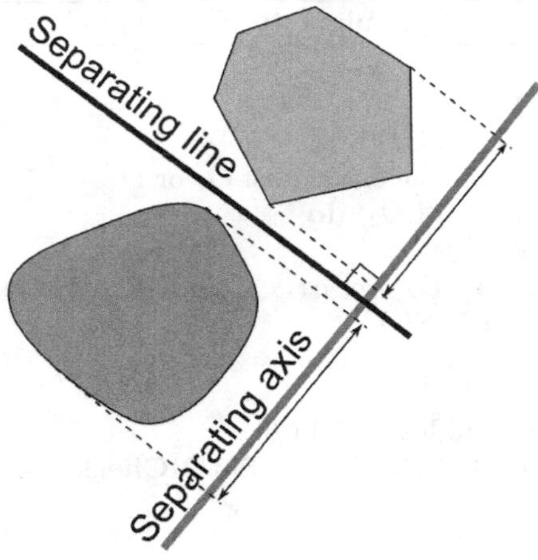

Fig. 4. Illustration of the Separating Axis Theorem.

Algorithm 3. SAT Naive Approach

SAT(O_1,O_2)
$state \leftarrow COLLISION$;
$\beta \leftarrow 0$;
while ($\beta < 360$) AND ($STATE = COLLISION$) **do**
  $axis \leftarrow$ line with β angle;
  if Projection of O_1 and O_2 over $axis$ = OVERLAP **then**
    $state \leftarrow NO\_COLLISION$;
    break;
  else
    $\beta \leftarrow \beta + \xi$;
  end if
end while

this modification is straightforward: assuming that two polygonal and convex objects do not collide, there is at least one line that passes between them without touching. Therefore, there is at least one line, parallel to the side of one of the objects, that also freely passes between them. Second, only non-parallel sides are considered when building the separating axis. As parallel sides result in the same separating axis, we shorten the number of searches for all of the parallel sides. Finally, because the algorithm verifies whether joint angles are feasible prior to the execution of the SAT, it is not necessary to check for self-collision for two subsequent links. Algorithm 4 summarizes the main modifications for a more efficient collision check among two generic, convex and bidimensional objects O_1 and O_2.

Figure 5 gives an overview of the organization of the planner structure. The proposed planner binds the concepts and modifications presented so far to

Algorithm 4. SAT for Space Manipulator Collision Check

SAT(O_1,O_2)

$state \leftarrow COLLISION$;
SetOfCheckedAngles $\leftarrow$ $EMPTY$;
$\beta \leftarrow$ angle from line containing one side of O_1 or O_2;
while ($\exists\beta$ to check in O_1 OR O_2) **do**
  $axis \leftarrow$ line with β angle;
  if Projection of O_1 and O_2 over $axis$ = OVERLAP **then**
    $state \leftarrow NO\_COLLISION$;
    break;
  else
    SetOfCheckedAngles $\leftarrow$ Add β;
    $\beta \leftarrow$ angle $\in$ (O_1 OR O_2) AND $\notin$ **SetOfCheckedAngles**;
  end if
end while

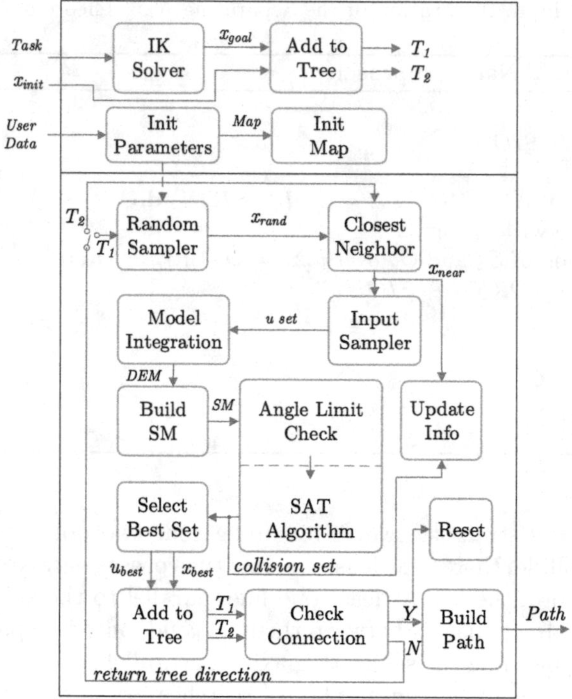

Fig. 5. Overview of the planner

successfully find a path between two configurations. The presented architecture has its main components and characteristics described as follows.

- **IK Solver:** Computes the inverse kinematic based on the end-effector goal position. Basically, the position of each DEM joint is computed iteratively, from last to first, based on the joint limitations and lengths of previous links. As infinite solutions may appear, one is randomly picked and converted back to the space manipulator (SM) model. If this configuration is assessed as collision-free, the IK solver has a valid solution.

- **Random Sampler:** Picks a valid random configuration. For that purpose, a sequence of angles is randomly chosen in order to build the DEM model. The SM is then build after the DEM. In the case some collision or joint limitation is detected, a new sample is computed. The same sampled configuration is used for the tree to grow in the opposite way.
- **Closest Neighbor:** Finds the closest neighbor from random configuration. [13] provides the pseudocode for this matter.
- **Input Sampler:** Samples an input set from the total input set (U_s from U). Consider Table 1 representing the discrete inputs. A number $m < k$ of samples is randomly chosen for each joint. Thus m^{n-1} set of inputs are sampled.
- **Model Integration:** Integrates the dynamic DEM model for every input in U_s.
- **Build SM:** Builds the free-floating manipulator from the DEM model. This is done by iteratively applying direct kinematics from end-effector to the base.
- **Collision Detection:** Checks every expansion made and identify the ones that are collision-free based on the SAT algorithm. This block also verifies if some joint has reached its opening limits.
- **Update Info:** Updates the constraint violation frequencies (CVFs) and marks inputs already evaluated.
- **Select Best Set:** Checks, among every expansion considered, the one that is collision-free and has the lowest cost of motion to the desired configuration.
- **Add to Tree:** Add new nodes and edges to the tree.
- **Check Connection:** Runs the routine of connection verification between two trees.
- **Build Path:** Builds the path between initial and final configurations after connecting two trees.
- **Init Parameters:** Loads all constant parameters of manipulator and algorithm. Initializes also variable parameters that are user-defined, such as task description and bias to goal.
- **Init Map:** Loads obstacles that compose the map. These must be described as poligonal objects.
- **Reset:** As observation of simulations, the convergence of the algorithm proposed in Fig. 5 is still jeopardized due to local minimas in some cases. It was noticed that the first 1000 samples suffice to provide a fair glimpse of how well the exploration would be conducted. Hence, this block restarts the planner if a reasonable approximation is not achieved during the first 1000 iterations. This enables the algorithm to spare effort in the search for a solution.

5 Results and Discussions

For the proposed task, the UARM-E robot was configured with two active joints in one single arm. The obstacle in the environment was considered to be fixed and rectangular. The task tries to find a path between two configurations without hitting the obstacle. For that purpose, a step size $\Delta T = 0.001\,\mathrm{s}$ is used for integrating the DEM, which is performed through Euler method. All of the results were obtained through simulations, runned in Matlab - version R2012a. The computer was powered with an *Intel®* *Core i*7, 3.40 GHz and 12 GB RAM. A tolerance region is created around the goal configuration. Aiming to provide an intuitive idea of the tree growth from initial to final configuration, all the nodes representing end-effector positions are plotted as points for both trees. Examples are given in Figs. 6 and 7.

If we allow the planner to continue the searching, it generally comes up with a better solution as the tree expands in free workspace. Figure 7 shows an expansion after 50000 iterations.

To achieve a convergence between two states, the metric was computed based on: Euclidean distance (x); Torques difference (τ); Velocities difference (v).

Consider two configurations a and b. In order to relate the parameters, variable x is made $x = -\sum |x_a - x_b|$, where x_a and x_b denote the positions of joints in a and b, respectively. Therefore, $x \to 0$ as the configurations get closer

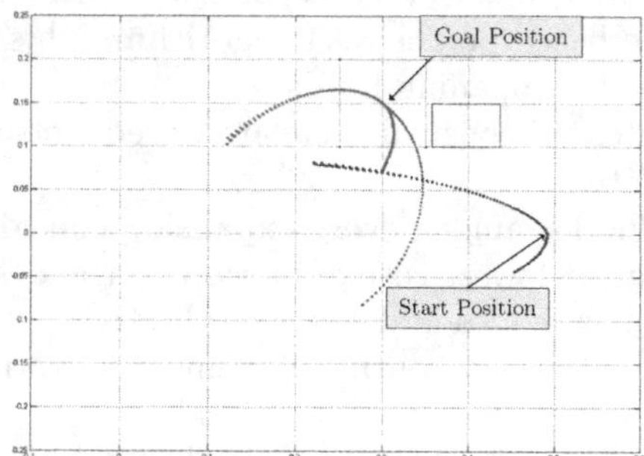

Fig. 6. Expansion after 1000 iterations. The green box stands for the tolerance region. The black box depicts the obstacle. The main tree is presented in blue, while the second tree, rooted at the goal configuration, is depicted in red. (Color figure online)

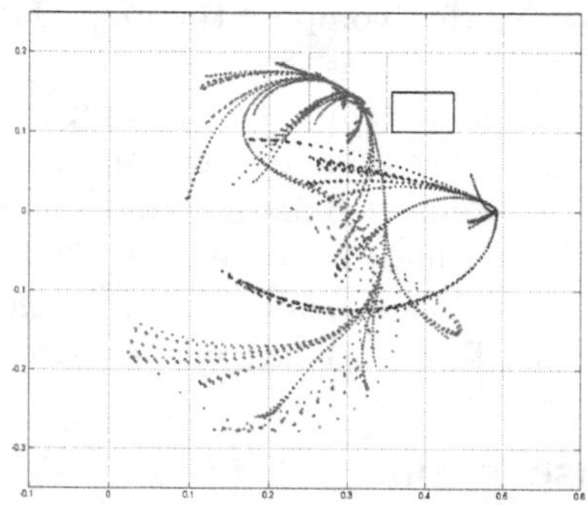

Fig. 7. Expansion after 50000 iterations

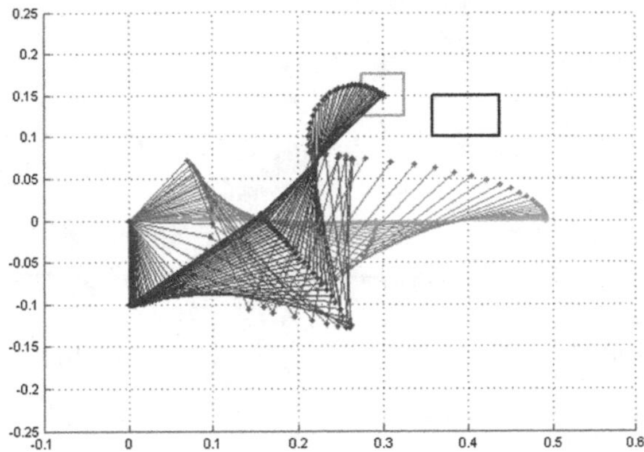

Fig. 8. Example of a computed path. The DEM representation of robot UARM-E gets darker as it approximates from the goal region.

to each other. Equation (4) defines the form of the metric used. Similarly, the difference of torques and velocities between all joints in a and b is computed as $\tau = \sum |\tau_a - \tau_b|$ and $\upsilon = \sum |\upsilon_a - \upsilon_b|$, respectively.

$$\rho(x) = k_1(x)x + k_2(x)\tau + k_3(x)\upsilon \tag{4}$$

Because variables τ and υ need to influence metric ρ more heavily as configurations get closer, coefficients k_i are represented as sigmoid functions. Equation (5) presents an example of these sigmoid functions after adjustment of the magnitudes of all constants. This function shape has also the advantage of achieving a smooth transition between configurations.

$$\begin{aligned} k_1 &= 0.8 - \frac{0.1}{1 + e^{(-5-100x)}}, \\ k_2 &= \frac{1.5 \cdot 10^{-3}}{1 + e^{(-6-50x)}}, \\ k_3 &= \frac{5 \cdot 10^{-6}}{1 + e^{(-5-120x)}}. \end{aligned} \tag{5}$$

A path was considered found after $|x| < \xi$, with $\xi < 10^{-5}$. With the goal of evaluating the planner performance, 100 evaluation tests were run. On average, the algorithm needed 3827 nodes to find a solution. There were 11 cases where the planner could not find a solution, even after 25000 nodes were grown. Among the results, the solution with faster convergence needed 1604 nodes only. On the other hand, the solution that took longer time was achieved after 16445 nodes. Because of its dependency on the environment's complexity, the computation time was not chosen as a measure of convergence speed. Figure 8 shows a typical solution of the computed path. Only the DEM is shown in this picture for a better visualization. Figure 9 shows other examples of solutions found for the same problem.

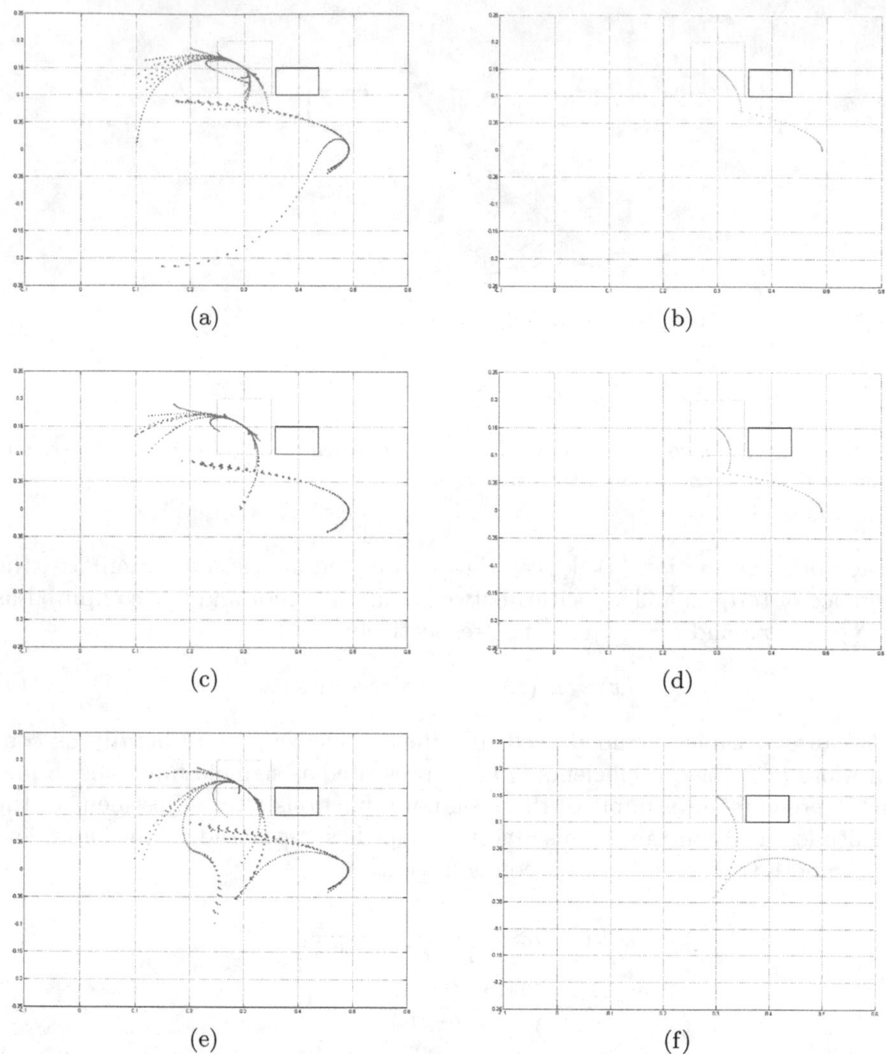

(a) (b) (c) (d) (e) (f)

Fig. 9. (a), (c) and (e) represent expansions with 16445, 7280 e 4916 nodes, respectively. Paths associated with these figures were connected by the planner using 90, 89 and 111 nodes, respectively. Path computed is represented by green nodes in figures (b), (d) (e) and (f) (Color figure online)

6 Conclusion

The main goal of this paper was to present a feasible approach for automatically planning a collision-free path for a free-floating manipulator. Our main contribution was to provide such method through cooperating the RRT with the complex dynamics of free-floating manipulators with the help of the dynamically equivalent manipulator. Furthermore, enhancements like growing bidirectional trees and reducing the metric sensitivity were improved in order to create a robust path planner. A straightforward collision-check for the space manipulator is also shortly presented in order to help for a fast implementation. The proposed methodology proved itself functional after several tests and different conditions. Future works aim to expand the planner to consider trajectories and evaluate the RRT* performance in a different programming environment, like C++ or Python. Finally, we plan to extend the algorithms to operate with two-arm manipulation as well.

Acknowledgment. This research was supported by grants from Fundação de Amparo à Pesquisa do Estado do Amazonas (FAPEAM) and the Fundação de Amparo à Pesquisa do Estado de São Paulo (FAPESP).

References

1. Yoshida, K.: Achievements in space robotics. IEEE Robot. Autom. Mag. **16**(4), 20–28 (2009)
2. Li, C., Liang, B., Xu, W.: Autonomous trajectory planning of free-floating robot for capturing space target. In: IEEE/RSJ International Conference on Intelligent Robots and Systems (IROS), pp. 1008–1013 (2006)
3. Papadopoulos, E., Dubowsky, S.: Dynamic singularities in free-floating space manipulators. ASME J. Dyn. Syst. Meas. Contr. **115**, 44–52 (1993)
4. Caccavale, F., Siciliano, B.: Quaternion-based kinematic control of redundant spacecraft/manipulator systems. In: IEEE International Conference on Robotics and Automation (ICRA), pp. 435–440 (2001)
5. Tortopidis, I., Papadopoulos, E.: Point-to-point planning: methodologies for under-actuated space robots. In: IEEE International Conference on Robotics and Automation (ICRA), pp. 3861–3866 (2006)
6. Nanos, K., Papadopoulos, E.: On cartesian motions with singularities avoidance for free-floating space robots. In: IEEE International Conference on Robotics and Automation (ICRA), pp. 5398–5403 (2012)
7. Pazelli, T.F.P.A.T.: Assembly and nonlinear h infinite control of free-floating base space manipulators. Ph.D. dissertation, EESC-USpP (2011)
8. Liang, B., Xu, Y., Bergerman, M.: Mapping a space manipulator to a dynamically equivalent manipulator. Robotics Institute, Pittsburgh, PA, Technical report, CMU-RI-TR-96-33, September 1996
9. Vafa, Z., Dubowsky, S.: On the dynamics of manipulators in space using the virtual manipulator approach. In: IEEE International Conference on Robotics and Automation (ICRA), pp. 579–585 (1987)
10. LaValle, S.M.: Rapidly-exploring random trees: a new tool for path planning. Computer Science Dept., Lowa State University, Technical report (1998)
11. LaValle, S.M., Kuffner Jr. J.J.: Randomized kinodynamic planning. In: IEEE International Conference on Robotics and Automation (ICRA), pp. 473–479 (1999)
12. Urmson, C., Simmons, R.: Approaches for heuristically biasing RRT growth. In: Proceedings of the IEEE/RSJ International Conference on Intelligent Robots and Systems (IROS), vol. 2, pp. 1178–1183, Outubro (2003)
13. Cheng, P., LaValle, S.: Reducing metric sensitivity in randomized trajectory design. In: IEEE/RSJ International Conference on Intelligent Robots and Systems (IROS), pp. 43–48 (2001)
14. Gottschalk, S.: Separating axis theorem. UNC Chapel Hill, Chapel Hill, NC, Technical report, TR96-024 (1996)

A Topological Descriptor of Forward Looking Sonar Images for Navigation and Mapping

Matheus Machado(✉), Guilherme Zaffari, Pedro Ballester, Paulo Drews-Jr, and Silvia Botelho

NAUTEC - Centro de Ciencias Computacionais, Universidade Federal do Rio Grande (FURG), Rio Grande, RS, Brazil
{matheusmachado,zaffari,paulodrews,silviacb}@furg.br

Abstract. The automation of the monitoring, inspection and underwater maintenance tasks by underwater robots require a mapping and localization system. One challenge of these systems is how to recognize previously visited place in sensory information. This paper proposes a extended version of a method to detect loop closure dealing with acoustic images acquired by a forward looking sonar (FLS). The method builds a graph of Gaussian probability density function. This structure represents both shape and topological relation. We improve the image segmentation step adding a local parameters adjustment regard to intensity peak analyze of acoustic beams and changed the graph matching metric. We evaluate the method in a real dataset acquired by a underwater vehicle performing navigation in a harbor area.

Keywords: Topological graph · Acoustic image · SLAM

1 Introduction

Over the past decades, Autonomous Underwater Vehicles (AUVs) have been deployed in a growing number of aquatic environments. It includes the observation of benthic habitats, shallow reefs, near-shore mangroves and marinas. The robots have been applied in monitoring and inspection tasks of underwater sites [11,19]. However, their operation involves a number of challenges due the characteristics of the environment that limit the use of cameras [3].

One of the most important challenges is related to the ability of self-localize and identify an unknown environment through its sensor readings. The methods to solve these problems are called Simultaneous Localization and Mapping (SLAM) [5]. One of the key issues to solve the SLAM problem is to detect previously visited areas, which are called loop closure. This detection reduces the displacement error on the path caused by the integration of sensory information, *i.e.* the drift of dead reckoning.

Typically, the detection is performed using sensors such as cameras, lasers, or sonars. The sonar is largely applied in underwater environments due to the

The use of robots in underwater exploration is increasing in the last years.

F. Santos Osório and R. Sales Gonçalves (Eds.): LARS 2015/SBR 2015, CCIS 619, pp. 120–134, 2016.
DOI: 10.1007/978-3-319-47247-8_8

limitation of the light propagation. The loop detection is basically the identification of the same place in an unknown environment, using sensory information captured from a distinct temporal and spatial condition. These places should be stable and stand out in the environment. Therefore, they need to be easily identified and their positions need to be relevant to the identification of the robot's location.

Most of the works on terrestrial SLAM found in the literature [10,12,13,17] close loops using corner extraction on optical images and describe the points of interest applying a local feature descriptor. The matching between descriptors is easily achieved by computing their similarity with Euclidean Distance.

Although the fact that these approaches present good performance on terrestrial environments, they are limited in underwater environments. The water turbidity reduces the contrast and, thus, reduces the capability to extract corners on optical images [2,4]. The acoustic imaging sensors do not suffer with this effect. However, they are unable to capture high resolution details like optical sensors. Thus, traditional approaches using local feature descriptors do not work well in a typical underwater condition [8]. Some studies described acoustic images in the space domain using shapes [1,9,14] or in the frequency domain [7]. Our work describes the shapes of segments in the space domain images using a set of Gaussian probability density function.

We also represent the topological relationship between these Gaussian distributions to reduce the number of outliers. This relationship is motivated by the application of navigation and mapping where outliers can degrade the robot's performance. A typical harbor area presents a semi-structured environment, composed of piers, boats, ships, see Fig. 8-a. Normally, these structures are static and their spatial relation can be explored. Here, we describe the topological relation using graphs. The graph is a well known data structure that allows us to represent the topology.

This method was first proposed in our previous work [16]. In this paper we present an extended version where we added a new image segmentation step based on the intensity peaks analysis to determine the segmentation thresholds in order to reduce the sensitivity of the method with parameters variation. We also changed the graph comparison metric to calculate the error between vectors instead of the weighted sum of errors eliminating the need to determine the weight of each error. The new approach was tested using the same dataset ARACATI 2014 [18] used in our previous work. This work also is direct related to a master thesis [15].

The remainder of the work is structured as follows. Initially, Sect. 2 describes the acoustic image generation process and the main problems associated with it. Section 3 presents a brief description of the proposed methodology. Section 4 shows the experimental results obtained in a real dataset acquired by an underwater vehicle. Finally, Sect. 5 summarizes the paper and discusses future directions.

2 Acoustic Image and Forward Looking Sonar

The forward looking sonars (FLS) are active devices which produce acoustic waves that propagate through the medium until they collide with an obstacle or be completely absorbed. When a wave collides with an obstacle, part of its energy is absorbed and the other part is reflected. The reflected portion which returns to the sensor is recorded using an array of hydrophones. The round trip of the wave is called *ping*.

The waves captured by the hydrophones are organized according to its return direction and their distance to the reflecting object. This information are estimated according to the capture time difference of each hydrophone to the same wave and the knowledge of the speed of sound in water. Acoustic returns from the same direction belong to the same beam. The returns recorded over time in a beam are called bin. A fan-shaped acoustic image $I(X, Y)$ is a way to represent the information recorded by the sonar for a certain period of time. In this image the pixels are associated with bins, and they are indexed according to their distance r and their azimuth direction θ from the sonar, see Fig. 1(a). Because of the way that the FLS is built, the height information of a bin are not captured and therefore the acoustic image is a 2D representation of the environment being observed.

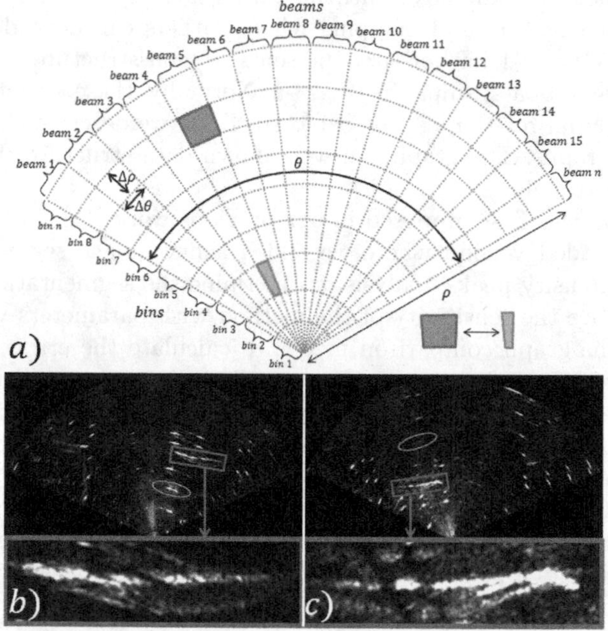

Fig. 1. Acoustic images: (a) example of a FLS acquisition, (b) and (c) image acquired in different positions, where the green boxes are the shape changes of the same object, and the yellow ellipses are the acoustic shadow effect. (Color figure online)

The acoustic images have some phenomena that are not found in optical images such as the inhomogeneous resolution. Thus, the amount of pixels used to represent a bin varies according to its distance r to the sonar. The intensity variations of each bin, that may be caused by the water attenuation or by changing in the sonar tilt. The intensity variations between the sonar beams caused due to the sensitivity differences between the hydrophones.

There are also phenomena related to the sonar activity, such as acoustic reverberation due to the capture of more than one acoustic returns from the same object producing duplicated objects in the image. This phenomenon is common in small environments or in cases where there are very close objects to the sonar.

The acoustic shadow is another effect produced by objects that block the passage of transmitted waves, producing a region without acoustic feedback after the blocking objects. These regions are characterized by a black spot in the image hiding a part of the scene.

A change on the incidence angle of the acoustic waves on the objects surface, produced by the sonar displacement, may change the bin intensity and the object's shape. Furthermore, the displacement of the sonar also causes the movement of the acoustic shadows which it may cause the occlusion or the appearing of objects in the scene.

Finally, the acoustic images suffers with various types of noise as the low signal-to-noise ratio where mutual interference of the sampled acoustic returns causes speckle noise.

Figures 1-b and c show the effect of the change in the viewpoint. The green box shows the shape of an object that changes due the point of view. The yellow ellipse in the Fig. 1-b shows an object that almost disappear in the Fig. 1-c. It occurs due to the acoustic shadow created by the object highlighted by the green box.

3 Methodology

The proposed descriptor for acoustic images is divided into three stages. Initially, the acoustic image is segmented by a peaks intensity search and a breadth first search to extract the neighbors pixels of the peaks as a new segment. After the image segmentation, the segments are described using probabilistic Gaussian functions. The last step of the method is the creation of a topological graph which establishes the relationships between each Gaussian function. After the description of acoustic images is provided a method to compare the descriptors and identify similar images.

3.1 Segmentation

Segmentation is performed based on the regions with higher acoustic intensity variations. Initially, we performed a search for intensity peaks for each beam of the acoustic image. After, each intensity peak found with intensity variation higher than H_{min} generates a new segment. The elements of the segment are searched using a breadth first search with a 8-connected neighborhood.

In Fig. 2 is shown an intensity profile of an acoustic beam, where the horizontal axis represents the bins and the vertical axis the intensity of each bin. The graph illustrates the intensity peak search of one acoustic beam. Each bin is analyzed and the variables I_{max}, I_{min}, H_{total} and $H_{current}$ are constantly updated, where I_{max} and I_{min} are the respective maximum and minimum intensities between the analyzed pixels, H_{total} is the largest intensity variation and $H_{current}$ is the intensity variation in the current bin. This variables are updated using Eqs. 1, 2, 3 and 4.

$$I_{min} = min[I_{min}, Bin(i)] \tag{1}$$

$$I_{max} = max[I_{max}, Bin(i)] \tag{2}$$

$$H_{current} = Bin(i) - I_{min} \tag{3}$$

$$H_{total} = I_{max} - I_{min} \tag{4}$$

During the beam analysis, one peak analysis is finished when a bin with lower intensity variation than H_{end} is found, *i.e.* $H_{current} < H_{end}$. The intensity variation H_{end} is a portion of the peak intensity variation calculated by Eq. 6. Many peaks are analyzed however, only the peaks with $H_{total} > H_{min}$ create a new segment.

$$\rho_{recursive} = I_{min} + H_{total} \cdot \pi_{recursive} \tag{5}$$

$$H_{end} = H_{total} \cdot \pi_{end} \tag{6}$$

The segment is extracted pixel by pixel by a breath first search regarding the 8-neighborhood constraint. Only the pixels with intensity higher than $\rho_{recursive}$ are visited and all visited pixels are included in the same segment. The intensity $\rho_{recursive}$, used as intensity threshold of our segmentation, is calculated by Eq. 5 and depends of the total intensity variation of the peak found. The intensity rates π_{end} and $\pi_{recursive}$ adjusts the sensitivity of the segmentation method. π_{end} is related with the amount of analyzed peaks, and $\pi_{recursive}$ is related with the size of the extracted segments. Figure 3 shows the peaks and segments analyzed in a single image beam. The complete segmentation is performed by analyzing all image beams.

The segment found in each search is a sample that is described using an Gaussian model. Each sample is defined by a row Y, a column X and a intensity I, and they are represented by a column vector as:

$$X = \begin{bmatrix} x_1 \\ x_2 \\ \dots \\ x_N \end{bmatrix}; Y = \begin{bmatrix} y_1 \\ y_2 \\ \dots \\ y_N \end{bmatrix}; I = \begin{bmatrix} i_1 \\ i_2 \\ \dots \\ i_N \end{bmatrix}. \tag{7}$$

3.2 Describing the Segments

One Gaussian is adjusted for each segment and is defined as $GA = (\mu X, \mu Y, \mu I, \sigma X, \sigma Y, \sigma I, \theta, N)$, where μX, μY and μI are the respective mean of X, Y and I.

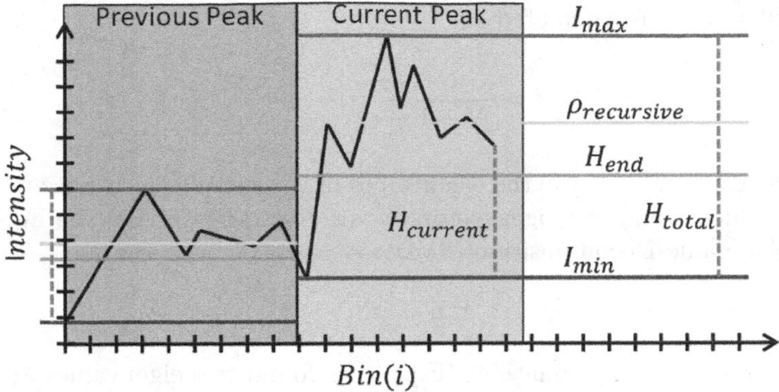

Fig. 2. Peak analysis of one beam. The blue line indicates the reference intensity of each peak, the red line indicates the maximum intensity of each peak, the green line indicates the minimum intensity variation of a bin required to complete close one peak analysis and start the next one, the yellow line indicates the intensity threshold used to extract a segment, the dotted orange line $H_{current}$ indicates the intensity variation of current bin and the orange dotted line H_{total} indicates the intensity variation of the peak. (Color figure online)

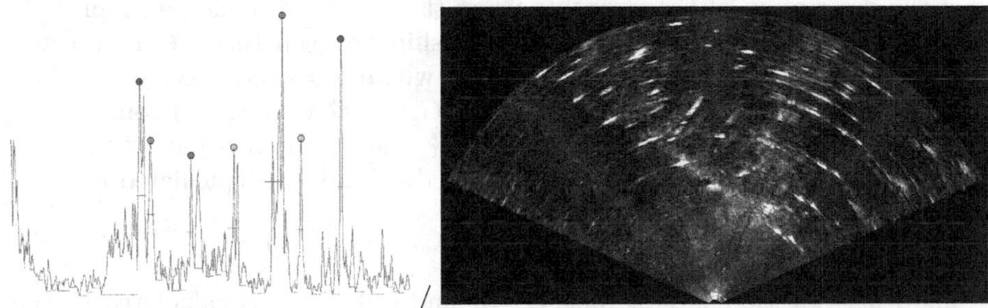

Fig. 3. Detected peaks on the intensity profile of an beam: (a) the graph with the intensity profile analysis where the horizontal axis represents the bins and the vertical axis represents the intensities; (b) the analyzed beam represented by the blue line within the extracted segments in the acoustic image. The detected peaks are identified by colored circles on both images. This result is obtained on a 16-bits image using $\pi_{end} = 0.6$, $\pi_{recursive} = 0.9$ and $H_{min} = 150$. The following $\rho_{recursive}$ were found: in the first blue peak 153, in the green peak 126, in the red peak 101, in the cyan peak 106, in the pink peak 173, in the yellow peak 104 and int the last blue peak 166. (Color figure online)

The variables σX, σY, σI, are the standard deviation of the variable X, Y and I, respectively. Finally, θ is the rotation angle with respect to the vertical image axis, and N is the sample size. Figure 5 illustrates the estimation.

The value of σI is calculated as:

$$\sigma I = \frac{1}{N-1}\sum_{i=1}^{N}(i_i - \mu i)^2. \tag{8}$$

The values of σX, σY are the eigenvalues of the sample covariance matrix Σ. The eigenvalues λ and the eigenvectors $\overrightarrow{v}$ are found by solving the Eq. 9 using the Singular Value Decomposition (SVD):

$$\Sigma\overrightarrow{v} = \lambda\overrightarrow{v}. \tag{9}$$

Solving the system of equations (Eq. 9), we found two eigenvalues λ_1, λ_2 and two eigenvectors $\overrightarrow{v_1}, \overrightarrow{v_2}$, where $\overrightarrow{v_1}$ is the unitary vector in the direction of the largest variance of the data, and $\sigma X = \lambda_1$. And $\overrightarrow{v_2}$ is the unitary vector in the direction of the second largest spread of the data, and $\lambda 2 = \sigma Y$. The rotation angle θ is the tangent of the eigenvector $\overrightarrow{v_1}$, $\theta = \arctan(\overrightarrow{v_1})$. Figure 5 depicts the vectors.

3.3 Establishing the Topological Relationship

After the description of the segments using the Gaussian model, a graph G is created to represent the topological relationship between them. G is directed and symmetric graph defined as $G = (V, E)$ where $V = \{GA_1, GA_2, ..., GA_N\}$ and $E = \{E_1, E_2, ..., E_M\}$. The edges $E_i = (GA_{src}, GA_{dest}, \theta_e, \rho_e)$ connect two vertices $\{GA_{src}, GA_{dest} \in V | GA_{src} \neq GA_{dest}\}$, where θ_e is the slope of the edge relative to the axis with largest spread of the data and θ_e is calculated by:

$$\theta_e = atan2(\mu Y_{dest} - \mu Y_{src}, \mu X_{dest} - \mu X_{src}) - \theta. \tag{10}$$

The length of the edges ρ_e links GA_{src} and GA_{dest}. It is calculated by the Euclidean distance between the means of Gaussian:

$$\rho_e = \sqrt{(\mu X_{src} - \mu X_{dest})^2 + (\mu Y_{src} - \mu Y_{dest})^2}. \tag{11}$$

The slope edges θ_e is calculated with reference to the axis of largest spread of the source Gaussian making them independent of the sonar viewpoint. See Fig. 5-a.

3.4 The Descriptors Matching Process

The last step is the detection of similar images based on comparison of topological graphs of each image. The graphs are compared vertex to vertex by the function $cmp(GA_m, GA_n)$, where the pair of vertex GA_m, GA_n belong to two different graphs. A pseudo code is shown int Fig. 7.

The function $nextAdjEdge(E_i, GA_j)$ is used to iterate through the adjacent edges of a vertex GA_j in ascending order of slope, accessing the next adjacent edge after E_i. The error between a pairs of edges of two vertices (GA_u, GA_v) are calculated by the Eq. 12. This equation finds the side of a scalene triangle that represent the error between two edges. This error is illustrated on Fig. 4.

$$\epsilon = \rho_{v1}^2 \cdot \rho_{u1}^2 - 2 \cdot \rho_{v1} \cdot \rho_{u1} \cdot cos(\theta_{eu} - \theta_{ev}). \tag{12}$$

The error between all pairs of edges is computed starting with edges with smallest slope until edges with greater slope. If the error ϵ between the current pair of edges is smaller than the maximum error ϵ_{max}, we add it to ϵ_{ACC}. The similarity edges counter Σ_{Hit} is increased and the next pair of edges is analyzed. Otherwise, the calculated error is discarded and the edge with the greatest slope is maintained, replacing only the edge with lower slope by its next adjacent edge. After the analysis of all adjacent edges between the two vertices, the average error between them is computed dividing the accumulated error ϵ_{ACC} by the amount of pairs of edges Σ_{Hit} and it is returned by the function as a comparison indicator between two vertices.

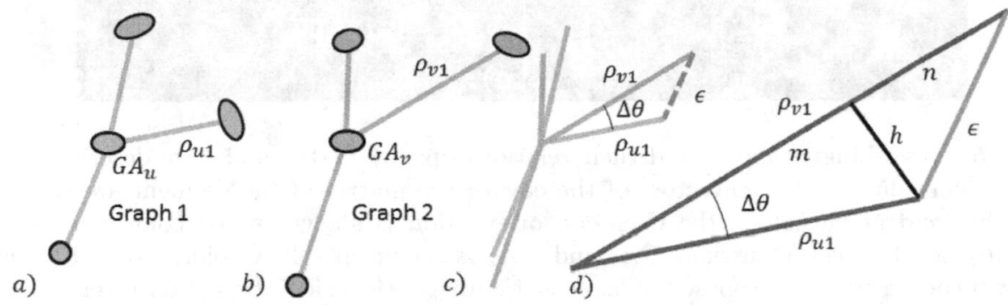

Fig. 4. Computing the error between edges: (a) and (b) show the vertices to be compared GA_u and GA_v, with their respective adjacency; (c) these two vertices centered in the same point, the error ϵ is depicted by the dotted blue line; (d) the variables ρ_{v1}, ρ_{u1} and $\Delta_\theta = \theta_{u1} - \theta_{v1}$ are used in Eq. 12. (Color figure online)

We find the similar vertices between two graphs estimating the error between all pairs of vertices using the function $cmp(GA_m, GA_n)$. A new bipartite graph G_{ERROR} is generated to associate each pairs of vertices and their respective errors, as is shown in Fig. 6-b. The pairs of vertices with significantly lower errors than the errors of their adjacent edges are considered similar vertices.

The similar acoustic image and a loop detection is defined depending of the similar vetices found between the graphs. The comparison method assumes that the environment is partially structured and partially static.

4 Experimental Results

The proposed method is evaluated in the dataset ARACATI2014 available from [18]. In this dataset, an underwater vehicle Seabotix LBV300-5 equipped with a forward looking sonar Teledyne BlueView P900-130 and a differential GPS

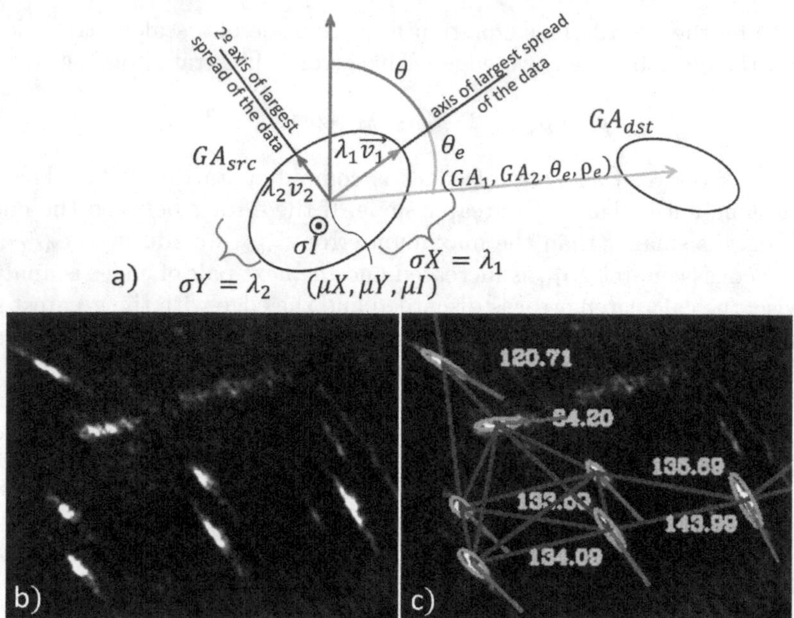

Fig. 5. Describing segments and their relationship: (a) Vertex and edge description - The eigenvalues and eigenvectors of the covariance matrix of the segment are shown in blue and green colors, the Gaussian orientation is shown in red color, the edge description between Gaussians GA_1 and GA_2 is shown in yellow color; (b) and (c) is shown the description proposed in an acoustic image, the orientation of each vertices is indicated by one read line and the edges are indicated by blue lines. Also the orientation of each Gaussian is indicated by cyan numbers in degrees. (Color figure online)

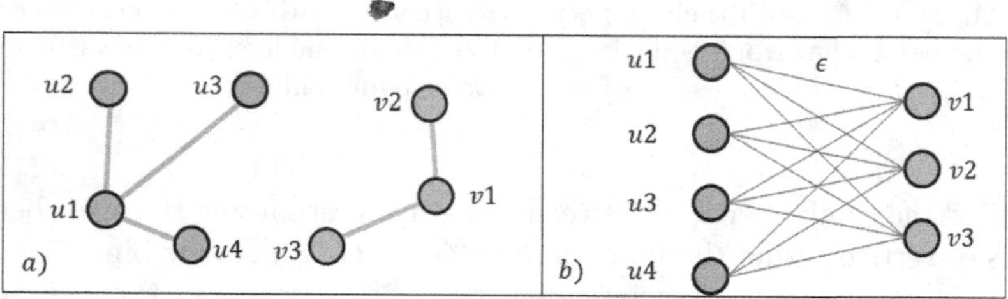

Fig. 6. Graphs comparison: In (a) is shown two graphs to be compared; In (b) is shown the bipartite graph G_{ERROR} generated with the error between all possible pairs of vertices.

location sensor are adopted. The robot travels about 300 m at a harbor in Rio Grande, Brazil. The environment is partially structured with boats, pier and wooden poles used to tie up the boats. In Fig. 8 is shown the marina and an example of acoustic image.

The comparison results of two image extracted from dataset ARACATI 2014 are shown in Fig. 10. In Figs. 10-a, b are shown the description of two acoustic

```
Function cmp(GA_n, GA_m)
Σhit ← 0
u ← firstEdge(u, GA_n)
v ← firstEdge(v, GA_m)
while u ≠ ∅ and v ≠ ∅ do
    Δθ ← ||u.θ_e − v.θ_e||
    ε ← ||u.ρ² · v.ρ² − 2 · u.ρ · v.ρ · cos(Δθ)||
    if ε < ε_max then
        ε_ACC ← ε_ACC + ε
        Σhit ← Σhit + 1
        u ← nextAdjEdge(u, GA_n)
        v ← nextAdjEdge(v, GA_m)
    else
        if u.θ_e < v.θ_e then
            u ← nextAdjEdge(u, GA_n)
        else
            v ← nextAdjEdge(v, GA_m)
        end if
    end if
end while
if Σhit > 0 then
    ε_ACC ← ε_ACC/Σhit
end if

return ε_ACC, Σhit
End Function
```

Fig. 7. Pseudo code of the function *cmp*. This function compute the similarity between two vertices of two distinct graphs, and returns the average compatibility error and the amount of compatibility pairs of edges between them.

images. In Figs. 10-c, d are shown the extracted segments. In Figs. 10-e, f is shown the adjusted Gaussian of each segment. Its orientation is indicated by a green line followed by a slop value in degree. In Figs. 10-g, h are shown the topological graphs that describe each image. Finally, in Fig. 10-i are shown the similar vertices found, indicated by red lines. The adopted parameters are defined on Table 1.

These parameters were manually adjusted by performing several tests. Regarding the acoustic images description steps, π_{end} parameter affects the number of segments extracted, *i.e.* the larger it is the more segments will be extracted, $\pi_{recursive}$ parameter affects the size of the segments, *i.e.* the lower it is the larger segments will be extracted, and r parameter affects the number of edges. The graph density influences the results in a way that more density graphs obtain better results but with an additional computational cost.

The proposed method was compared with the same ground truth used in [16], where a set of 35 pair of images were manually segmented and established their similarity. The Table 2 shows the achieved results. Approximately 24 % of

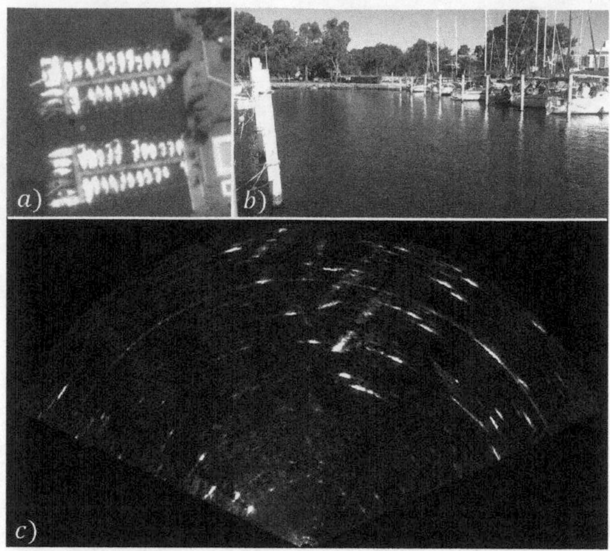

Fig. 8. Dataset ARACATI2014 from [18]: In (a) the trajectory is shown in red, where the green and blue points are the start and end points; In (b) is shown a picture of the marina; In (c) is shown an example of acoustic image. (Color figure online)

Table 1. Parameters adopted to obtain the experimental results.

Parameter	Value
r	300 pixel
π_{end}	0.6
$\pi_{recursive}$	0.98
minSampleSize	20 pixel
maxSampleSize	12000 pixel
$\rho_{similar}$	4.0

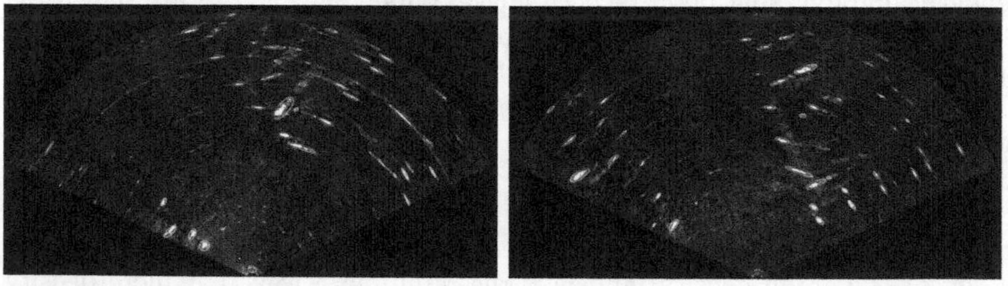

Fig. 9. Comparison of image segmentation methods. In both images, the ellipses in red were created by segmentation without the peaks intensity analysis used in [16], the green ellipses were created using the peaks intensity analysis. The blue lines are intersections between red and green ellipses. (Color figure online)

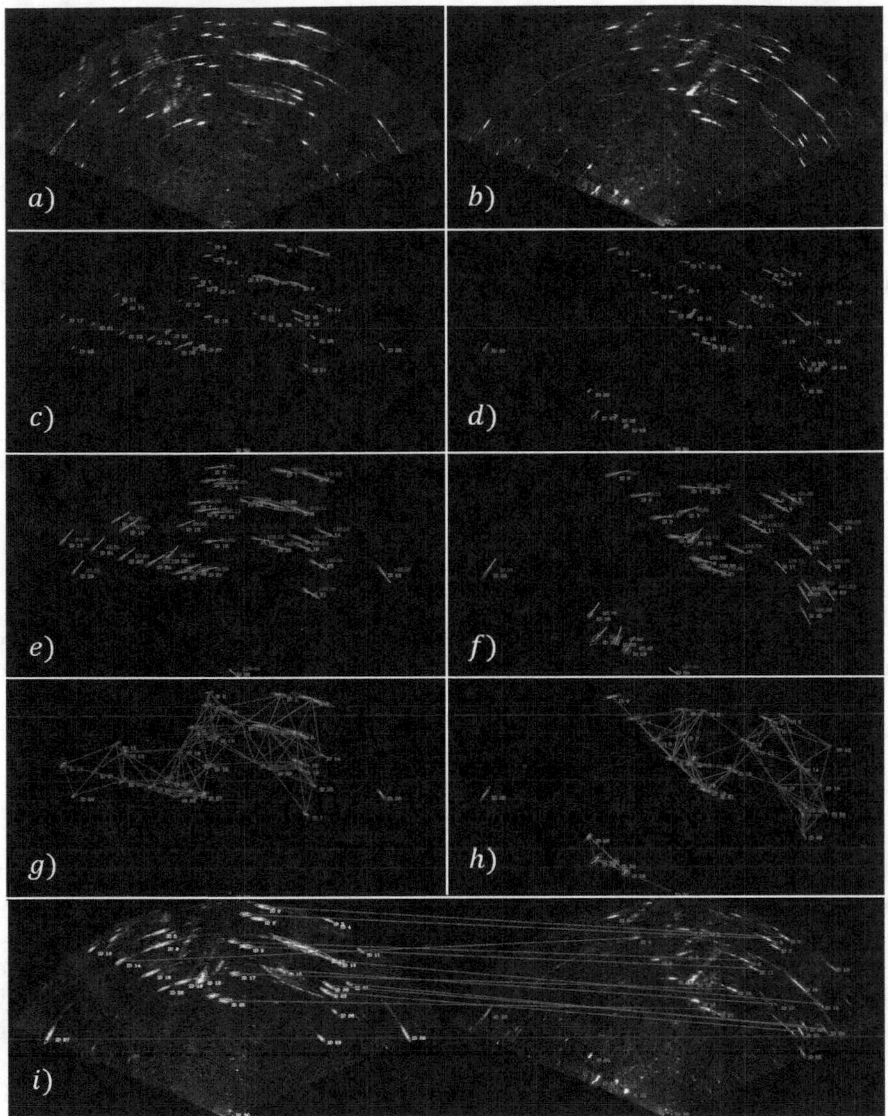

Fig. 10. Process of description and comparison of images: (a) and (b) the original acoustic images; (c) and (d) the extracted segments; (e) and (f) show the adjusted Gaussian to the extracted regions of interest; (g) and (h) the built graph with the relations between the extracted regions of interest; (i) the similar vertices found between two acoustic images, in red color. (Color figure online)

the matches were found by the proposed method, 19 pairs had at least 5 matches found and 24 pairs had at least 3 matches found. Although the limited amount of matches found, it was able to correspond a pair of images using our method.

With regard to previous results presented in [16] the proposed improvements results in a better performance for the same tests using fewer parameters. Regarding to segmentation step, a comparison between a segmentation using the proposed local parameters adjustment and without local adjustments was performed using the same image, the results are shown in Fig. 9. Some segments that were extracted using the peaks analysis were not extracted using segmentation with fixed threshold.

Table 2. Comparison of our results in relation to the manually obtained ground truth.

Results	
Total tests	35 pairs of img.
Correct matchs found	24.93 %
5+ correct match	19 pairs
3+ correct match	24 pairs
No match found	4 pairs

The comparison method operates in partially structured environments since the shape of the segments remain constant. The approach faces difficulties to find matches when large rotations occur in the sonar field of view because its causes changes on segments shape and direct affect the Gaussian orientation.

5 Conclusions

This paper presents an extended version of our method [16] to describe and detect similar acoustic images considering the shape of objects and their topological relationship.

In order to reduce the sensitivity of the method due to parameter changes, a new image segmentation step, that adapts the segmentation parameters locally according to the intensity peaks of each acoustic beam, was proposed. Resulting in a more robust segmentation once the search threshold varies according to the peak intensity variation. The second proposed improvement change the comparison metric between graphs eliminating the need of parameters in the comparison step. The results show a better performance using fewer parameters.

The presented description and detection of similar acoustic images is limited to partially structured environments and the acoustic images of forward looking sonars. The detection can be used to improve the autonomous navigation in a SLAM problem of underwater vehicles.

The next steps will be focused in evaluate the method in other datasets and make comparisons with other approaches found in the literature. Furthermore, we intent to use the method to obtain an estimation of visual odometry and evaluate its performance in terms of loop closure detection inside a SLAM system like DolphinSLAM [6,18].

Acknowledgment. The authors thank to colleagues of the NAUTEC-FURG. This research is partly supported of CNPq, CAPES, FAPERGS and PRH-27 FURG-ANP/MCT. This paper is a contribution of the INCT-Mar COI funded by CNPq Grant Number 610012/2011-8.

References

1. Aykin, M.D., Negahdaripour, S.: On feature matching and image registration for two-dimensional forward-scan sonar imaging. J. Field Robot. **30**(4), 602–623 (2013)
2. Codevilla, F., Gaya, J.O., Duarte, N., Botelho, S.: Achieving turbidity robustness on underwater images local feature detection. In: British Machine Vision Conference (BMVC) (2015)
3. Drews-Jr, P., Nascimento, E., Campos, M., Elfes, A.: Automatic restoration of underwater monocular sequences of images. In: IEEE/RSJ International Conference on Intelligent Robots and Systems (IROS), pp. 1058–1064 (2015)
4. Drews-Jr, P., Nascimento, E., Codevilla, F., Botelho, S., Campos, M.: Transmission estimation in underwater single images. In: IEEE International Conference on Computer Vision Workshops (ICCVW), pp. 825–830 (2013)
5. Durrant-Whyte, H., Bailey, T.: Simultaneous localization and mapping: part I. IEEE Robot. Autom. Mag. **13**(2), 99–110 (2006)
6. Guth, F., Silveira, L., Botelho, S., Drews-Jr, P., Ballester, P.: Underwater SLAM: challenges, state of the art, algorithms and a new biologically-inspired approach. In: IEEE RAS EMBS International Conference on Biomedical Robotics and Biomechatronics (BioRob), pp. 981–986 (2014)
7. Hurtos, N., Cufí, X., Petillot, Y., Salvi, J.: Fourier-based registrations for two-dimensional forward-looking sonar image mosaicing. In: IEEE/RSJ International Conference on Intelligent Robots and Systems (IROS), pp. 5298–5305 (2012)
8. Hurtos, N., Nagappa, S., Cufí, X., Petillot, Y., Salvi, J.: Evaluation of registration methods on two-dimensional forward-looking sonar imagery. In: MTS/IEEE OCEANS, pp. 1–8 (2013)
9. Johannsson, H., Kaess, M., Englot, B., Hover, F., Leonard, J.: Imaging sonar-aided navigation for autonomous underwater harbor surveillance. In: IEEE/RSJ International Conference on Intelligent Robots and Systems (IROS), pp. 4396–4403 (2010)
10. Konolige, K., Agrawal, M.: FrameSLAM: from bundle adjustment to real-time visual mapping. IEEE Trans. Robot. **24**(5), 1066–1077 (2008)
11. Kuhn, V.N., Drews-Jr, P., Gomes, S., Cunha, M., Botelho, S.: Automatic control of a ROV for inspection of underwater structures using a low-cost sensing. J. Braz. Soc. Mech. Sci. Eng. **37**(1), 361–374 (2014)
12. Maddern, W., Milford, M., Wyeth, G.: CAT-SLAM: probabilistic localisation and mapping using a continuous appearance-based trajectory. IJRR **31**(4), 429–451 (2012)
13. Milford, M., Wyeth, G.: Mapping a suburb with a single camera using a biologically inspired SLAM system. IEEE Trans. Robot. **24**(5), 1038–1053 (2008)
14. Ribas, D., Ridao, P., Neira, J., Tardos, J.: SLAM using an imaging sonar for partially structured underwater environments. In: IEEE/RSJ International Conference on Intelligent Robots and Systems (IROS), pp. 5040–5045, October 2006
15. dos Santos, M.M.: Descrição e detecção de regiões subaquáticas parcialmente estruturadas em imagens acústicas adquiridas por um sonar de imageamento frontal. Master's thesis, Universidade Federal do Rio Grande - FURG (Appr 2016)

16. dos Santos, M.M., Ballester, P., Zaffari, G.B., Drews-Jr, P., Botelho, S.: A topological descriptor of acoustic images for navigation and mapping. In: 2015 12th Latin American Robotics Symposium and 2015 3rd Brazilian Symposium on Robotics (LARS-SBR), pp. 289–294, October 2015
17. Sibley, G., Mei, C., Reid, I., Newman, P.: Vast-scale outdoor navigation using adaptive relative bundle adjustment. Int. J. Robot. Res. **29**(8), 958–980 (2010). http://ijr.sagepub.com/content/29/8/958.abstract
18. Silveira, L., Guth, F., Drews-Jr, P., Ballester, P., Machado, M., Codevilla, F., Duarte, N., Botelho, S.: An open-source bio-inspired solution to underwater SLAM. In: IFAC Workshop on Navigation, Guidance and Control of Underwater Vehicles (NGCUV) (2015)
19. Williams, S., Mahon, I.: Simultaneous localisation and mapping on the great barrier reef. IEEE Int. Conf. Robot. Autom. **2**, 1771–1776 (2004)

Non-stationary VFD Evaluation Kit: Dataset and Metrics to Fuel Video-Based Fire Detection Development

Cristiano Rafael Steffens(✉), Ricardo Nagel Rodrigues, and Silvia Silva da Costa Botelho

Center of Computer Science, Federal University of Rio Grande, Avenida Itália – Km. 8, Rio Grande, Brazil
{cristianosteffens,ricardonagel,silviacb}@furg.br
http://c3.furg.br

Abstract. Datasets play a major role in the advance of computer vision techniques nowadays. Open, complete and challenging ground truth data, combined with standardized metrics are essential to push the development and allow the proper evaluation of computer vision algorithms. Even though a significant amount of work on VFD (video-based fire detection) systems has been developed, compare different algorithms is a laborious task due to the lack of common evaluation schemes and evaluation datasets. We address both of these issues by presenting a dataset of fire videos along with frame by frame annotations to be used for non-stationary fire detection algorithms training and validation. By the time, this is the largest dataset released on this subject matter. Standard video file formats and open markup languages where used to allow compatibility and convenient integration with the most popular computer vision libraries. The dataset includes hand-held, robot attached and drone attached footages and aims to boost the development of fully autonomous firefighter robots. The presented ground truth and metrics adapt to the majority of the state-of-the-art techniques and provides a reliable and unbiased solution to compare them. The dataset, example source-code and documentation are publicly available under the Creative Commons 3.0 license on GitHub.

Keywords: Dataset · Database · Evaluation · Validation · Fire

1 Introduction

The video-based fire detection has been a research topic longer than two decades now. The first research in this area dates back to 1993, when Healey *et al.* [24] presented a real-time system for automatic fire detection using color video input from stationary cameras. The physical properties of fire are considered for the

C.R. Steffens—Acknowledges the financial support of CNPq-Brazil, Finep and CAPES.

F. Santos Osório and R. Sales Gonçalves (Eds.): LARS 2015/SBR 2015, CCIS 619, pp. 135–151, 2016.
DOI: 10.1007/978-3-319-47247-8_9

development of algorithms that explore the spectral, spatial, and temporal properties of fire events. Latter, in 1996 Foo [16] proposed a knowledge based system based on heuristics of statistical measures such as mean, median, standard deviation, and first-order moment derived from the histogram and image subtraction analyses of successive image frames. Also in 1996, Plumb and Richards [36] present a prototype VFD system capable of determining the location and heat release rate of the fire employing transient temperatures using temperature-sensitive, color-changing sensors.

As research kept improving the results of VFD systems and video acquisition hardware became ubiquitous, some techniques have made their way to become commercial products. Only to name a few, we can list SIGNIFIRE video flame, smoke and intrusion detection system [15], AlarmEye AE3000 PC Based Video Fire Detection System [25], and FireVu Video Smoke Detection [1], which promise early and effective detection. Stipaničev *et al.* [40] lists some commercial VFDs for forest fire detection, showing their strengths and weaknesses, and presents IPNAS, a complete terrestrial tower based structure featuring cameras, wireless communication and software for forest fire surveillance. Despite the maturity that these solutions have reached, while all vendors advocate towards the capabilities of their VFD systems, it remains difficult to compare them.

On academic research, ideally, algorithms should be published with sufficient details to replicate the claimed results or, at least, with an executable binary and datasets. In this sense, the present work aims to offer a public test database and common evaluation metrics, allowing researchers to test and establish clear comparisons and consequently better benchmarks. Proprietary ground truth data is a barrier to independent evaluation of metrics and algorithms. The interested parties should be able to duplicate the metrics produced by various types of algorithms, validating them against the ground truth data and so comparing the results.

Current publications in the Video-based fire detection field can be grouped in two main categories according to their input data: stationary – where the camera is fixed to a tower or building – or non-stationary – where the camera is carried by a person or any moving equipment such as robots, cars or drones. The majority of the current research focuses on stationary systems, which are, usually, aimed to forest and outdoor surveillance. Usually, the proposed solutions are tested using a set of videos provided by the Bilkent VisiFire[1] Sample Video Clips as is the case for Celik *et al.* [4], Toreyin *et al.* [42–45], Habiboglu *et al.* [20] or by KMU Fire & Smoke[2] database, as is the case for Park *et al.* [35], Kwak *et al.* [28] and Shidik *et al.* [38]. More recently, Gunay *et al.* [19], Jin *et al.* [26] and Kong *et al.* [27] used the ICV database[3]. Other recent research such as Labati *et al.* [29] does not mention nor provide further access to the dataset. All the aforementioned datasets are incomplete, in the sense that they do not provide annotations for a standardized evaluation. Once the mentioned datasets

[1] Available at http://signal.ee.bilkent.edu.tr/VisiFire/Demo/SampleClips.html.
[2] Available at http://cvpr.kmu.ac.kr/Dataset/Dataset.htm.
[3] Available at http://vision.inha.ac.kr. Password protected.

lack essential information, such as the amount of negative and positive samples, it is impossible to compute the accuracy, specificity, fall-out and false negative rate among other relevant statistics.

Non-stationary VFDs are more recent and therefore there are only a few published researches targeting this problem. As far as we know, there is no publicly available fire detection dataset for tests on non-stationary videos. For instance, the dataset used in Borges *et al.* [2] is not available under the provided link. In personal contact the authors claimed the project was executed with private partners and neither code nor datasets where release. On the other hand, Chenebert *et al.* [9] do not provide any further information about the test data that has been used.

In fact, the lack of a standardization for evaluating the output of detection algorithms results in a situation where the reported performance results are strongly dependant on the definition of what is a correct detection. Since the evaluation process is not standardized, a comparison between two different algorithms remains difficult. Back in 2015, Tolouse *et al.* [46] presented a first attempt to bring some common metrics proposing a dataset and framework for benchmarking wildland fire segmentation algorithms. The dataset, which is composed of 100 RGB images acquired from the internet and specialized researchers, contains images of wildland (outdoor vegetation) fire in different contexts, such as fuel, background, luminosity, and smoke. All images of the dataset are characterised according to the principal colour of the fire, the luminosity, and the presence of smoke in the fire area. With this characterisation, the authors claim it to be possible to determine on which kind of images each algorithm is efficient. Although the authors do not provide a downloadable file, the dataset can be used via Octave/Matlab code trough their site[4].

The first contribution of this work, which extends a paper previously published by Steffens *et al.* [39], is the creation of a new non-stationary dataset composed by 20 annotated videos, featuring a wide range of difficulties including occlusion, different scales, camera vibration and a variety of bright and contrast conditions. As a second contribution we address the evaluation problem by presenting a new evaluation scheme composed by the following elements:

- An algorithm to find correspondences between a fire detector output and the annotated fire regions;
- Two separated rigorous and precise methods for evaluating any algorithm's performance on the proposed dataset. These two methods are intended for two different applications: fire location and frame-by-frame classification;
- C++ source code that implements these procedures.

The publicly available dataset and evaluation scheme proposed provide an straightforward manner to compare the performance of different algorithms. With an easy to use software library, it allows to develop machine learning based approaches once the training data can be easily generated. In this sense, researchers can focus solely on their methods, which will further prompt

[4] Available at http://firetest.cs.wits.ac.za/benchmark.

researchers to work on more difficult versions of the mobile robot based fire detection problem.

2 Dataset

In order to propose a new dataset and evaluation metrics the first step is to study how researchers tested their prior work. The non-stationary video based fire detection was first introduced by Borges and Izquierdo [2], which presents a method that analyzes the frame-to-frame changes of specific low-level features describing potential fire regions in order to classify newscast videos as containing or not hydrocarbon flames. The algorithm is structured in two main steps: color based classification and texture based classification. A statistical based equation is proposed for the color classification. In a second step color, area size, surface coarseness, boundary roughness and skewness of the R channel within the estimated fire regions are used as descriptors. The final classification is given by a Bayes classifier. Borges and Izquierdo [2] evaluate their detector by checking the classification of an entire video, disregarding the detection time and the location on the frame.

In 2011, Chenebert et al. [9] proposed a non-temporal texture driven approach for fire detection, combining a color-threshold equation previously proposed in [8]. The main idea of this work was to use supervised learning classifiers such as neural networks and regression trees. As texture descriptors they used ten bins histograms on the hue and saturation channels from the HSV color space. The authors also propose the use of the Gray-level Covariance Matrix [23], energy, entropy, contrast, homogeneity and correlation to find these parameters. Based on the descriptors, the authors claim that they were able to process 12 frames per second with an average precision close to 87.83 % using classification trees. The results are given considering exclusively if each frame was classified as fire or non-fire.

Given the wide range of possible applications for the proposed dataset, we release it as an unique wide pack that developers can use in different ways. In machine learning based fire detection techniques researchers may break it in training, test and evaluation sets while other approaches may possibly use the whole dataset only for evaluation. Therefore, we provide functions that allow software developers to access a specific frame and its associated annotations, making it easy to implement sample splitting techniques such as cross-validation or bootstrapping. Another important property of the state of the art detectors that leads us to release the dataset as an unique package is that some techniques, such as the ones proposed in [7,8,33,34], are dependant on a video sequence, using the flickering frequency and optical flow in order to determine the position and location of the fire region.

2.1 Video Properties

The dataset was created using 28022 frames distributed in 24 videos and published on the internet under the Creative Commons 3.0 license. It features a total of 14397 fire frames, which stand for 51.37 %, while the amount of 13625 non-fire frames represents the remaining 48.62 %. Considering that some fire frames present more than one annotated fire region, the dataset is composed by 17917 annotations. In order to provide a complete and challenging non-stationary video set, enabling a true evaluation of the detection algorithms, the files present the following properties:

- Variety of fire sources: different liquid and solid fuels produce different flames. The fuel and oxidizer may have a negligible influence on the flickering according to Hamins et al. [22] but still affect the color and shape.
- Uneven illumination: the videos were recorded in different and uncontrolled light conditions. It is worth to mention that fire is, by itself, a source of light that affects the surrounding objects. The dataset also features a large contrast and brightness range.
- Camera movement: the videos that compose the dataset where recorded using either hand-held or robot-attached cameras presenting forward and back, up and down, left and right movement and roll, pitch and yaw rotation.
- Different color accuracy settings: Different cameras were used resulting in a heterogeneous set of videos.
- Clutter: the fire flames can be obscured by surrounding objects, affecting them.
- Partial Occlusion: occlusion has been reported to be one of the biggest challenges to prior approaches specially when using flame contour and optical flow.
- Motion blur: camera shaking is very common when using a robot-attached camera due to the relative motion between the camera and the scene while the shutter is open.
- Scale and projection: fire does not have a specific size, scale or view point. The distances vary from less than 1 m to nearly 15 m meters.
- Reflection: as fire is a light source it may affect the objects that surround it, which will produce reflection and/or any other optical phenomenon.

2.2 Annotations

Fire can often assume random shapes, colors and transparency characteristics, which may have a direct impact on the detected area. For some image regions, deciding whether or not it represents a fire region can be a challenge. Besides the intrinsic flame properties, several factors, such as the image low resolution, scale and occlusion may turn this determination ambiguous. Due the lack of an objective criterion for including (or excluding) a fire region we resort to human judgment for this decision.

Sometimes the fire flickering process results in small fire flame regions that are completely separated from the main fire source. As fire flickers between 2 Hz and 10 Hz it is almost impossible to annotate the exact contour of each single fire flame. Therefore, the annotations are given by a rectangle that embraces the whole fire region. In cases where the fire flames are distant enough to be separated in rectangles without intersection our approach is to annotate them

Fig. 1. Approach for the validation process.

as distinct regions, even if they are actually related to the same source. The main reason to separate them is to keep the data clean and allow it to be used with machine learning approaches. Small fire sparkles are left out. For every frame that presents visible fire one or more annotations are made. A example is shown in the Fig. 1.

The annotations are released as XML files. The XML file format was chosen once it is a W3C standard, endorsed by software industry market leaders and easy to read and understand. Every video is provided with its own annotation file, making it simple to extend the dataset or separate it in smaller parts for training, testing and evaluation steps.

The heat map presented in Fig. 2 gives an idea of the fire region placement throughout the dataset. Regions with cold colors present less fire occurrences. Hot colors represent a higher number of occurrences. It can be noted that the majority of the ground truth annotations occur at the center of the image, endorsing some assumptions that have been made in [2]. The average area of a annotation is 61512 pixels (aprox. 250 × 250 px square). In proportion, fire regions size stands for 8.92 % of the frame size. Figure 3 shows a few excerpts of the videos that compose the dataset.

2.3 Software Tools

To use the ground truth data available in XML format along with the corresponding video files, we also released relevant software artifacts. An OpenCV [3] based

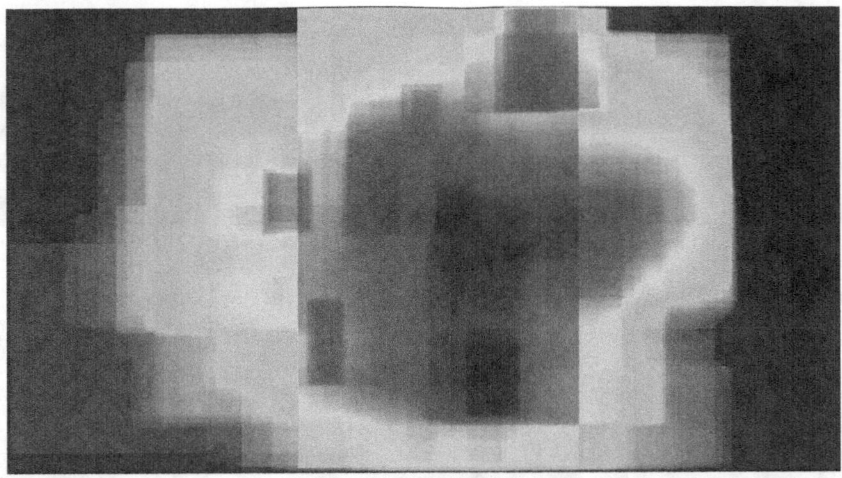

Fig. 2. Heat map showing the density of the annotated regions throughout the dataset.

file manipulation library is provided so that developers can easily access the dataset content calling standard methods. We also provide the software implementation and documentation to enable researchers to compare their algorithms against the ground truth data using the methodology presented in Sect. 3. The interface of the annotation tool is shown in Fig. 4. Quick commands, usage tips and current frame information is presented on the bottom of the frame.

A ground truth visualization tool is provided, allowing the user to load a video file and the corresponding XML file into an application, displaying the annotations overlaid on the image. It implements some features that allow the user to get pertinent data such as image statistics and histograms.

There will be occasions where there is new ground truth data to add or existing data to modify. Therefore a ground truth editor is provided that allows the users to do it graphically. As the annotations are delivered in a XML format the user can also edit the file directly using any text editor.

3 Evaluation Criteria

One challenge in comparing fire detection systems is the lack of agreement on the desired output. The reported performance results are highly dependant on the definition of what is a correct detection result. Therefore we propose three different evaluation approaches: frame based, where it becomes a classification problem, location based, where it becomes a location problem and time based, where the detection delay is considered. We do not consider the approach presented by [2] where they only take in account the classification as a whole video containing or not fire.

On the frame based evaluation, each frame is an instance as in a binary classifier. This approach is the most commonly used to evaluate fire detection

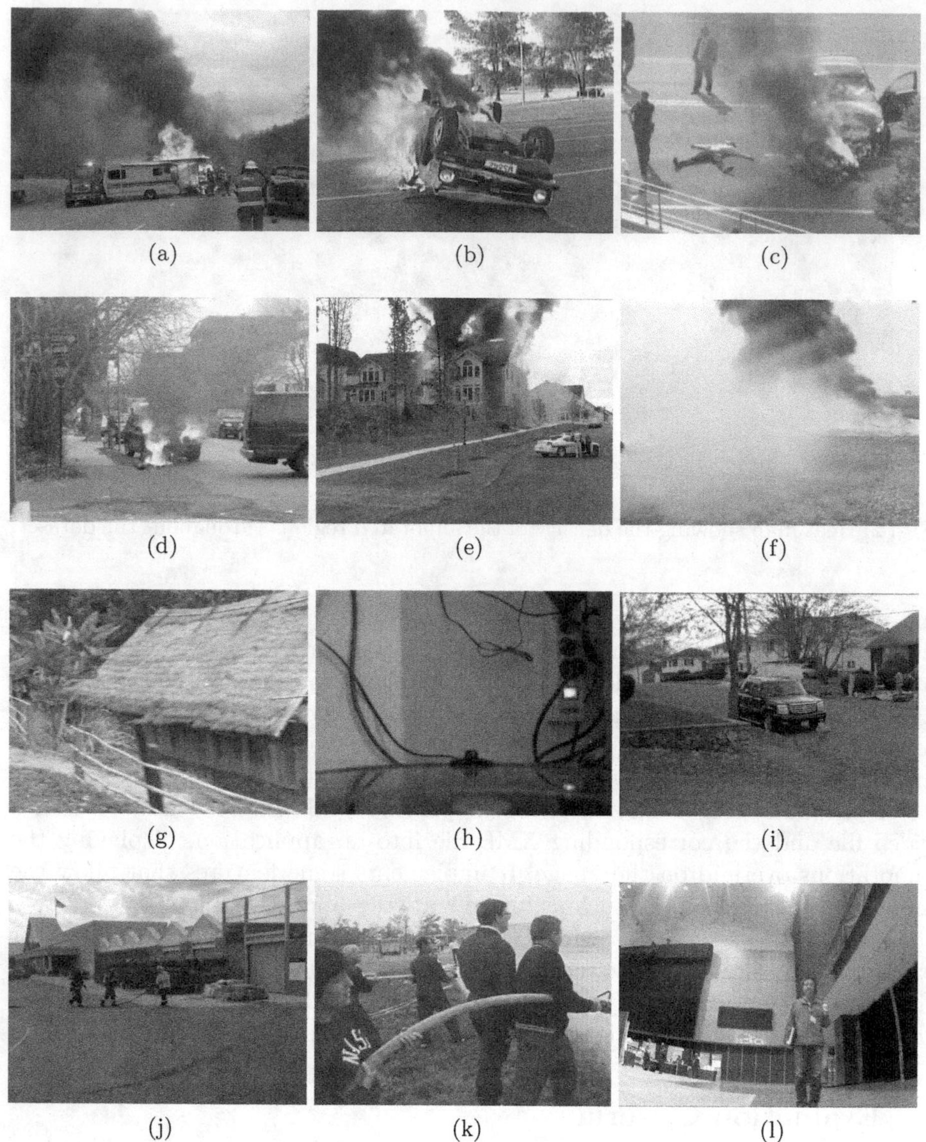

Fig. 3. Sample frames for some of the videos in the dataset. The two top rows show some videos that present fire flames, while the bottom rows show some negative samples.

systems. Even though the authors did not fully explore the potential of the approach, it can be assumed that it has been used before in [5,6,9], where the recall has been labeled as Detection Rate and the precision has been labeled as False Alarm Rate. Considering the whole frame as fire or non-fire makes it

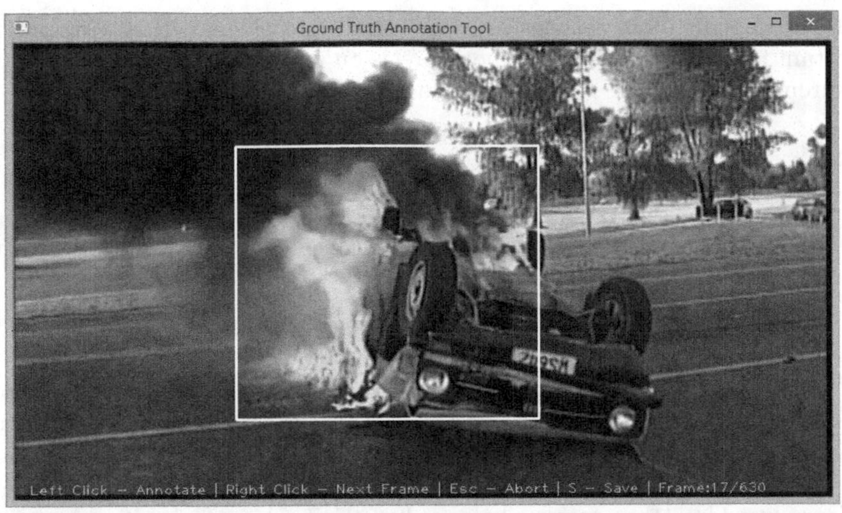

Fig. 4. Ground truth annotation tool interface.

possible to compute the true positive rate TPR (a.k.a. Recall or Hit Rate), true negative rate TNR (a.k.a. Sensitivity), positive predictive value PPV (a.k.a. precision), negative prediction value NPV, false positive rate FPR (a.k.a Fall-Out), false discovery rate FDR, and false negative rate FNR. The corresponding equations are presented in Eqs. 1, 2, 3, 4, 5 and 6 where P and N represent respectively the positive and negative samples count in the dataset and T_P and T_N are the number of true positive and true negative detections.

$$TPR = \frac{T_P}{P} \tag{1}$$

$$SPC = \frac{T_N}{N} \tag{2}$$

$$PPV = \frac{T_P}{T_P + F_P} \tag{3}$$

$$NPV = \frac{T_N}{T_N + F_N} \tag{4}$$

$$FPR = \frac{F_P}{N} \tag{5}$$

$$FDR = \frac{F_P}{F_P + T_P} \tag{6}$$

Considering the fire detection problem as a frame-by-frame binary classification task also makes it possible to compute the accuracy ACC (Eq. 7), F_1 score (Eq. 8) and Matthews correlation coefficient MCC (Eq. 9). In so far as we know, the first time MCC and F Score were used used to compare fire detectors points back to Collumeau *et al.* [11]. Latter, they have also been used in

[39, 46]. Although these metrics are not usually considered, they can provide important insights by considering both precision and recall, resulting in a better measurement to evaluate the quality of a detector.

$$ACC = \frac{T_P + T_N}{P + N} \tag{7}$$

$$F_1 = \frac{2 \times PPV \times TPR}{PPV + TPR} \tag{8}$$

$$MCC = \frac{T_P T_N - F_P F_N}{\sqrt{(T_P + F_P)(T_P + F_N)(T_N + F_P)(T_N + F_N)}} \tag{9}$$

The F-Measure was first introduced by Chinchor [10] as a measure that combines both precision and recall in one single metric through the harmonic mean. When the recall and precision values are considered to have the same weight that measure is named as F1-Score. Yet the Matthews correlation coefficient, was introduced in [31] as another balanced metric for binary classifiers that considers true and false positives and negatives. It can be used even if the classes are of different sizes. The MCC scale goes from -1 up to $+1$, where -1 indicates a total disagreement between prediction and observation, 0 indicates a random prediction and $+1$ indicates that the classifier output perfectly matches the ground truth labels.

While the frame based approach may be appropriated for most cases, it may also be interesting to evaluate the location of the detection. Many researchers have yet proposed benchmarking metrics for face and object detection. Most of them, however, assume that there will be only one detection that matches the ground truth annotation which is not appropriated for the fire detection problem. As fire does not have a fixed shape and color and can separate into many flame sparks it is better to use a many to one approach.

Detections are considered true or false positives based on the area overlap with the ground truth rectangles. The similarity function S is given by the Eq. 10 where d_i is the detector output and gt_i is the annotation. A detection is considered as correct when $S > 0.5$.

$$S = \frac{d_i \cap gt_i}{d_i} \tag{10}$$

Once the location overlap based approach does not have the negative sample count we can only compute metrics that do not rely on the negative data. Using the positive data and considering that by definition the precision is the fraction of retrieved instances that are relevant, we are able to compute it using the Eq. 11. The recall, which is defined as the fraction of relevant instances that are retrieved, is given by Eq. 12. D represents the number of detections while GT represents the number of annotations in the evaluation dataset.

$$PPV = \frac{T_P}{D} \tag{11}$$

$$TPR = \frac{GT - F_N}{GT} \tag{12}$$

Another fundamental information when evaluating fire detection systems is the time gap between the moment the fire starts to be detected and the first time it appears in the ground truth. The latency Lt is given by the Eq. 13 where d_i is the frame in which the first detection occurs and gt_i is the first time fire appears in the ground truth annotations. Usually the fire detectors present a intrinsic latency because they use the flame flickering as input.

$$Lt = min(d_i) - min(gt_j) \tag{13}$$

One last important metric when evaluating the quality of a non-stationary video-based fire detection system is the accuracy of the detection location. The average similarity $\overline{S}$ may be an useful information to direct the development of algorithms for active vision and mobile robot systems.

4 Experimental Setup

The dataset has been used to compare three different state-of-the-art techniques. The source code for each solution was implemented in C++ and the default parameters were used. The methods proposed by Celik [4] and Zhou [47], are both based on temporal evaluation. Chenebert's [9] approach is texture based and non-temporal. The results for this experimental setup are given in Tables 1 and 2, which respectively show the performance considering the frame by frame classification and the location based metrics.

Table 1. Frame by frame results.

Metrics	Better	Celik [4]	Zhou [47]	Chenebert [9]
TPR	↑	0.739	0.987	0.990
SPC	↑	0.317	0.022	0.724
PPV	↑	0.654	0.638	0.857
NPV	↑	0.410	0.501	0.979
FPR	↓	0.682	0.977	0.275
FDR	↓	0.345	0.361	0.142
FNR	↓	0.260	0.012	0.009
ACC	↑	0.585	0.635	0.890
F_1 Score	↑	0.694	0.775	0.919
MCC	↑	0.060	0.036	0.773

While the temporal based methods show a high recall and could make us think that the results are satisfactory, the balanced metrics show that, in fact,

Table 2. Location based results

Metrics	Better	Celik [4]	Zhou [47]	Chenebert [9]
L_{PPV}	↑	0.251	0.019	0.832
L_{TPR}	↑	0.732	0.440	0.979
F_1 Score	↑	0.384	0.037	0.902
$\overline{S}$	↑	0.250	0.020	0.801

the predictions are almost random, having a small correlation with the expected outputs. On the same hand, when the location based metrics are considered, Table 2 proves that the Chenebert's method outperforms both Celik's and Zhou's methods, being the only one to report a high L_{PPV} and L_{TPR}. The mean similarity $\overline{S}$ also shows that the stationary camera based systems do not present a good detection/ground truth area intersection. The Borges *et al.* method was not tested once the code is not publicly available and it requires some threshold values that are not presented on the original paper.

5 Additional Considerations

While the proposed evaluation kit presents a complete set of videos, annotations and software artifacts, we also have to acknowledge some of its limitations. First and most important, as has been mentioned in the Sect. 2.2 the annotations do not provide the exact contours of the fire regions, which in turn does not enable developers to compute the Hafiane's criterion [21]. The Hafiane's criterion (a.k.a. Hafiane quality index) is a supervised evaluation criterion for region based segmentation methods. It considers the position, shape, and size of the segmented regions and has been used before for the evaluation of VFD systems in [11,46]. While we think the Hafiane's criterion could be an useful metric, we have to consider that, in turn, the aforementioned publications used datasets with only a few hundred images or less.

Also regarding the annotations, we reinforce they are based on OpenCV XML format and have not been tested with other computer vision and deep learning libraries. However, we believe this wont pose a major problem, once the format is a standard in the software development industry. In comparison to the framework for benchmarking of wildland fire segmentation algorithms from [46] the proposed evaluation kit lacks relevant information from the context, such as climate conditions, burning material or camera pose. That could be an important improvement for future updates in the dataset.

The videos are in the MPEG-4 file format, which requires specific codecs. On the other hand, mp4 codecs are available for all mainstream operational systems. By default the videos are stored in the YUV color format, with 8 bits bit depth. As they were obtained from different and converted to this format latter for the sake of standardization, this might result in an overall non-significant quality loss. When using OpenCv library it will, by default, convert the frames to BGR (blue, green, red) color space to simplify manipulation.

Aside from being the first dataset aimed to evaluate and validate non-stationary fire detection systems, the presented dataset is also larger than others which have been previously proposed. For instance, [11] used 76 pictures, [46] used 100 images. When it comes to datasets used in other publications, but not publicly available or annotated in details, it is important to notice that the presented kit has almost balanced fire and non-fire samples. Labati *et al.* [29] used 72852 frames in total, but the data was very unbalanced, with only 1328 smoke samples. Similarly Toreyin *et al.* [43] used a dataset with 83745 frames in 61 sequences, but only 19 of them contained fire regions. That is an important remark, given that many machine learning and data mining algorithms do not work properly on imbalanced datasets.

Although, it is aimed to be used in non-stationary systems, it can also come in hand to evaluate stationary VFD systems. Stationary approaches can be divided in two large groups: (I) systems based on single images and (II) systems based on multiple frames. The latter ones usually depend on fire flickering features extracted from videos and therefore they can not be trained or evaluated with a shaking camera. The first ones, on the other hand, depend only on color, boundaries or texture clues extracted from a single frame, and therefore they can be trained using the presented dataset.

The proposed dataset has been integrated to CvWorks – Computer Vision Framework[5] and all parts of the kit are available on Github[6]. The dataset is downloadable and can be easily extended due its annotation model. Public datasets play an important role for the development and evaluation of computer vision solutions. Datasets such as Image-net [12], Pascal VOC [13] and Calltech [14,18] had a great impact on the development of object recognition. So had Kitty [17] and TUM-RGB-D [41] to the development of autonomous vehicles, and AM-FED [32] and CK+ [30] for the development of human action and emotion recognition. In the same way, our dataset can be used for the development of machine learning approaches such as deep neural networks, which are a current trend in this field.

For methods that need to compute a threshold value, we recommend to follow the same procedure as Rudz *et al.* [37]. In this approach, thresholds are found using a third of the images of the dataset taken randomly. The value of the threshold that maximises the F_1 Score (Eq. 8) for these images is estimated with a direct pattern search algorithm. Once the threshold value is found it should be used for all testing steps. Otherwise, adjusting the threshold and parameters during the process would result in over-fitting, invalidating the whole results.

[5] Available at http://www.cvworks.c3.furg.br.

[6] Available at https://github.com/steffensbola/furg-fire-dataset.

6 Conclusion

The validation kit presented, featuring a dataset of videos annotated at the frame level, evaluation metrics, along with the software artifacts, enable researchers to accurately evaluate their video-based fire detection methods. The dataset is publicly available under a non-restrictive copyright license which allows it to be freely shared and redistributed. Open file formats are used to provide large compatibility and easy use. Example implementations with source code are provided using the OpenCv computer vision library, which currently is a *de facto* standard for computer vision research. For researches who intend to use the provided kit with another programming language we provide evaluation metrics that are unambiguous and straightforward to implement.

In this work, as a demonstration, three state-of-the-art VFD algorithms were implemented and their performances analysed. The obtained results justify the effort to build the presented dataset and evaluation scheme, once they show there is still room for further improvements in non-stationary VFD systems. The dataset is a convenient resource to support the development of machine learning approaches and benchmark tools, enabling the researchers to focus on new algorithms and improvements rather than implementing their own metrics and creating their own datasets from scratch. Aside from being a standardized validation for non-stationary VFD systems, the proposed dataset can boost research and push the development of autonomous fire hazard combat alternatives such as early stage fire alarm, automatic fire extinguishers and firefighting robots.

Our evaluation kit addresses a computer vision application that did not have a publicly available evaluation scheme yet. Therefore, the importance of the presented dataset for fire detection algorithms can be compared to what ground truth data represented in fields like face, object detection and tracking decades ago, fostering their development to reach a level where they have become reliable enough to be used in real life applications.

References

1. AD Group: Firevu video smoke detection (2016). http://www.firevu.com/
2. Borges, P.V.K., Izquierdo, E.: A probabilistic approach for vision-based fire detection in videos. IEEE Trans. Circ. Syst. Video Technol. **20**(5), 721–731 (2010)
3. Bradski, G.: The OpenCV library. Dr. Dobb's J. Softw. Tools **25**(11), 120–126 (2000)
4. Celik, T.: Fast and efficient method for fire detection using image processing. ETRI J. **32**(6), 881–890 (2010)
5. Celik, T., Demirel, H.: Fire detection in video sequences using a generic color model. Fire Saf. J. **44**(2), 147–158 (2009)
6. Celik, T., Ozkaramanli, H., Demirel, H.: Fire and smoke detection without sensors: image processing-based approach. In: 15th European Signal Processing Conference, EUSIPCO, pp. 147–158 (2007)
7. Chen, J., He, Y., Wang, J.: Multi-feature fusion based fast video flame detection. Build. Environ. **45**(5), 1113–1122 (2010)

8. Chen, T.H., Wu, P.H., Chiou, Y.C.: An early fire-detection method based on image processing. In: 2004 International Conference on Image Processing, ICIP 2004, vol. 3, pp. 1707–1710. IEEE (2004)
9. Chenebert, A., Breckon, T.P., Gaszczak, A.: A non-temporal texture driven approach to real-time fire detection. In: 2011 18th IEEE International Conference on Image Processing (ICIP), pp. 1741–1744. IEEE (2011)
10. Chinchor, N.: MUC-4 evaluation metrics. In: Proceedings of the 4th Conference on Message Understanding, MUC4 1992, pp. 22–29. Association for Computational Linguistics, Stroudsburg, PA, USA (1992). http://dx.doi.org/10.3115/1072064.1072067
11. Collumeau, J.F., Laurent, H., Hafiane, A., Chetehouna, K.: Fire scene segmentations for forest fire characterization: a comparative study. In: 2011 18th IEEE International Conference on Image Processing (ICIP), pp. 2973–2976. IEEE (2011)
12. Deng, J., Dong, W., Socher, R., Li, L.J., Li, K., Fei-Fei, L.: Imagenet: a large-scale hierarchical image database. In: 2009 IEEE Conference on Computer Vision and Pattern Recognition, CVPR 2009, pp. 248–255. IEEE (2009)
13. Everingham, M., Van Gool, L., Williams, C.K.I., Winn, J., Zisserman, A.: The PASCAL visual object classes (VOC) challenge. Int. J. Comput. Vis. **88**(2), 303–338 (2010)
14. Fei-Fei, L., Fergus, R., Perona, P.: Learning generative visual models from few training examples: an incremental bayesian approach tested on 101 object categories. Comput. Vis. Image Underst. **106**(1), 59–70 (2007)
15. Fike Corporation: Signifire video flame, smoke and intrusion detection system, January 2016. http://www.fike.com/products/signifire-video-flame-smoke-intrusion-detection-system/
16. Foo, S.Y.: A rule-based machine vision system for fire detection in aircraft dry bays and engine compartments. Knowl.-Based Syst. **9**(8), 531–540 (1996)
17. Geiger, A., Lenz, P., Stiller, C., Urtasun, R.: Vision meets robotics: the kitti dataset. Int. J. Robot. Res. **32**(11), 1231–1237 (2013). 0278364913491297
18. Griffin, G., Holub, A., Perona, P.: Caltech-256 object category dataset (7694) (2007). http://authors.library.caltech.edu/7694
19. Gunay, O., Cetin, A.E.: Real-time dynamic texture recognition using random sampling and dimension reduction. In: 2015 IEEE International Conference on Image Processing (ICIP), pp. 3087–3091. IEEE (2015)
20. Habiboglu, Y.H., Gunay, O., Cetin, A.E.: Real-time wildfire detection using correlation descriptors. In: 19th European Signal Processing Conference (EUSIPCO 2011), Special Session on Signal Processing for Disaster Management and Prevention, pp. 894–898 (2011)
21. Hafiane, A., Chabrier, S., Rosenberger, C., Laurent, H.: A new supervised evaluation criterion for region based segmentation methods. In: Blanc-Talon, J., Philips, W., Popescu, D., Scheunders, P. (eds.) ACIVS 2007. LNCS, vol. 4678, pp. 439–448. Springer, Heidelberg (2007). doi:10.1007/978-3-540-74607-2_40
22. Hamins, A., Yang, J., Kashiwagi, T.: An experimental investigation of the pulsation frequency of flames. Symp. (Int.) Combust. **24**(1), 1695–1702 (1992). Elsevier
23. Haralick, R.M.: Ridges and valleys on digital images. Comput. Vis. Graph. Image Process. **22**(1), 28–38 (1983)
24. Healey, G., Slater, D., Lin, T., Drda, B., et al.: A system for real-time fire detection. In: 1993 IEEE Computer Society Conference on Computer Vision and Pattern Recognition Proceedings, CVPR 1993, pp. 605–606. IEEE (1993)
25. InnoSys Industries Inc: Alarmeye ae3000 pc based video fire detection system (2016). http://www.innosys-ind.com/

26. Jin, D., Li, S., Kim, H.: Robust fire detection using logistic regression and randomness testing for real-time video surveillance. In: 2015 IEEE 10th Conference on Industrial Electronics and Applications (ICIEA), pp. 608–613. IEEE (2015)
27. Kong, S.G., Jin, D., Li, S., Kim, H.: Fast fire flame detection in surveillance video using logistic regression and temporal smoothing. Fire Saf. J. **79**, 37–43 (2016)
28. Kwak, J., Ko, B.C., Nam, J.Y.: Forest smoke detection using CCD camera and spatial-temporal variation of smoke visual patterns. In: 2011 Eighth International Conference on Computer Graphics, Imaging and Visualization (CGIV), pp. 141–144. IEEE (2011)
29. Labati, D., Genovese, A., Piuri, V., Scotti, F.: Wildfire smoke detection using computational intelligence techniques enhanced with synthetic smoke plume generation. IEEE Trans. Syst. Man Cybern. Syst. **43**(4), 1003–1012 (2013)
30. Lucey, P., Cohn, J.F., Kanade, T., Saragih, J., Ambadar, Z., Matthews, I.: The extended Cohn-Kanade dataset (CK+): a complete dataset for action unit and emotion-specified expression. In: 2010 IEEE Computer Society Conference on Computer Vision and Pattern Recognition Workshops (CVPRW), pp. 94–101. IEEE (2010)
31. Matthews, B.W.: Comparison of the predicted and observed secondary structure of t4 phage lysozyme. Biochim. Biophys. Acta (BBA)-Protein Struct. **405**(2), 442–451 (1975)
32. McDuff, D., El Kaliouby, R., Senechal, T., Amr, M., Cohn, J.F., Picard, R.: Affectiva-MIT facial expression dataset (AM-FED): naturalistic and spontaneous facial expressions collected in-the-wild. In: 2013 IEEE Conference on Computer Vision and Pattern Recognition Workshops (CVPRW), pp. 881–888. IEEE (2013)
33. Mueller, M., Karasev, P., Kolesov, I., Tannenbaum, A.: Optical flow estimation for flame detection in videos. IEEE Trans. Image Process. **22**(7), 2786–2797 (2013)
34. Nguyen-Ti, T., Nguyen-Phuc, T., Do-Hong, T.: Fire detection based on video processing method. In: 2013 International Conference on Advanced Technologies for Communications (ATC), pp. 106–110. IEEE (2013)
35. Park, J., Ko, B., Nam, J.Y., Kwak, S.: Wildfire smoke detection using spatiotemporal bag-of-features of smoke. In: 2013 IEEE Workshop on Applications of Computer Vision (WACV), pp. 200–205. IEEE (2013)
36. Plumb, O., Richards, R.: Development of an economical video based fire detection and location system. US Department of Commerce, Technology Administration (1996)
37. Rudz, S., Chetehouna, K., Hafiane, A., Laurent, H., et al.: Investigation of a novel image segmentation method dedicated to forest fire applications. Meas. Sci. Technol. **24**(7), 075403 (2013)
38. Shidik, G.F., Adnan, F.N., Supriyanto, C., Pramunendar, R.A., Andono, P.N.: Multi color feature, background subtraction and time frame selection for fire detection. In: Robotics, Biomimetics, and Intelligent Computational Systems (ROBIONETICS), 2013 IEEE International Conference on, pp. 115–120 (2013). doi:10.1109/ROBIONETICS.2013.6743589
39. Steffens, C.R., Rodrigues, R.N., da Costa Botelho, S.S.: An unconstrained dataset for non-stationary video based fire detection. In: IEEE LARS/SBR 2015. Uberlândia, October 2015
40. Stipaničev, D., Vuko, T., Krstinić, D., Štula, M., Bodrožić, L.: Forest fire protection by advanced video detection system-croatian experiences. In: Third TIEMS Workshop-Improvement of Disaster Management System, pp. 26–27. Citeseer (2006)

41. Sturm, J., Engelhard, N., Endres, F., Burgard, W., Cremers, D.: A benchmark for the evaluation of RGB-D SLAM systems. In: 2012 IEEE/RSJ International Conference on Intelligent Robots and Systems (IROS), pp. 573–580. IEEE (2012)
42. Toreyin, B.U., Cetin, A.E.: Wildfire detection using LMS based active learning. In: 2009 IEEE International Conference on Acoustics, Speech and Signal Processing, ICASSP 2009, pp. 1461–1464. IEEE (2009)
43. Töreyin, B.U., Dedeoğlu, Y., Güdükbay, U., Cetin, A.E.: Computer vision based method for real-time fire and flame detection. Pattern Recogn. Lett. **27**(1), 49–58 (2006)
44. Toreyin, B., Dedeoglu, Y., Cetin, A.: Wavelet based real-time smoke detection in video. In: 13th European Signal Processing Conference, pp. 1–4, September 2005
45. Toreyin, B., Dedeoglu, Y., Cetin, A.: Contour based smoke detection in video using wavelets. In: 14th European Signal Processing Conference, pp. 1–5, September 2006
46. Toulouse, T., Rossi, L., Akhloufi, M., Celik, T., Maldague, X.: Benchmarking of wildland fire colour segmentation algorithms. IET Image Process. **9**(12), 1064–1072 (2015)
47. Zhou, X.L., Yu, F.X., Wen, Y.C., Lu, Z.M., Song, G.H.: Early fire detection based on flame contours in video. Inf. Technol. J. **9**(5), 899–908 (2010)

GPU-Services: GPU Based Real-Time Processing of 3D Point Clouds Applied to Robotic Systems and Intelligent Vehicles

Leonardo Christino(✉) and Fernando Osório(✉)

Mobile Robotics Lab (LRM), Center for Robotics (CRob/USP-SC), ICMC, University of São Paulo, Trabalhador São-carlense, 400, São Carlos, São Paulo 13566-590, Brazil
{leomilho,fosorio}@icmc.usp.br

Abstract. The GPU-Services project fits into the context of research and development of methods for data processing of three-dimensional sensors data applied to mobile robotics and intelligent vehicles. The implemented methods are called services on this project, which provide 3D point clouds pre-processing algorithms, such as, data alignment, segmentation of safe/unsafe navigable zones (e.g. separating ground from obstacles and borders/curbs) and elements of interest detection. Due to the large amount of data provided by the sensors to be processed in a very short time, these services use the GPU (NVidia CUDA) to perform partial or complete parallel processing of these data. The project aims to provide data processing services to an autonomous car, forcing the services to approach real-time processing, which is defined as completing all data processing routines before the arrival of the sensor's next frame. This work was implemented considering 3D data acquired from a LIDAR, more specifically from a Velodyne HDL-32. The sensor data is structured in the form of a cloud of three-dimensional points, allowing for great parallel processing. However, the major challenge is the high rate of data received from this sensor (around 700,000 points/sec or 70.000 points/frame at 10 Hz), which gives the motivation of this project: to use the full potential of sensor and to efficiently use the parallelism of GPU programming. The GPU services are divided into four steps: The first step is an intelligent extraction, reorganization and spacial correction of the data provided by the Velodyne multi-layer laser sensor; The second stage is the segmentation of planar data; The third stage is object segmentation; The fourth stage is to develop a methodology that unite the results from the previous steps in order to better detect the curbs. The services were implemented and the performance was evaluated using traditional sequential data processing (CPU data processing) and parallel data processing (GPU CUDA implementations). Besides that, different NVidia GPUs were also tested, allowing us to process the acquired data much faster than using the CPUs, and in some cases faster than it was provided by the Velodyne sensor.

Keywords: 3D point cloud · Sensor data processing · Services · GPU · Parallel processing · LIDAR

F. Santos Osório and R. Sales Gonçalves (Eds.): LARS 2015/SBR 2015, CCIS 619, pp. 152–171, 2016.
DOI: 10.1007/978-3-319-47247-8_10

1 Introduction

The mobile robots research is presented as an area in constant and rapid expansion. Today and in the near future, mobile robots are being applied in various areas, from the household to terrestrial and aerial vehicles, with civilian, agricultural, industrial and military applications. Because of this wide range of areas, mobile robotics has become an area of extensive research and knowledge requires various fields such as, for example, Mechanical, Electrical and Computer Engineering.

Within the mobile robotics area, there are those able to perform their task assigned without much supervision and with little intervention and/or coordination by humans, called Autonomous/Unmanned Mobile Robots (UGVs and UAVs). These robots perform tasks such as mapping environments, self-localization, path planning, action planning and execution of the navigation (trajectory tracking and obstacle avoidance). To do these and other mobile robots tasks need to receive information from the environment (through sensors), make decisions and plan actions (by decision-making algorithms), and perform actions on the environment (through actuators).

The area of autonomous mobile robotics is relatively recent and the biggest challenge development is to create robots capable of interacting with the environment, learn and make decisions right to complete their tasks successfully. A direct application of this technology is the development of intelligent autonomous vehicles. This area was created within the scope of a system of autonomous highways in the 60s [4], and now contributes to increase safety on roads and improve traffic flow in a road network, thus reducing the number of traffic accidents, decreasing jams and helping the disabled and elderly transportation. In intelligent vehicles applications for urban spaces, the DARPA Urban Challenge [2] was one of the most important initiatives to promote the development of these vehicles, resulting in projects like the Google Self-Driving car, and where the project CaRINA [5] from USP-ICMC stands out in Latin-America. There are also a lot of interest in vehicles automation applications, like people and cargo transportation, as well as in applications in mining and agriculture.

In the development of algorithms for an autonomous vehicle, many areas of computation are exercised. A prime example are the areas of control and performance in robotics, involving speed control, acceleration and torque for executing a mission. Areas related to planning, such as the creation of the missions of the vehicle, trace the route he should perform, keep tracking of the vehicle position in the path/map (and its destination), are also highly important. Finally, areas of sensing, such as identification of safe zones for navigation (e.g. streets, roads and flat surfaces), obstacle detection, pedestrian identification and objects tracking in the scene, also show how complex can be an autonomous mobile robotic system.

A serious problem and limitation with this type of application is its response time. Since it is a critical application, it has some real-time environment constraints to properly control the vehicle and prevent traffic accidents. Besides that, sensors provide lots of critical data, but usually it is very hard to extract useful information from these data. The present project, GPU-Services, looks to develop faster methods for data processing to achieve this goal. The processing speed becomes even more critical because of the need for various algorithms to run simultaneously (e.g. mapping, auto-localization, path planning, and obstacle avoidance). Parallelism exploitation methodologies, such as cluster processing, grid computing and other custom techniques have been sought in order to solve these problems, but each of them also presents limitations, as communication costs, energy costs and may still have high financial costs. These approaches are of limited application inside a vehicle, so it is preferred to adopt an embedded parallel processing technology with lower cost and lower space and energy consumption.

The development of parallel algorithms technology for super-computing have used much of the potential of commonly used video cards, also known as GPUs (Graphical Processing Units), with the advent of an extension of the C/C++ language developed by NVidia called CUDA [16]. Since the data received by the sensor is structured in 3D point clouds, mostly relying on local spatial information processing, which enables the exploitation of parallelistic algorithms. The use of GPUs is a good way to process 3D sensor data, execute some usual tasks, solve common problems related to mobile robots, and detect obstacles (danger) fast enough. Another feature of the sensor adopted in this project (Velodyne HDL-32 sensor, a Laser based LIDAR, described in Sect. 3.2) is the high rate of incoming sensor data (about 700,000 points per second), which completes the motivation for this project: to use efficiently all the sensors potential by using the GPUs parallelism to obtain a reasonable response time (ideally, processing the data as fast as it can be generated and acquired from the sensor).

This project is part of the autonomous vehicles research under development at the LRM Laboratory (Mobile Robotics Lab. - ICMC/USP), related also to the CaRINA Project. It aims to develop services for three-dimensional data processing applied to mobile robotics sensors, and more particularly, applied to the Velodyne sensor adopted in our autonomous vehicles. These services use elements of parallel processing through the super-computing power of GPUs. The focus of this project is to provide sensory data processing services for an autonomous vehicle, forcing the services to approach a real-time system. The services will be divided into stages which seeks faster processing times by exploiting parallelism. The GPU services are divided into four steps: The first step is an intelligent extraction, reorganization and spacial correction of the data provided by the Velodyne multi-layer laser sensor; The second stage is the segmentation of planar data; The third stage is object segmentation; The fourth stage is to develop a methodology that unite the results from the previous steps in order to better detect the curbs. The results achieved with this work have several applications to researches underway in the laboratory which it was developed, but can also be applied to other problems that require faster 3D point clouds data processing.

2 Related Works

Given the current strong interest of car companies and research centers in this area, several projects of autonomous vehicles have emerged in recent years [2] [13]. Almost a decade after the pioneering project: ALVINN, three competitions sponsored by DARPA in 2004, 2005 [19] and 2007 [11,14], defined important milestones in the development of autonomous vehicles. In the first two editions of the DARPA Challenge, participants were required to develop vehicles capable of traveling long distances in the desert autonomously, and in its latest edition vehicles were required to complete a route on streets (urban environment) respecting all applicable traffic laws. After the DARPA challenges, competitions and other challenges have been created, such as European initiative ELROB (The European Land-Robot Trial) which is carried out regularly, and GCDC (Grand Cooperative Driving Challenge). Several companies and universities also dedicated whole laboratories for the development of this technology [13].

2.1 DARPA Challenges

The DARPA Grand Challenge was proposed in 2003 to encourage research and progress in the area of autonomous vehicles. With a prize of 1 million in its first edition that occurred in 2004, its main goal was to create a vehicle capable of travel approximately 220 km of rough terrain in the desert in a maximum of ten hours. Although 107 teams have registered and 15 started the route, neither team was able to travel more than 5 % of the proposed path, a result that clearly shows the difficulty involved in the challenge. A second edition of the challenge was held in 2005, with a prize of two million dollars. In this edition 195 teams registered and 23 were qualified and started the route, with only five autonomous vehicles finishing it. With time 6 h 53 min, Stanley vehicle from Stanford University (Fig. 1(a)) won the competition. The strategy used by the Stanford team was to create a 3D map in real time ground ahead of the vehicle (obtained from LIDARs), dodging obstacles and possible depressions. Also video cameras were used to identify changes in the pattern of ground. In addition to the sensors, were used in the car six computers for processing information and decision-making. It is important to note that the GPS points describing with considerable precision the trajectory that the vehicle should follow was previously provided to the teams, reducing the problem to navigation and bypass local obstacles [19].

In November of 2007, DARPA conducted the 2007 Urban Challenge on an Air Force Base in California. In it, the competitors developed autonomous vehicles to navigate through an urban environment, moving to the middle of traffic vehicles, negotiating intersections and avoiding obstacles while respecting the traffic laws of State of California [11]. In this edition, 89 teams were initially accepted to participate, 11 passed through qualifications and participated in the challenge and of these only six completed the 96 km path. The Boss vehicle shown in Fig. 1(b), from the Tartan Racing team led by Carnegie Mellon University, was awarded first place to complete the route in 4 h and 10 min with the Stanford team placing second with their Junior car. The Boss used a navigation algorithm able to locate and estimate the track format, detecting the street borders and main path references using a Velodyne (HDL-64 model) as the main sensor to estimate this. Besides that, a system for detecting stationary and moving obstacles over time [14] was also implemented using the 3D perceived data.

(a) Junior (b) Boss

Fig. 1. Winners of the two DARPA challenges

2.2 Current Projects

The successful project at Stanford University in DARPA competitions was conduced by Sebastian Thrun and his group. After that, he was hired by Google in 2007 to create the "Google X" research and advanced projects department, being responsible for research and development of fully innovative projects, conducing the R&D of the Google Self-Driving Car. The Self-Driving Car Team from Google X developed the most robust autonomous car available today [7]. This project gave start of a lot of research on autonomous vehicles development by car makers, as Audi, Mercedes, and Tesla Motors, among others. Also start-ups entered the race as AuroBots, Peleoton, GetCruise and ber. There are also companies developing equipment and algorithms for autonomous vehicles, such as Bosch, Delphi, Continental, TRW and even NVidia. This project is then well placed among the technologies used today in this research field.

In Brazil, the research related to the development of this theme is recent. We can highlight the SENA project of the School of Engineering of São Carlos, the CADU project from Federal University of Minas Gerais (UFMG), the project from UNIFEI (Drive4U), the VERO project (Robotic Vehicle) from CTI/Campinas, the autonomous vehicle project LCAD-UFES and the project CaRINA (Intelligent Robotic Car for Autonomous Navigation) from the Mobile Robotics Laboratory (LRM) of ICMC (São Carlos), University of São Paulo (Fig. 2). Among these projects, those that showed better results so far, considering the implementation of an autonomous and urban vehicle, were the CADU and the Carina projects (see [5]). Some relevant researches in this project CaRINA were: the work from [15] which aims to develop an intelligent navigation system for autonomous vehicles based on landmarks and topological maps of the environment, Alberto Hata [9,10], Danilo Habermann [8] and Patrick Shinzato [17], which developed different works based on information obtained from the Velodyne sensor for safe navigation.

(a) Carina 1

(b) Carina 2

Fig. 2. The two versions of the CaRINA project

3 Conceptual Basis

3.1 ROS

ROS (Robot Operating System) is an operating system that provides libraries and tools to help software developers to create applications in robotics [20]. It runs on well-known platforms such as Linux and provides an abstraction of the hardware and drivers usually used in this research area. The communication between entities (called nodes) has a core of "subscribe/post" of topics (robotic providing services), and is exemplified by the diagram in Fig. 3. A node is similar to an independent program, but containing the ROS abstraction facilities. They can compile, start and stop independently of other nodes. Topics (or services) make the communication between them by using network-layer packets, and they are constructed from standard data types, such as integers, floats or chars, or also structures composed by a data set of standard data types.

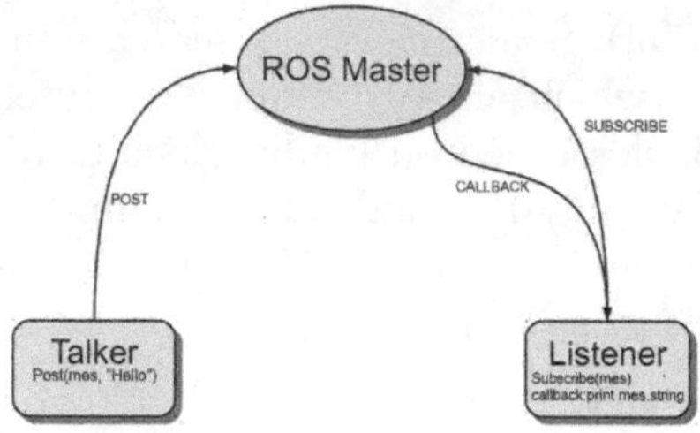

Fig. 3. ROS communication architecture

3.2 Velodyne

The Velodyne (Fig. 4) is an omnidirectional 3D sensor (360-degrees perception of the surroundings of the vehicle) that uses multiple beams of laser sensors (LIDAR type device), which have multiple focused beams with an vertical angular aperture and which rotates in the horizontal plane, thus allowing 360-degrees observation of the environment at different heights (multilayer/environmental plans 3D clippings). The lasers are then connected to a controller card that for a given horizontal and vertical scanning angle it decodes at which distance (x, y, z) the obstacles are, and measures the reflection intensity of the point of incidence. The sensor controller sends the data of the 3D point cloud to the computer, which is running the core ROS process, which then converts it into a ROS message. The composition of this message (PointCloud2 type) is an array of points (x, y, z, intensity, ring).

Its biggest advantage is also its biggest drawback, since the rotation of the sensor occurs at 10 Hz, it provides about 70,000 points per rotation. Another problem is that those points that have an effect within the limits of distance of the sensor are sent. The input format on the computer is then a large unorganized vector points due to the data rate provided by sensor (70,000 points/rotation, 10 rotations/second), there is a need for a fast program to be able to use all the quality and quantity of sensor data without losing data acquired from the next spin. For that, the solution was the NVidia video cards with CUDA.

3.3 CUDA

CUDA is a parallel computing platform and programming model created by NVIDIA. It allows significant increases in computing performance by harnessing the power of the Graphics Processing Unit (GPU) and its parallelism by the parallel execution of programs distributed on a large number of processing cores. Examples of its usage are image processing to identify hidden plaques in arteries, traffic flow analysis, and fluid molecules visualization [3,12]. It is observed that the use of 3D point clouds, like those obtained from the Velodyne sensor data, involves large amounts of data. The team in NVidia and the international scientific community have been developing materials and products to encourage the use of both technologies together [6]. Programming done in CUDA is based

(a) Velodyne Sensor (b) Data Sample

Fig. 4. The Velodyne-32E LIDAR sensor

on the popular languages: C and C++, and NVidia has developed some CUDA extensions and libraries which enables the CPU to delegate tasks to the GPU. The biggest difficulty of developing for GPU is its parallelism (non-sequential and non-standard programming as usually done in CPUs). There has been many upgrades and different architectures of GPUs, but one in specific demonstrates the appeal of CUDA developing for autonomous vehicles: the NVidia's vehicle developer platform Drive PX (Fig. 5(b)). Schematics of GPU CUDA architecture can be seen in Fig. 5.

4 GPU Services

Aiming to unite the technologies and techniques previously mentioned, the project aims to create an environment that makes efficient and effective use of the large amount of data supplied by providing for other projects fast processed data. The applicability of data processing point cloud is not limited to the Velodyne sensor or to autonomous cars, allowing a good scope for using the results here described, but the main application and focus of this work to be described is for use in that configuration. It is expected the following services (Fig. 6): data structure correction and spatial correction of the point cloud and detection relevant elements for vehicle navigation (road borders, curbs and obstacles). This project used the Velodyne HDL-32E sensor model, ROS system version Jade, parallel processing on GPU video cards with NVidia CUDA Toolkit version 7.5 (or 6.5 in case of the Jetson configuration), Linux Ubuntu version 15.4, Python 2.7, NumPy 1.9.227 and PyCUDA 2015.3, whose hardware is detailed in Table 1.

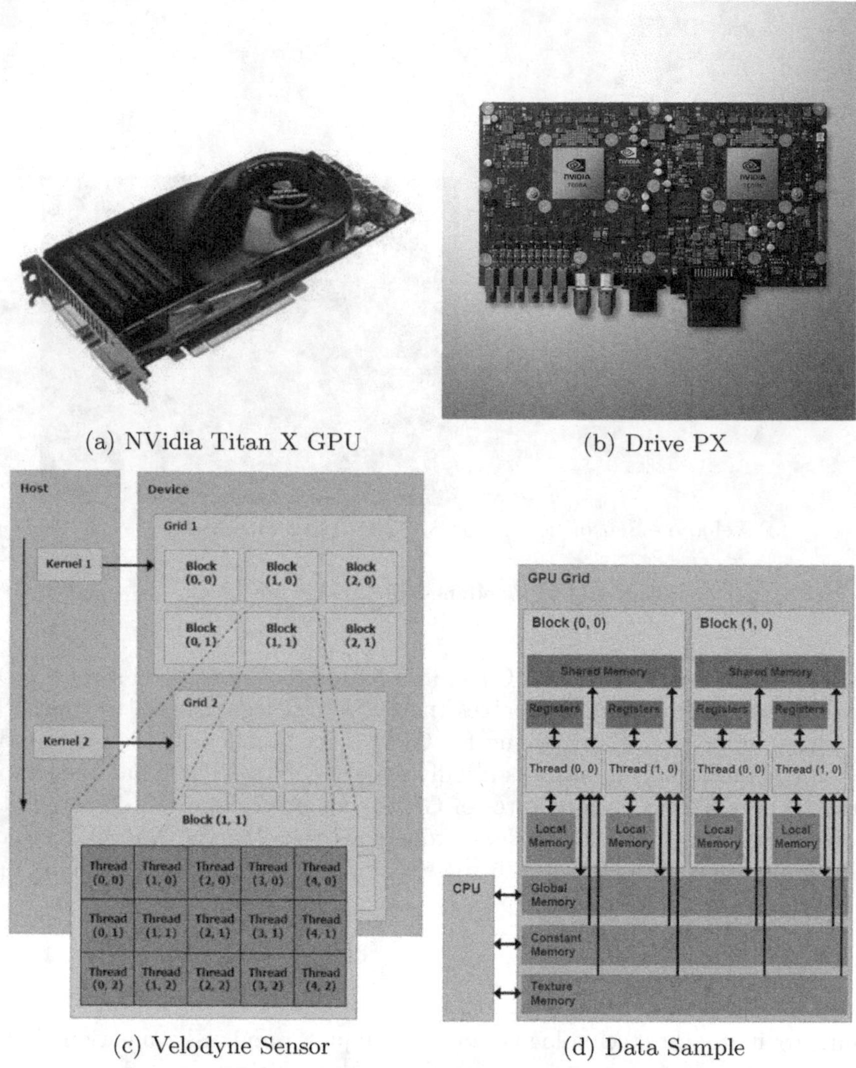

(a) NVidia Titan X GPU

(b) Drive PX

(c) Velodyne Sensor

(d) Data Sample

Fig. 5. CUDA enabled NVidia GPU examples and communication schematics

Table 1. Computer hardware table

Name	CPU	GPU	CUDA Cores	RAM
Car-PC	i5 2.9GHz	620GT	96	6GB
Common-PC	i3 3.3GHz	450GTX	384	8GB
Server-PC	i7 4.0GHz	TitanX	3072	16GB
Embedded-PC	A15 2.3GHz	TegraK1	192	2GB

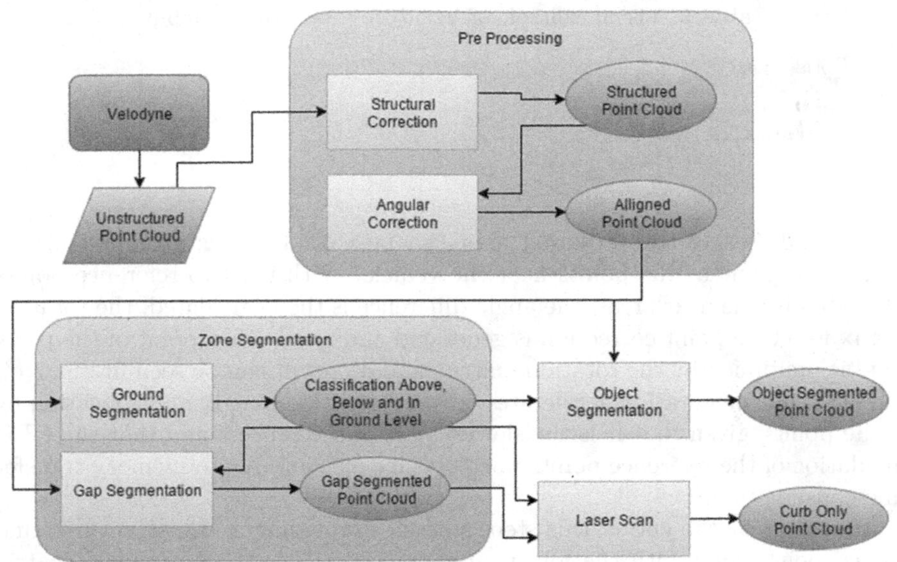

Fig. 6. Services communication diagram

4.1 Pre-processing

The first service is responsible for remodeling the point cloud provided by the Velodyne sensor. It has two steps: the first to generate a matrix out of the unstructured and unordered points arriving from the sensor, therefore allowing quick access and cropping of specific matrix regions, and the second step to provide angular correction of the planar distortion created by the difference from the ground plane to the sensors fixing base plane, simplifying subsequent services.

The first step utilizes data mapping methods based in trigonometry math, namely arc-tangent, and constructs a structured point cloud of 32 rows and 1800 columns. The matrix can be changed and customized, but these values were chosen because they are well divisible by the number of multiprocessor cores of the video cards and also because it is the most similar format to the structure internally generated by the sensor (32 lasers beams with throughput approximately 70 thousand points per frame, which is to a scan of a full 360-degrees, meaning a 0.2° separation per column). Table 2 exemplifies this need for a reorganization step. The CPU code used Python code with NumPy for its execution, and the GPU implementation has one thread per point.

The second step has an algorithm which corrects any angular distortion of a given point cloud. The necessity of this step can be seen in Fig. 7(a). The service calculates two reference points in the near front and rear of the vehicle (two arrays 900×6 representing a 90 opening and 6 rings of depth, namely the 3rd ring to 8th ring) as shown in Fig. 7(a). It is considered that the sensor is in a typical vehicular environment, where the vehicle is initially on a plane/flat

Table 2. Visual example of Velodyne's unorganized input

Packet Arrival Order	1	1	1	2	2	2	3	3	3	...	7	7	7
Matrix Element (ring x column)	1,1	1,2	1,3	1,14	1,15	1,16	1,6	1,7	1,8	...	1,32	2,1	2,2

surface and, hence, the proposed area can be considered as ground truth of the ground plane as the points near the vehicle. With the two reference points calculated by the near areas, the angle difference is then calculated, the rotation matrix for cloud point correction is generated and finally each point of the point cloud is multiplied by the rotation matrix, and its result can be seen in Fig. 7(c). The GPU implementation accelerates only the rotation matrix multiplication by all the points, given that it is the slowest part of the process and that the GPU calculation of the reference points was less time efficient due to memory transfer time consumption.

Therefore, at the end of this step, a service providing a well-structured and aligned point cloud with the following characteristics is then provided: matrix correction with each line as a ring of the point cloud, each column as a reading angle from the point cloud and readings not received from the sensor have a default value of 0, indicating no available Distance Reading for that point; and Planar Alignment which all points are adjusted (corrected) based on the planar distortion estimation, which is caused by the angular difference of the sensors fixation frame related to the ground plane.

4.2 Zone Segmentation

The next step is to provide services that can split the large data set for local detection elements. The data is divided into smaller regions allowing specialization and acceleration of other services. The zone segmentation service is divided into two steps: Ground Segmentation and Gap Segmentation.

The first step aims to separate the ground plane from the point cloud. By using the previous service, this service uses a height parameter and applies a threshold filter, classifying the points into three categories: under, above and in the ground plane. For this service, the height of the sensor is treated as 1.8m, with about 20 cm around that mark as threshold to be considered inside the plane (Fig. 8(a)). The result can be seen in Fig. 8(b) and the GPU algorithm is composed of one thread per point.

For the second step, it is known that in normal traffic situations there are usually two parallel continuous guides that follow the direction of vehicular movement. Towards detecting the curbs, different methods were tested based on success stories of the Darpa Challenge: ring compression detection method [2,9], used by the autonomous car JUNIOR, who participated in the DARPA Urban Challenge, and a method based on detection of planar regions used by the autonomous car BOSS [2], winner of the Darpa Urban Challenge. The first method shows that given a flat area, such as a street, an object on top of the area

(a) CARINA's angular distortion

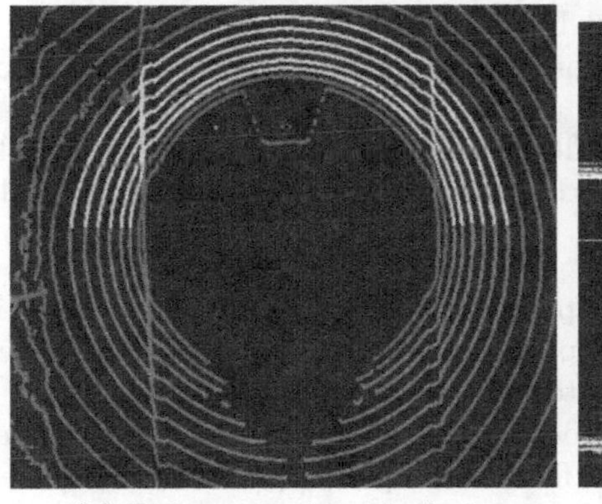

(b) Reference point calculation

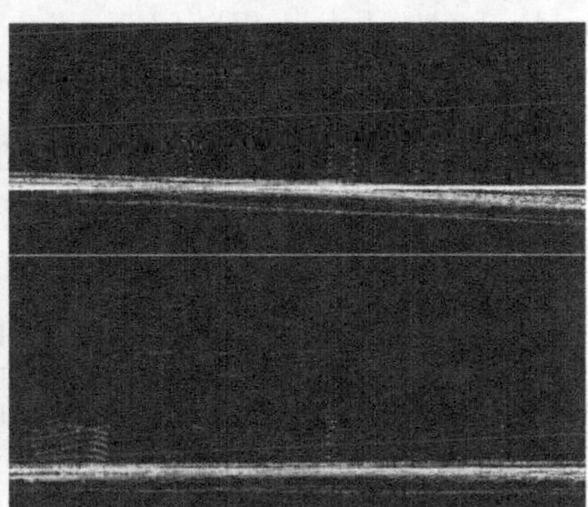

(c) Angular distortion correction

Fig. 7. Angular distortion service

(a) Velodyne's height

(b) Ground segmentation result

Fig. 8. Ground segmentation service

can be identified by the compression factor of the laser rings. This is detected by comparing a ring to its adjacent rings. The second method is based on the use of reflection intensity of the received convolutions of the sensor, then a step function is applied to the result. The lateral guides (street curbs) usually have different reflective index compared to asphalt, which is then detected. The method is applied both to the position (x, y, z) as the intensity in this design. The developed methods were:

- Compression rings detection by the hypotenuse of points between two radials (Fig. 9(a)). This method uses planar distances (x and y) of each point to calculate the distance from the origin (hypotenuse), and then it is normalized by the expected average height of the entire ring or the region around the point.
- Detection of angle variation between two radials by law of sines, which is done by detecting the difference between h and h2, which are calculated through d2 respectively. This method uses the points angles from planar distances (X and Y), normalized by the ring index (symbolizing a hypotenuse) and then divides by the height (Z) of the point by the sine of the normalized value.
- Reflection intensity variation detection in relation to the median reflection intensity of the ring.
- Convolution detector using the Sobel filter. This method uses the Sobel masks technique [18], a well-known image processing method, and applies it to the Point Cloud matrix.

The four above solutions were developed in CPU and then were compared in order to evaluate their performance, computational loads and, in particular, their accuracy. The method that gets the better efficiency, effectiveness and which is better adjusted to the parallelization process was then selected to be developed in GPU. The method adopted was the first technique (hypotenuse) while also applying a median filter in each ring to reduce noise and a ranking threshold for removal of values known to be wrong due to interference of other sensors coupled near the Velodyne. The result can be seen in Fig. 9(b) and the GPU algorithm is composed of one single kernel of one thread per point with three different steps: thread cooperation with shared memory to calculate a median matrix of the size of the original point cloud matrix, normalization and classification.

4.3 Object Segmentation

The next GPU service method developed is the object segmentation, which allows external projects to use classification techniques and further methodologies to detect pedestrians, cars and other objects that usually compose the scene around a vehicle. This service used the DBScan [1] technique, which partitions the space by local voxel adjacency of cluster of points. To achieve the goal, each point recursively classifies its close neighbors as one class until all points are classified. The parallelization process involved separating the technique into two parts:

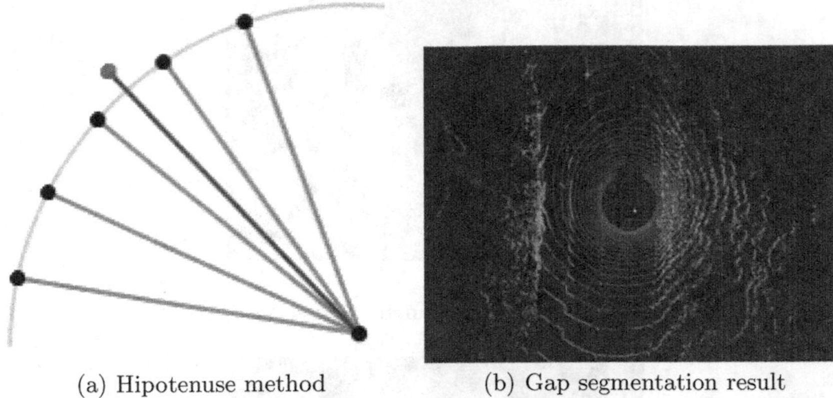

(a) Hipotenuse method (b) Gap segmentation result

Fig. 9. Gap segmentation service

- Identification: This step inserts a unique identifier for each point in the cloud and produces a matrix of limited size (due to hardware limitations) identifying the neighboring points of each point. This is run by one thread per point.
- Classification: This step is a recursive stage where the matrix of the identified neighboring points are used to set the segmented value (Blob Id) to the minimum value of the group. Therefore, due to successive runs, eventually all points will be segmented. This is run by one thread per row of neighboring points.

Thus it was possible to parallelize and run the DBSCAN algorithm in the GPU. The size values of the matrix of neighboring pixels and the number of repetitions are set depending on the size of the cloud points and the number of different objects to be segmented is expected to each frame following Eq. 1, where alpha denotes the level of precision. Depending on the system capacity, different configurations are recommended, since a large array involves large memory usage and time spending on memory access. Through preliminary tests, it was observed that an accuracy of 0.1 is enough to identify the location of objects in the environment with good accuracy. In this case, a point cloud with 1000 points and 10 objects would have a neighbor vector of size 5 and 2 repetitions with a precision level 0.1. After the execution, a data cleaning procedure removes groups with few elements, removing then any errors. The result of this service can be seen in Fig. 10.

$$\frac{Cloud\,Size}{Objects\,Quantity} = \frac{Neighbor\,Vector\,Size * Repetitions}{\alpha} \tag{1}$$

The aim of this service was to use GPU computing power to extract objects in the scene without worrying about the semantics associated with the objects. To improve the use of the GPU, the intermediate data between the two stages and between repetitions of the second stage were not transferred back to the CPU, allowing an efficient performance. Since the service uses such a versatile

Fig. 10. Object segmentation service

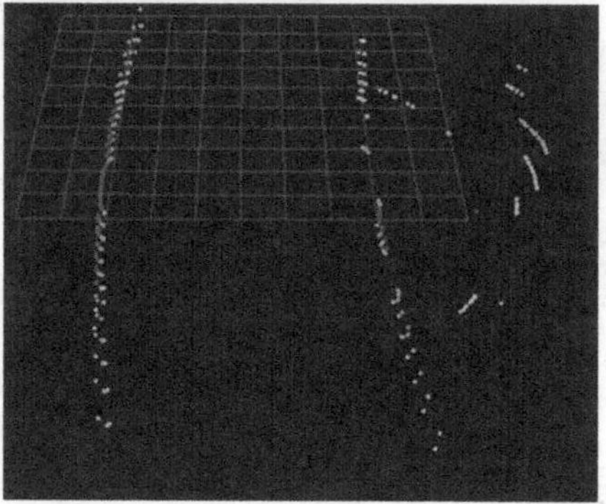

Fig. 11. Laser scan segmentation service

algorithm as the DBSCAN, it allows any point cloud in any context as input for object segmentation.

4.4 Laser Scan

The final service combines many of the previous services in order to identify correctly street curbs. For this, starting from point cloud with zone segmentation, we use a similar strategy employed in the BOSS vehicle, described in Sect. 2.1, which was later developed and perfected by Patrick Shinzato (SHINZATO and Wolf, 2015). Laser scan reading is a data reduction transformation of the point cloud to a single layer laser sensor. For this, it uses the result of the zone segmentation service (Sect. 4.2), and excludes the points that are not part of the curbs by saving the nearest point of origin for each radial angle reading. For the parallelization process, each angle was processed by a different thread (making that 1800 threads) and its result can be seen in Fig. 11.

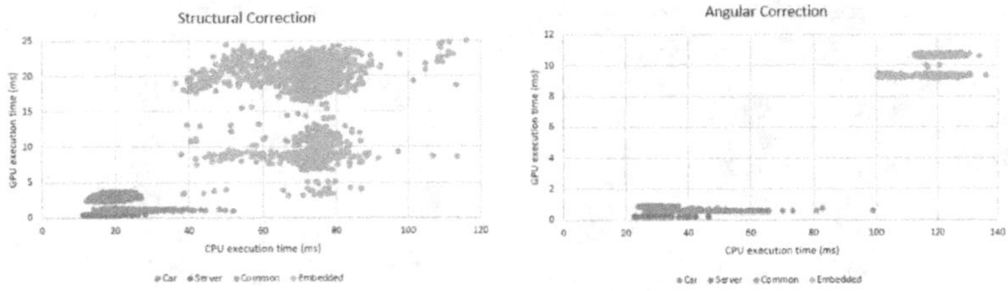

(a) Pre-processing (structural correction) time comparison

(b) Pre-processing (angular correction) time comparison

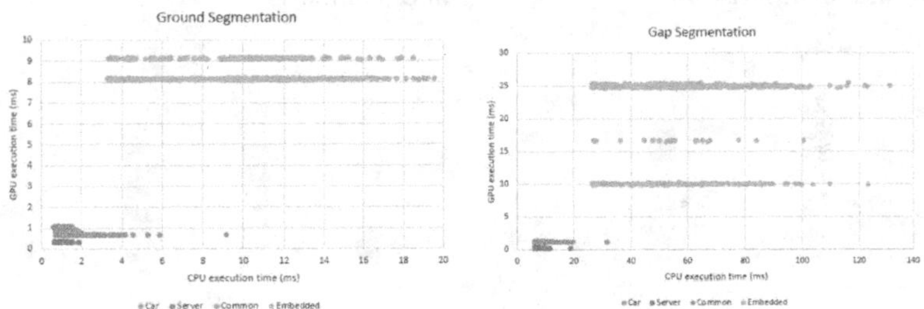

(c) Zone segmentation (ground) time comparison

(d) Zone segmentation (gap) time comparison

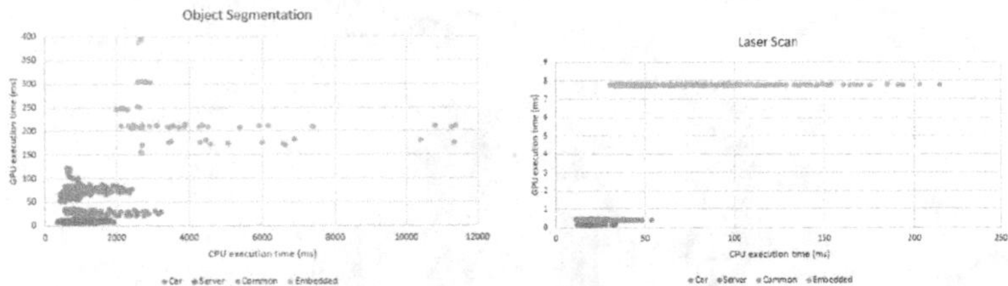

(e) Object segmentation time comparison

(f) Laser scan time comparison

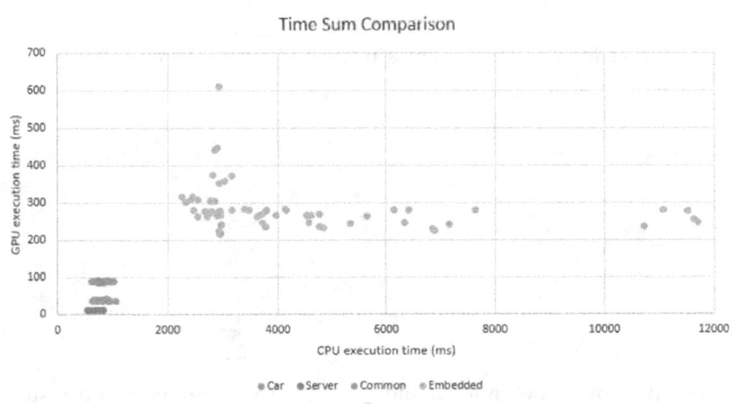

(g) Full time comparison

Fig. 12. Time graphs comparisons of services

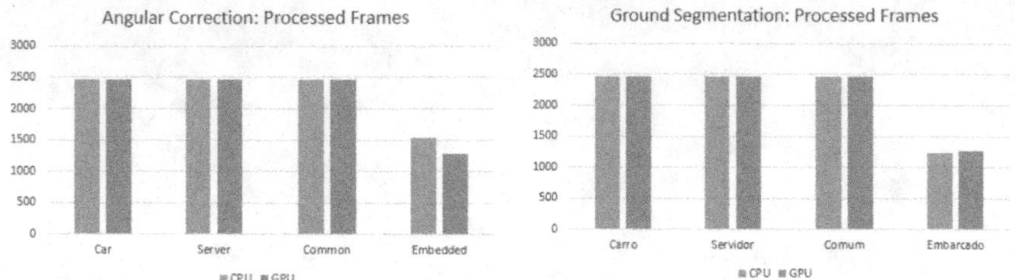

(a) Pre-processing (angular correction) total frames comparison

(b) Zone segmentation (ground) total frames comparison

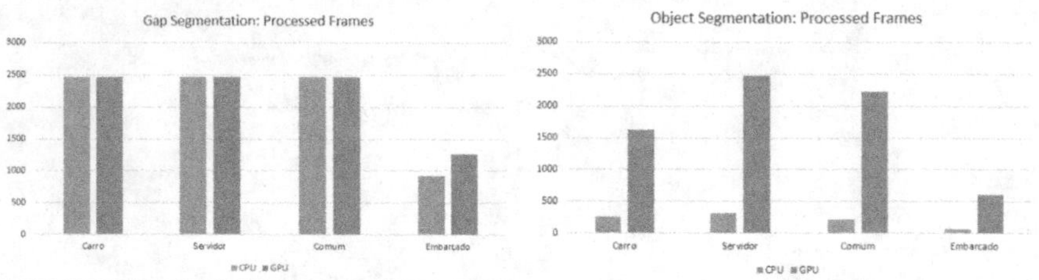

(c) Zone segmentation (gap) total frames comparison

(d) Object segmentation total frames comparison

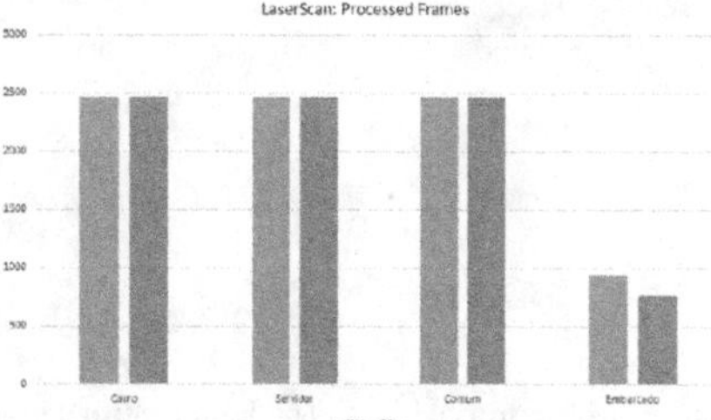

(e) Laser scan total frames comparison

Fig. 13. Total frames processed comparison

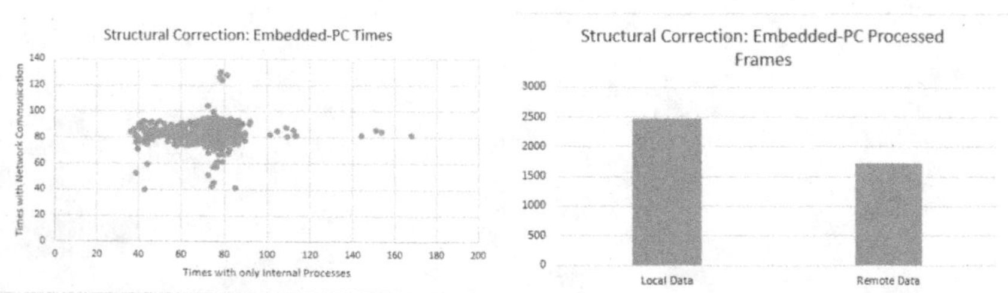

(a) Time comparison between local and remote driver

(b) Gap segmentation result

Fig. 14. Gap segmentation service

5 Conclusion

In this paper it is described services developed for point cloud processing obtained from the Velodyne sensor. It provides data processing methods that are important and very useful to the implementation of driver support systems (ADAS) and aid in the autonomous driving vehicles. The results of various implementations and different hardware is shown in Fig. 12, where each axis is one type of execution (GPU time X CPU time). Each test was made of the same 2470 frames for correct comparison. Some important notes are:

- The times in the CPU have greater variability (X axis) than the GPU time (Y axis);
- In general, CPU times are greater to, or at least equal to, the GPU time;
- The embedded-pc has GPU times roughly as good as the server's CPU time;
- Although only times are compared, power issues are also relevant, leaving for architect engineers to use the hardware most adequate for his project's purpose;
- In special, the Object Segmentation Service shows the capability of the GPU acceleration process due to the complexity of the algorithm and its consequent time speedup;
- The full time comparison is shown in Fig. 12(g), which is very similar to the Object Segmentation service result in Fig. 12(e) because of the service's big time consumption.

Depending on the configuration of the hardware and the services used, the processing times many times exceed the 100 ms mark of the sensor's frame rate. Therefore, the Fig. 13 shows the comparison of the amount of frames processed in each service.

Finally, another comparison made in Fig. 14 show the results of using the sensor driver locally or remotely, receiving the data over the network. In both cases, the CPU is roughly equally used as seen in Fig. 14(a), however Fig. 14(b) shows that receiving the data remotely renders less frames processed due to I/O activities. This comparison was made using the CPU version of the pre-processing service (step 1) in the embedded architecture. Because of the result, the remote method is not used in the other services.

In addition to the methods that were implemented through services, they were also presented and discussed the results of the performance of such methods, tested on different platforms (CPU and NVIDIA-CUDA GPU different models). It was possible to demonstrate that the great gain in terms of obtained run-time when implemented exploiting parallelism and run on GPUs.

In this paper we present GPU-powered solutions called services that provide pre-processing and pre-segmentation of sensorial data with large throughput rate and good speedups compared to the sequential implementations. Also, the GPU services allows reducing the CPU processing load, allowing its use for other higher-level robot tasks. This project is also able to achieve its real-time mark since when summing up all GPU execution times we have a mean GPU execution of about 45ms, much lower than the 100ms mark, providing therefore a real-time system to the Velodyne sensor data processing for the vehicle. It is expected for future work to unite all information into a single topological map provided as a service for semantic gathering of all procedures used until now, which was described in this article as the last service.

Acknowledgment. This project was financially supported by Fundação de Amparo à Pesquisa do Estado de São Paulo (FAPESP), Process #2013/13880-9. The CaRINA Project (Autonomous Vehicle and Sensors) was also financially supported by CNPq, FAPESP and INCT-SEC. We also acknowledge the support granted by the LRM Laboratory - ICMC/USP.

References

1. Birant, D., Kut, A.: ST-DBSCAN: an algorithm for clustering spatial-temporal data. Data Knowl. Eng. **60**, 208–221 (2007)
2. Buehler, M., Iagnemma, K., Singh, S.: The DARPA Urban Challenge: Autonomous Vehicles in City Traffic. Springer, Berlin (2009)
3. Dias, S., Bora, K., Gomes, A.: CUDA-based triangulations of convolution molecular surfaces. In: Proceedings of the 19th ACM International Symposium on High Performance Distributed Computing, HPDC 2010, pp. 531–540. ACM, New York (2010). http://doi.acm.org/10.1145/1851476.1851553. ISBN: 978-1-60558-942-8
4. Fenton, R.E., Cosgriff, R.L., Olson, K., Blackwell, L.M.: One approach to highway automation. Proc. IEEE **56**, 556–566 (1968)
5. Fernandes, L., Souza, J., Pessin, G., Shinzato, P., Sales, D., Mendes, C., Prado, M., Klaser, R., Magalhães, A.C., Hata, A., Pigatto, D., Branco, K.C., Grassi Jr., V., Osorio, F.S., Wolf, D.: CaRINA intelligent robotic car: architectural design and applications. J. Syst. Archit. **60**, 372–392 (2014)
6. Biermeyer, J.O., Templeton, T.R., Berger, C., Gonzalez, H., Naikal, N., Rumpe, B., Shankar, S.S.: Rapid integration and calibration of new sensors using the Berkeley Aachen Robotics Toolkit (BART). In: AAET - Automatisierungssysteme, Assistenzsysteme und eingebettete Systeme für Transportmittel: Beiträge zum gleichnamigen 11. Braunschweiger Symposium vom 10. und 11. Februar 2010, Deutsches Zentrum für Luft- und Raumfahrt e.V. am Forschungsflughafen, Braunschweig/Intelligente Transport- und Verkehrssysteme und - dienste Niedersachsen e.V. (Hrsg.), ITS Niedersachsen, Braunschweig (2010). http://publications.rwth-aachen.de/record/126284. Pages 17 S
7. Guizzo, E.: How google's self-driving car works, October 2011
8. Habermann, D., Silva, R., Wolf, D., Osorio, F.: Detecção e classificação de objetos com uso de sensor laser para aplicações em veículos autônomos terrestres. In: XV Simpósio de Aplicações Operacionais em Áreas de Defesa (SIGE), pp. 55–59. DCTA-ITA (2013)
9. Hata, A.Y., Habermann, D., Osorio, F.S., Wolf, D.F.: Road geometry classification using ANN. In: 2014 IEEE Intelligent Vehicles Symposium Proceedings, pp. 1319–1324. IEEE (2014)
10. Hata, A.Y., Osorio, F.S., Wolf, D.F.: Robust curb detection and vehicle localization in urban environments. In: 2014 IEEE Intelligent Vehicles Symposium Proceedings, pp. 1257–1262. IEEE (2014)

11. Henderson, T.C., Minor, M., Drake, S., Quist, J., Roberts, J., Sani, H., Rasmussen, M., Collins, A., Sun, Y., Fan, X., Louis, St., Mikuriya, S., Dean, K.: Robust autonomous vehicles DARPA urban challenge. DARPA Grand Challenge Tech Papers (2007)
12. Langdon, W.B.: Performing with CUDA. ACM (2011)
13. Luettel, T., Himmelsbach, M., Wuensche, H.J.: Autonomous ground vehicles: concepts and a path to the future. Proc. IEEE **100**, 1831–1839 (2012)
14. Montemerlo, M., Becker, J., Bhat, S., Dahlkamp, H., Dolgov, D., Ettinger, S., Haehnel, D., Hilden, T., Hoffmann, G., Huhnke, B., Johnston, D., Langer, D., Lev, A., Levinson, J., Marcil, J., Orenstein, D., Paefgen, J., Penny, I., Petrovskaya, A., Pflueger, M., Stanek, G., Stavens, D., Vogt, A., Thrun, S.: Junior: the stanford entry in the urban challenge. In: Buehler, M., Iagnemma, K., Singh, S. (eds.) The DARPA Urban Challenge. Springer, Heidelberg (2007)
15. Sales, D.O., Correa, D.O., Fernandes, L.C., Wolf, D.F., Osório, F.S.: Adaptive finite state machine based visual autonomous navigation system. Eng. Appl. Artif. Intell. **29**, 152–162 (2014). doi:10.1016/j.engappai.2013.12.006
16. Shane, R., Rodrigues, C.I., Baghsorkhi, S.S., Stone, S.S., Kirk, D.B., Hwu, W.W.: Optimization principles and application performance evaluation of a multithreaded GPU using CUDA. In: Proceedings of the 13th ACM SIGPLAN Symposium on Principles and Practice of Parallel Programming, PPoPP 2008, pp. 73–82. ACM, New York, NY, USA (2008). http://doi.acm.org/10.1145/1345206.1345220. ISBN: 978-1-59593-795-7
17. Shinzato, P.: Estimação de obstáculos e área de pista com pontos 3D esparsos. CCMC-ICMC-USP, USP São Carlos, Brazil (2015)
18. Russ, J.C.: Image Processing Handbook, 6th edn. CRC Press, Inc., Boca Raton (2016). ISBN: 978-1-4398-4063-4 (Ebook-PDF)
19. Thrun, S., Montemerl, M., Dahlkamp, H., Stavens, D., Aron, A., Diebel, J., Fong, P., Gale, J., Halpenny, M.: The robot that won the darpa grand challenge. J. Field Robot. **23**, 661–692 (2006)
20. Troniak, D.M.: PR2 rides the elevator - a problem in vision-based localization (2012)

Collaborative Object Transportation Using Heterogeneous Robots

Ramon S. Melo[(✉)], Douglas G. Macharet, and Mario FernandoM. Campos

Computer Vision and Robotics Laboratory, Department of Computer Science, Universidade Federal de Minas Gerais, Belo Horizonte, MG, Brazil
{ramonmelo,doug,mario}@dcc.ufmg.br

Abstract. The use of multi-robot systems can be seen in many different contexts in recent years. One of them is the object transportation problem, which has many applications, such as simple moving objects as well as in more complex scenarios, like tasks typically involved in building sites and structures assembling. Despite the fact that much effort has been focused on what may apparently be a relatively simple task, several facets of the problem still remain open and need to be tackled. In this work, we propose a complete methodology which encompasses all related stages of the problem (*i.e.* path planning, task allocation and control). Several experiments with simulated robots and with real ground robots were conducted in order to provide a thorough evaluation and validation of the methodology.

Keywords: Cooperative transport · Object manipulation · Task allocation

1 Introduction

The use of mobile robots in many different contexts and applications has increased significantly in recent years. It is notorious the use of these autonomous systems in activities such as surveillance, search and rescue, object transportation, among others.

All aforementioned activities can be executed both individually as well collaboratively, in this case called Multi-Robot Systems (MRS). The use of multiple robots presents several advantages like increased robustness and, in most cases, time reduction to accomplish a task. However, the use of such systems also brings many challenges, such as robot localization, path planning, task allocation and control. A remarkable advantage in the use of such systems is the potential to perform difficult activities without the need of specialized agents, for example, use simple robots to transport an object instead of a complex manipulator.

The accomplishment of any task using a MRSinvolves many subproblems that must be considered. Among these problems, we can highlight the localization of the robots, path planning, coordination and the task allocation. The localization problem is related with agent itself as well the environment where the robots are

F. Santos Osório and R. Sales Gonçalves (Eds.): LARS 2015/SBR 2015, CCIS 619, pp. 172–191, 2016.
DOI: 10.1007/978-3-319-47247-8_11

working and any other object that they interact.The path planning problem can be addressed in many ways, in some cases to minimize the total time spent or the overall traveled distance for example. Regarding the team coordination, that involves task allocation among the agents, as well a study of how combine the available resources to accomplish the activity respecting all possible restrictions.

One of the main activities in which robotic systems are employed is the transportation and manipulation of objects. These tasks, although simple, are fundamental in numerous activities of our daily life such as the transport action, as well as the manipulation of objects and tools in the execution of tasks. The transport actions can be broadly classified into two categories: (i) prehensile manipulation, in which the object is grasped by the robot, and (ii) non-prehensile manipulation, in this case the robot uses actions like throwing, rolling or pushing to accomplish the transportation [5,12,13,16].

Although robots still do not show the same dexterity as humans for specific tasks, robots may attain better efficiency and effectiveness. Furthermore, robots can be readily deployed to operate in harsh and inhospitable environments such as fire, deep ocean, and nuclear accidents. Another example is a collapsed mine, which may become totally inaccessible for humans, but where a robot agent could probably gain access by using its manipulator to clear the way into tunnels and cavities. A service robots improve their manipulation capabilities, their use in homes have been steadily increasing, and much investigation is underway on the use of robots for in home assistance to the elderly and patients [9].

The main contribution of this paper is a framework for heterogeneous groups of robots to collaboratively transport a set of objects using both prehensile and non-prehensile manipulation, with emphasis on the task allocation phase. We present a complete framework which encompasses all related stages of the problem. The methodology was evaluated in both simulated and real environments, demonstrating the effectiveness and flexibility of the technique in respect to the task needs, such as minimizing execution time or energy consumption.

The remainder of the paper is organized as follows. Next section discusses related works in the literature regarding different aspects on the autonomous transportation problem. Section 3 presents our methodology, describing the mechanisms used for planning the path for each object, decomposition into segments and selection by the robot. Experiments considering a simulated and real environments are presented in Sect. 4. Finally, Sect. 5 discusses the results and indicate future directions of investigation.

2 Related Work

The object transportation using a group of robots is composed by many steps, such as the path planning, task allocation and team control and coordination. This work addresses these three aspects, however has as main focus the task allocation phase.

The applicability of robots capable of manipulation and transportation of objects has grown, mainly seeking to improve the efficiency and gain in different tasks. Practical examples of this use can be seen in scenarios like: withdrawal of mine sediments [15], execution of surgeries [10], object transportation [13], handling of goods in a warehouse [8], construction of goods and structures [2, 11,18,19].

Many works demonstrate real applications of object transportation using the model of MRS, considering prehensile and non-prehensile techniques. A cooperative transportation using terrestrial robots is shown in [5], where the agents surround the object and move it avoiding obstacles. Another example is the Kiva System [8], a solution used in warehouses to manipulate the commodities mostly used by companies like *Amazon*[1]. Using a team of quadrotors, it is possible to build structures based on preexisting segments [11], or using blocks [2].

An example of construction using autonomous robot agents is the Aerial Robotic Construction (ARC), which defines a change in the paradigms of design and manufacture of items. In the research, it is exposed the characteristics of this type of system, which are: (i) the construction of the support structures does not require frames to be performed, (ii) the digital model can be complex, and the agents are explicitly guided by this, and (iii) the system is scalable in the sense of not being tied to the environment [25].

In many cases, centralized control facilitates decision making and task allocation in problems involving multiple agents. The authors of [14] present a technique using the concept of *task instance*, which is characterized by the work a single agent must execute in a period of time. In transportation, it is possible to define task instances with actions like search, removal of obstacles and object movement. Together with a centralized control, the use of potential fields is a simple and intuitive way of planning, which can be refined with the addition of dipolar function [22], or be segmented using a control which addresses the following two questions: (i) how to navigate through the environment and (ii) how to manipulate the object [26].

Another kind of coordination is based on the leader-follower approach. An example is the work presented in [6], which exemplifies this technique applying the task allocation algorithm MURDOCH [7], in which an agent has the task of a *watcher* that informs to the *pusher* agents the velocities that should be impressed upon the object so that it reaches its destination. In this case, the leader does not directly participate in the transportation task. In contrast, another work presents the case where the leader informs other robots in the team its trajectory so that they follow it during transport [21].

The *constrain-move* method by [1], is one example of a fully distributed coordination, in which a team of robots is divided into two subgroups: one restricts the object's movement, while the other drives the motion. Both teams work together and simultaneously to take the object to its destination. Using the same idea of motion restriction, there is also the use of potential fields, by applying this technique in three steps: (i) the agents are attracted by the object, (ii) which are then disposed equally around it, and finally, (iii) move to the final position taking the object [20].

---

[1] https://www.amazon.com/.

The Task Allocation phase can be described as a set of methods to distribute task among a team of agent and manage the coordination aspect until the mission is complete. The work presented in [4] tackle this problem transforming the task allocation problem into the graph coloring problem, where each agent represents a color and each vertex is a location that needs to be visited. Similarly, [23] shows a multi-agent control system that solves a *Binary Integer Linear Programming* (BIP) problem and creates an allocation matrix.

To explore an environment satisfactorily, the authors of [17] propose an auction-based approach, considering not only the bids of the agents, but the path planning to decide which agent will perform the task.

To inspect a ship using a team of agents, the authors of [3] describe the environment as a graph, where each node is a room that must be visited, and is described by its characteristics, like if it is submerged or have obstacles. Based on the available resources, like the agents types and capabilities, it is possible to allocate sub-teams to specific areas besides presents a sub-optimal total distance traveled solution.

A comparison between two methods of task allocation is presented in [24], where agents must transport a collection of packages. The first approach is a centralized planner that has information about all aspects of the environment, like production frequency and tasks allocated to each agent. Based on these information, it determines if a particular agent should perform a task or stay, for example, in standby mode. The other approach is a decentralized one. In this case, each agent has information only of its surrounds, the source position and the place of deposition. In order to decide to change from task execution to standby mode, the agent deliberates based on specific, predefined rules.

3 Methodology

The object transportation problem using a group of robots may be decomposed into many steps, such as path planning, task allocation and team control and coordination. In this work we address all of these aspects, whereas the main contribution of this work is the task allocation and coordination phase.

3.1 Problem Statement

Given a known environment, we define by $\mathcal{W} \subset \mathbb{R}^3$ a convex polygon as the workspace. Let $\mathcal{R} = \{r_1, r_2, ..., r_k\}$ composed by k robots, each with a type defined in $\mathcal{T} = $ {ground, aerial}, where <ground> type can only push objects and <aerial> type can grasp and fly with the object. We define $\mathcal{O} = \{o_1, o_2, ..., o_y\}$ the set of y objects that must be transported. Likewise, we define the set $\mathcal{B} = \{b_1, b_2, ..., b_x\}$ of x blocking objects that are immovable and will restrict the motion of the robots and the objects.

The transportation problem is then defined as moving an object $o_i \in \mathcal{O}$ from its initial position S to a final position S_d, both in $\mathcal{W}$. A subgroup of robots from $\mathcal{R}$ will be responsible to move each object minimizing the total cost of transportation (time and energy) and considering the constraints imposed by $\mathcal{B}$. The main objective is to describe a set of plans $\mathcal{P}$, that can be performed by the agents, and culminates in the transportation of all objects. The planning process in most works is mainly focused on how the robots must navigate, in this work however, we focus on the object's trajectory, which will guide the robots' paths.

The problem at hand was approached considering two sub-problems: (1) Object path planning – a path is created for the objects given an utility function (Sect. 3.2), which will be used to guide the robots in its transportation, and (2) Task allocation and Coordination – the agents are evaluated based on their performance in accomplishing the task, and then are coordinated to execute it.

Figure 1 presents a flowchart of the framework, in which are highlighted the steps involved in the process of transportation: (i) workspace definition, (ii) action planning to be executed during the transportation process, (iii) task allocation among the agents, (iv) coordination and execution of the task. Each phase will be described in the following subsections.

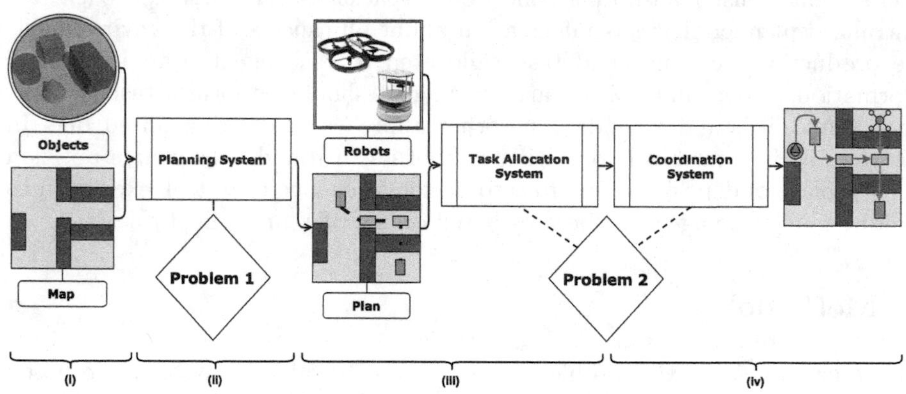

Fig. 1. Framework diagram, showing the process steps to transport objects using a group of robots. The total process was divided into five steps to better understanding and development.

3.2 Utility Function

In the path planning process, of both objects and agents, in order to consider the heterogeneity of robots, certain characteristics are analyzed related to the type of robot in order to find the best type that meets the system needs.

In this sense, two parameters are evaluated for each agent type, which describe the usefulness of performing a certain action a_i within a plan $\mathcal{P}$, using an utility function in order to quantify this action in such way that can be compared. The following features are used in the evaluation:

- **Dislocation Time:** Time required to move between cells in $\mathcal{W}$. This aspect is related to the average speed that a robot can perform a given displacement. Described by the function $\Upsilon(\cdot)$;
- **Dislocation Cost:** Cost to move between cells. For each type of robot used in transportation, the average energy expenditure that it has during displacement is calculated. Described by the function $\Psi(\cdot)$.

Formally, the Eq. 1 describes the utility function $\Theta(\cdot)$, which is evaluated for a certain robot type, resulting in a punctuation based on its characteristics. The generated value of this analysis is used to guide the path planning process of the objects (as described in Sect. 3.3).

$$\Theta(t_i, a_i) = \alpha \cdot \Upsilon(t_i) + (1 - \alpha) \cdot \Psi(t_i). \tag{1}$$

The constant defined by $\alpha \leftarrow \{\alpha \in \mathbb{R} | 0 \leq \alpha \leq 1\}$, is used as a weighting factor between both considered aspects to reflect the prioritization of certain activities, such as execute the task in the shortest time possible or with the lowest energy expenditure. This model is flexible because it can be reconfigured to meet diverse needs, adding to the planning particularities of each agent. The resulting value of this function is used as part of the heuristics used in the search of the best plan of movement.

The overall utility of a plan $\mathcal{P}$ is calculated as the sum of the utility values of each structure n in the plan, given by:

$$\Theta_p(\mathcal{P}) = \sum_{i=1}^{q} \Theta(k(n_i)). \tag{2}$$

Considering that the utility function is based on the type t_i of an agent and a certain action a_i, we define Eq. 3, which extracts such information directly from n, as defined below:

$$k : n_{\mapsto} t_i, a_i. \tag{3}$$

Equation 2 is used to determine, among several movement plans, which is the best, in other words, the one with the lowest total cost to carry out the actions. Its use will be further detailed in Sect. 3.5.

3.3 Path Planning - Object

The first step to transport a object is the creation of a motion plan, this process is sub-divided into two stages: (i) path planning and (ii) path segmentation. The later will serve as the baseline for the robot's path planning.

The proposed method is a variation classical tree search algorithms based on the creation of a search graph. First, we define the set $\mathcal{F}$, which describes all nodes available during the search expansion, and the set $\mathcal{SN}$, with all explored structures n.

The set $\mathcal{F}$ contains the states that were created in the expansion phase of a n. The structure n with the lowest utility cost is selected for expansion. In the start of the execution, a new node n is created, describing the initial position of the object, being added to $\mathcal{F}$. Algorithm 1 describes the tree node expansion process.

Algorithm 1. Search Expansion ($\mathcal{M}$, $\mathcal{F}$, $\mathcal{SN}$, $\mathcal{W}$)

```
n ← select_state_from(F)
SN ← SN ∪ {n}
S ← get_position(n)
for all a_i ∈ M do
    S' ← apply_action(a_i, S)
    S_r ← is_executable(a_i, S')
    if S' ∈ SN or is_colliding(W, S') or not S_r then
        SN ← SN ∪ {S'}
        continue
    end if
    n' ← create_node(S', S_r, a_i)
    F ← F ∪ {n'}
end for
```

The function *is_colliding* verifies if the object is colliding with any $b_i \in \mathcal{B}$, and the function *is_executable* considers, first, if exists an agent available of a type t that can perform the action a_i and then this action is evaluated if it is feasible to be executed, in this case the position S_r in which the robot must be to accomplish the movement is calculated, and if this position passes by the collision test it is a valid motion.

The search algorithm ends in two cases: (i) when $\mathcal{F} = \emptyset$, indicating that there is no path to the destination state S_d, or (ii) when the selected n is in the neighborhood of the final state, in this case, a new node n is created and added to the plan – this node defines the movement to state S_d.

The segmentation step executes the division of the plan $\mathcal{P}$ created by the search algorithm, generating the set $\mathcal{S}$ as result. This set will be used during the robot's planning phase and thereafter to task allocation.

To perform the segmentation, an auxiliary set $\mathcal{S}_p$, with references to the n structures used as base to segments, must be created. Algorithm 2 describes the set creation process.

A segment set $\mathcal{S}$ is created using all segmentation points $sp_i \in \mathcal{S}_p$. Each segment is a section of the original plan $\mathcal{P}$, with a sequence of node structures. Algorithm 3 presents the segment creation step.

After this process, the set $\mathcal{S}$ is defined with e segments, each one is classified as having a movement type of the set $\mathcal{T}$. Figure 2(a) illustrates a sample planning for one object within a particular environment.

The segmentation has its importance specially on terrestrial transportation. Considering that <ground> robots can only push the objects, whenever there is a change of direction, the robot must push the object from another side, and at this moment, there is the possibility that another agent would be more capable

Algorithm 2. Segment Point Set Creation ($\boldsymbol{\mathcal{P}}$, $\boldsymbol{\mathcal{S}}_p$)

```
for n_i, n_{i+1} ∈ P do
    if is_aerial(n_i) and is_aerial(n_{i+1}) then
        continue
    end if
    a_i^{n_i} ← get_action(n_i)
    a_i^{n_{i+1}} ← get_action(n_{i+1})
    if a_i^{n_i} != a_i^{n_{i+1}} then
        sp_i ← create_segment_point(n_i, n_{i+1})
        S_p ← S_p ∪ {sp_i}
    end if
end for
```

Algorithm 3. Segment Set Creation ($\boldsymbol{\mathcal{S}}_p$, $\boldsymbol{\mathcal{S}}$)

```
for all sp_i, sp_{i+1} ∈ S_p do
    if S = ∅ then
        s_i ← create_segment(S_o, sp_i^1)
    else
        if is_last(S_p, sp_i) then
            s_i ← create_segment(sp_i^2, S_d)
        else
            s_i ← create_segment(sp_i^2, sp_{i+1}^1)
        end if
    end if
    S ← S ∪ {s_i}
end for
```

to accomplish the action with a smaller cost than the agent which is currently running the task.

Furthermore, the aerial motion may not need to be segmented, even if the direction changes several times. This restriction is applied because, from the moment the agent start an aerial motion, another agent may take upon moving that object only after it has been deposited, in other words, when the aerial transportation has been completed.

Each segment, s_i, formed by a sequence of n structures is similar to a plan, since it also has states S_o and S_d, defined. These states are used in the robot's path planning.

3.4 Path Planning - Agents

After the path planning and segmentation for the object movement phase ends, it begins the path planning phase for the agents involved in the task. These plans will be used in the task allocation phase.

The path planning algorithm used in this step is similar to Algorithm 1 (Sect. 3.3), with the following modifications:

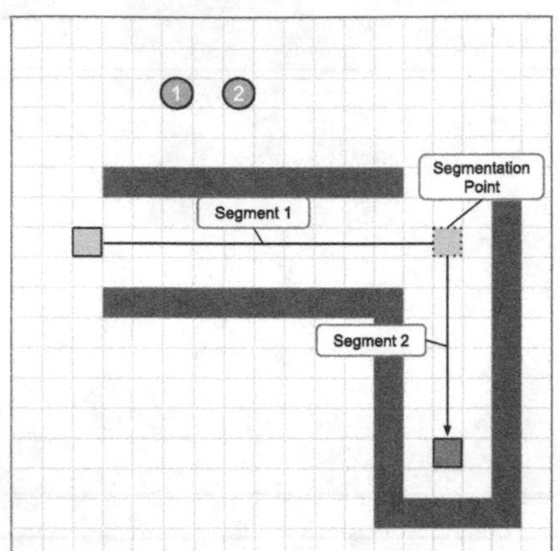

(a) Sample environment with one object and two agents for transportation, with two segments in the object plan.

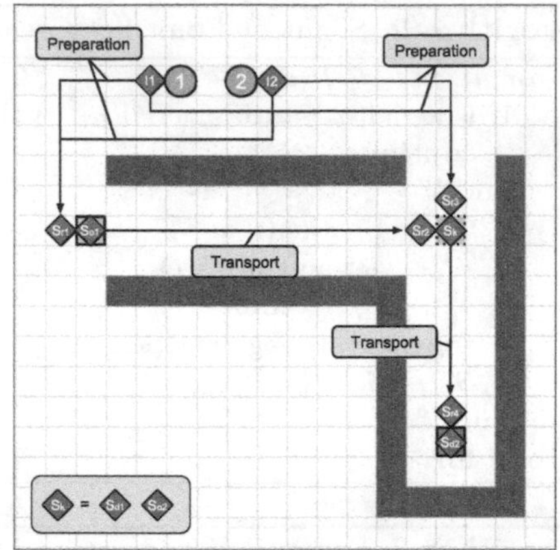

(b) Sample environment showing all the possible paths that the agents can execute and it's respective types. In this case the object's plan has two segments.

Fig. 2. Example workspace with one object and two agents. (a) shows the object plan to me executed, and (b) show the plans for the agents.

- The collision test using the function *is_colliding* uses the robot's body as reference;
- The function *is_executable* is not executed, since it only applies to the object.

In addition to the states S_o and S_d of each segment, a new state type I is created and defined as the initial state of each robot. From these states the path planning for the agents are performed. The created plans are classified by both types of set $\mathcal{TP}$ = {initial, transition, movement} and set $\mathcal{TM}$ = {pre-transport, transport}, as described below:

- **Initial:** Plan from the state I to state S_r described in the structure S_o of a segment. This plan ensures the arrival of the agent to state where the transportation will begin. It is classified as type <pre-transport>;
- **Transition:** Plan from the position S_r of the state S_d of a segment s_i to the position S_r of S_o of segment s_{i+1}. Through this plan, an agent can continue the transport in another segment. It is classified as type <pre-transport>;
- **Movement:** Plan from the position S_r of S_o to the position S_r of S_d of a segment. In this plan, the object properly transported across the segment. It is classified as type <transport>.

All created plans are added to set $\mathcal{RP} = \{rp_1, rp_2, ..., rp_j\}$. These values are used to execute the task allocation for all agents (Sect. 3.5).

Figure 2(b) demonstrates all generated plans that the agents can perform to execute the transportation task and their respective types, out of their starting positions I, going to each state S_o or S_d of segments. For differentiation, states S_r are enumerated.

The plan classification by the types of the set $\mathcal{TM}$ is explained by the fact that, in order to perform the transportation of an object on a given segment s_i, the agent must execute a movement of type <pre-transport> first, and then perform a movement of type <transport>.

3.5 Task Allocation

In the task allocation step, the goal is to determine a set of movements that once performed culminate in the object transportation to the final desired position, whilst minimizing the total utility function.

To perform the object transportation, it is necessary to create a set $\mathcal{EP} = \{ep_1, ep_2, ..., ep_h\}$, called *Execution Plan*, containing a sequence of h plans rp_i from the set $\mathcal{RP}$, that are concatenated to enable task completion. One way to create the set $\mathcal{EP}$ is explained by Algorithm 4.

Algorithm 4. Execution Plan Set Creation ($\mathcal{S}$)

for all $s_i \in \mathcal{S}$ **do**
  $rp_i \leftarrow$ get_plan(s_i, <pre-transport>)
  $r_i \leftarrow$ get_robot(rp_i)
  $rp_j \leftarrow$ get_plan(s_i, <transport>, r_i)
  $\mathcal{EP} \leftarrow \mathcal{EP} \cup \{rp_i, rp_j\}$
end for

In Algorithm 4, the function *get_plan* retrieves a random plan rp from a segment based on the type passed as argument, or can retrieve a plan also based on a determined robot. The function *get_robot* just get the responsible agent by that plan. To calculate the cost value of the created set *Execution Plan*, the cost of each plan ep_i are summarized, as follows:

$$\Theta_t(\mathcal{EP}) = \sum_{i=1}^{h} \Theta_p(ep_i). \tag{4}$$

With *Execution Plan* created by Algorithm 4, it is possible to accomplish the object's transportation, however, the objective is to obtain the set $\mathcal{EP}$ with the lowest cost possible, named $\mathcal{EP}*$, formally described as:

$$arg\,min\ \Theta_t(\mathcal{EP}*) = min \sum_{i=1}^{h} \Theta_p(ep_i). \tag{5}$$

The process of creating the $\mathcal{EP}*$ set was developed using a graph representation with the application of a Minimum Spanning Tree (MST) algorithm. The graph creation method used to perform the task allocation is shown in Algorithm 5. Following these steps it is possible to create the graph $\mathcal{AG}$ demonstrated in

Algorithm 5. Task Allocation Graph Creation ($\boldsymbol{\mathcal{S}}$, $\boldsymbol{\mathcal{R}}$, $\boldsymbol{\mathcal{RP}}$)

for all $s_i \in \boldsymbol{\mathcal{S}}$ **do**
    $S_o \leftarrow$ get_origin(s_i)
    $S_d \leftarrow$ get_target(s_i)
    create_vertex(S_o, $\mathcal{AG}$)
    create_vertex(S_d, $\mathcal{AG}$)
end for
for all $r_i \in \boldsymbol{\mathcal{R}}$ **do**
    $I \leftarrow$ get_initial_position(r_i)
    create_vertex(I, $\mathcal{AG}$)
end for
for all $rp_i \in \boldsymbol{\mathcal{RP}}$ **do**
    create_edge(rp_i, $\mathcal{AG}$)
end for

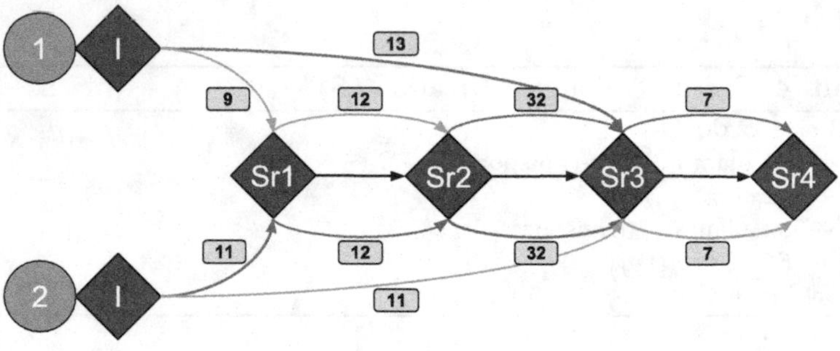

Fig. 3. Representation of the Allocation Graph $\mathcal{AG}$ used to select the best robot for each segment during the Task Allocation phase. Selected plans are represented in green, red are canceled. (Color figure online)

Fig. 3. The figure exemplifies the cost values for each edge, calculated using Θ_p (Eq. 2).

The graph creation algorithm uses several functions: *get_origin* gets the initial node from the segment; *get_target* retrieves the final node of the segment; *create_vertex* creates a new vertex on the graph; *get_initial_position* gets the initial position of the agent; and *create_edge* creates an edge between two vertexes.

In order to execute the MST algorithm in an optimized way, it is possible to make an reduction of $\mathcal{AG}$ graph, demanding fewer edges. This is accomplished based on the statements of Sect. 3.4, where each pair of edges that must be considered in sequence are grouped into a single edge, having as weight the sum of the original edges. Figure 4 exemplifies the same graph $\mathcal{AG}$, but in its reduced form, called $\mathcal{AG}_c$.

The graph $\mathcal{AG}_c$ has the same utility values as graph $\mathcal{AG}$, this way any operation done in the compacted version can be cast onto the original graph. Additionally, the edges from graph $\mathcal{AG}_c$ now have two types of $tp_i \in \boldsymbol{\mathcal{TP}}$, the instances <initial, movement> and <transition, movement>.

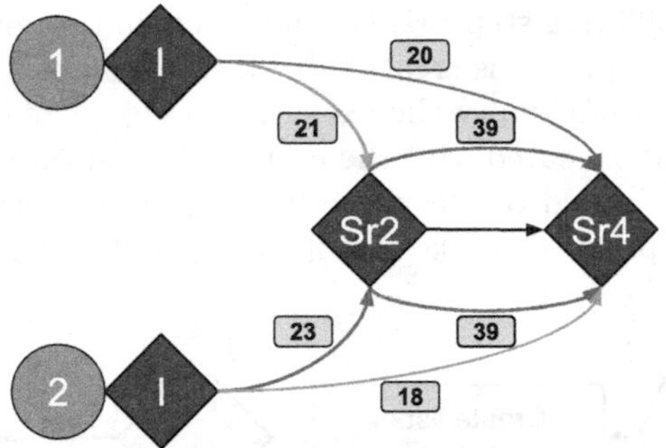

Fig. 4. Compact version of Allocation Graph ($\mathcal{AG}$) used as input for the Kruskal Algorithm on Task Allocation phase, named $\mathcal{AG}_c$. In this graph, each edge represents two edges from the original graph $\mathcal{AG}$, facilitating the application of the MST Algorithm.

To select the best plans and accomplish the task allocation among the agents, from graph $\mathcal{AG}_c$, the Kruskal's algorithm is applied following the rules below:

1. The edge selection follow the same sequence of segments in the set $\boldsymbol{\mathcal{S}}$;
2. After a edge selection, the edge of type <initial, movement> from the selected robot will be removed from the next selection phase;
3. And, all edges of type <transition, movement> from the others robots will also be removed.

The rule 1 is used because the transportation of each object must be executed in a predetermined order following the earlier created plan; the rule 2 is necessary because the robot executing the transport through the edge of that type will no longer be at its start position and cannot perform this type of movement. And considering the rule 3, no other robot can execute this type of movement, since only one agent could accomplish it, namely the one that was previous selected.

After the execution of the MST algorithm, the graph $\mathcal{AG}_c$ will discriminate the set of edges that minimizes the total utility value. Figure 4 demonstrates this, in which the edges in green were selected and the red ones were discarded.

Recalling that each edge in the graph $\mathcal{AG}_c$ represents two edges in graph $\mathcal{AG}$, the edge selection can be expanded to the complete graph. Based on this, the graph $\mathcal{AG}$ with the edges selected and discarded is presented in Fig. 3.

Through the construction of the graph $\mathcal{AG}$, it is possible to allocate the segments among the agents, thus minimizing the total utility value.

3.6 Coordination

The coordination phase is the process of controlling the agents to accomplish the overall task. To perform the object's transportation by a team, each robot must execute its task in a given order and in a coordinated manner.

The proposed methodology to coordinate the agents, using the before mentioned algorithms, and transport a set of objects is described in Fig. 5, and can be divided into the following steps: (i) for all objects in the environment, a list of all objects not yet transported is created; (ii) if the list is not empty, one of then is arbitrary selected, if it's empty the execution is completed; (iii) the planning phase for the agents is executed; (iv) the cost of all plans are calculated and the task allocation is performed to select the best agents to transport that object; (v) the coordination process to take the object until its final position occurs.

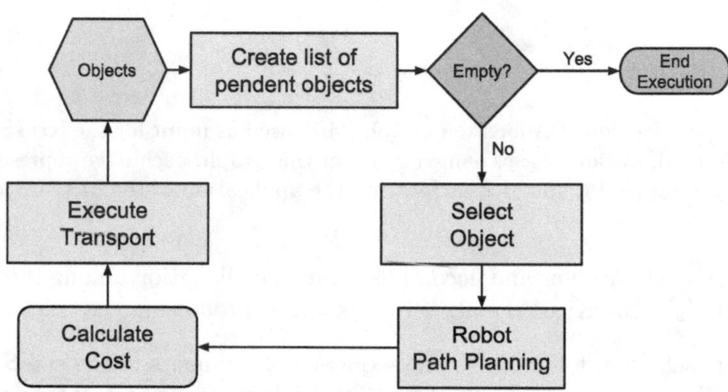

Fig. 5. Graphical representation of the Coordination Process to Object Transportation task. From the object list, those not yet transported are evaluated by the agents to execute their own plans to complete the movementation for all items.

As an assumption considered during the development of this methodology, only one agent can manipulate an object at a time, so only one is authorized to perform a task at a given time. The authorization control is done by the exchange of a *token* $\in \mathcal{TO}$, that is shared by the robots. The *token* has a property of type $tm_i \in \mathcal{TM}$, which represents the type of movement that is allowed.

As mentioned in Sect. 3.4, each task in the system represents a plan that must be covered to execute the transportation. Before executing the control mechanism, the set $\mathcal{TO}$ is populated with one *token* of type <transport> and *token* of type <pre-transport> in equal quantity of e, and then the Algorithm 6 is executed by each agent.

The creation of only one *token* of type <transport> restricts the quantity of agents performing plans of that type, because the object is handled by only one agent at a time. In order to speed up the whole process, the agents responsible for the next segments receives a *token* of type <pre-transport>, so that they can perform the preparation step, and then wait for the current agent executing the plan of type <transport>.

Using the aforementioned method, the agents are capable of accomplishing their tasks in an orderly fashion and under consensus, in addition to using the step before the transportation to gain time during the whole process. This task execution phase is performed through the exchange of information between the

Algorithm 6. Coordination Mechanism ($\mathcal{S}$)

```
for all s_i ∈ S do
    r_i ← get_robot(s_i)
    give_token(<pre-transport>, TO, r_i)
    if is_current_robot(r_i) then
        give_token(<transport>, TO, r_i)
        execute_task(r_i)
        retrieve_token(<pre-transport>, TO, r_i)
        retrieve_token(<transport>, TO, r_i)
    else
        execute_task(r_i)
        retrieve_token(<pre-transport>, TO, r_i)
        repeat
            wait()
        until has_transport_token(TO)
    end if
end for
```

agents, such as the current state of their actions and thc interchange of *token* among them.

4 Experiments

This section presents the experiments performed using the method described by this work showing the obtained results. The experiments were divided into two categories: (i) qualitative evaluation, considering simulations with several robots and (ii) demonstrative experiment, experiments with real agents.

4.1 Simulations

The presented methodology is capable of generating different kinds of plans to accomplish the transportation task. This is possible by the use of the utility function and the changing the value of the constant α, showing that the method is flexible in order to meet a certain demands, like realize the task trying to minimize the total time or the energy used.

Initially, we present an experiment that illustrates how the methodology behaves by using different values of α. Figure 6 shows the workspace used in the experiment and the resulting paths of each variation of the constant. Table 1 presents the values used in each experiment used to weight each characteristic, time and energy. As it can be seen, the methodology is suitable for different types of vehicles and it may take into account the specific task needs.

It is possible to observe that, with different values of α, the generated paths are totally different, when the weight tends to the Time aspect, the path is aerial, since it is the fastest way. However, when the Energy aspect is the objective, the terrestrial agent is used in the whole transport, considering its low energy use.

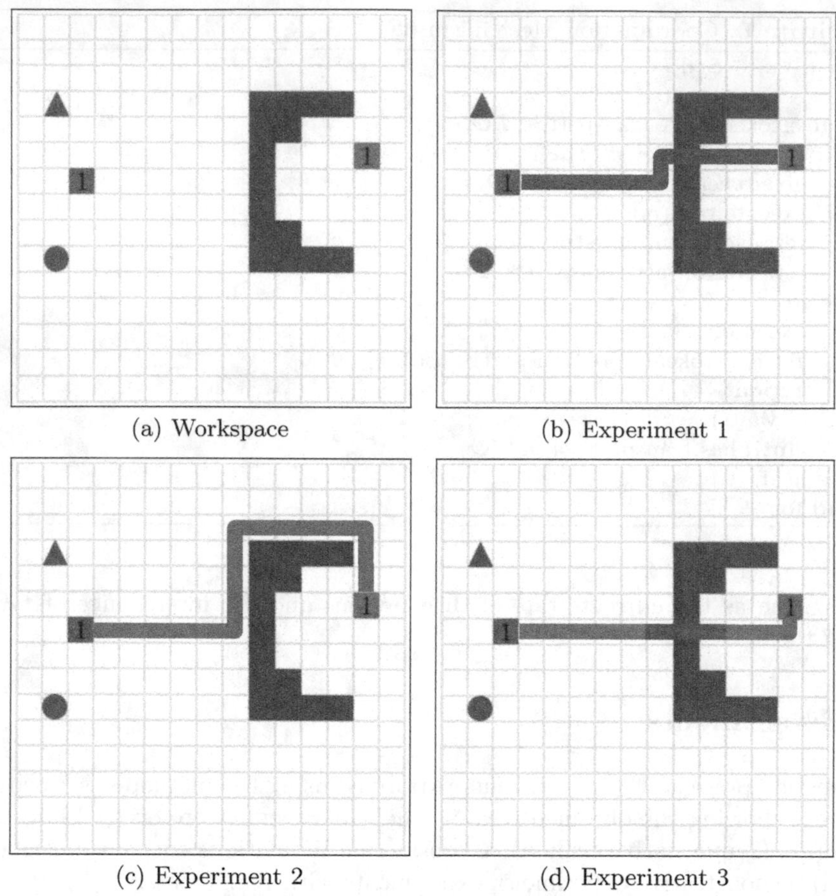

(a) Workspace (b) Experiment 1

(c) Experiment 2 (d) Experiment 3

Fig. 6. Different paths obtained accordingly to different values of α. Two vehicles are available, a ground (circle, red path) and an aerial (triangle, blue path). (Color figure online)

Finally, in the case with a medium value, both agents are used to transport the object.

The next experiment was conducted in order to empirically assess the asymptotic behavior of the methodology as the number of available robots increases. The framework was implemented in *Python* and all experiments were conducted in a computer with a Intel Core i7 2.4 GHz processor, 8 GB of RAM and Ubuntu 14.04 64-bit OS.

Figure 7 shows that the method presents a linear behavior with the number of robots. It is also possible to observe that as the initial and final configurations get farther away, there is also an increase in the computation time. This behavior was expected considering that all the agents must calculate its own paths to transport the object and then all of them will be considered during the task allocation phase.

Table 1. Different weighting constants used in the experiment.

Experiment	α	Time	Energy	Fig.
1	1	1	0	6(b)
2	0	0	1	6(c)
3	0.5	0.5	0.5	6(d)

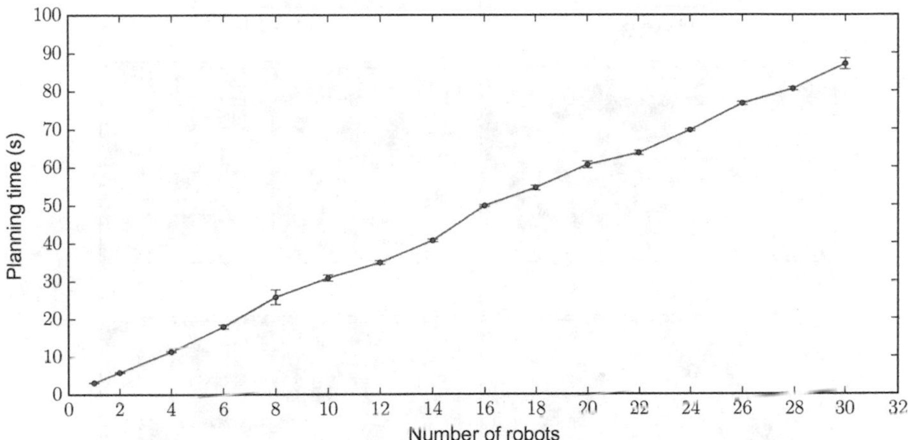

Fig. 7. Empirical asymptotic behavior of the methodology. The experiments showed that the increase of the number of agents implies linearly in the total time needed to plan and allocate tasks among the team.

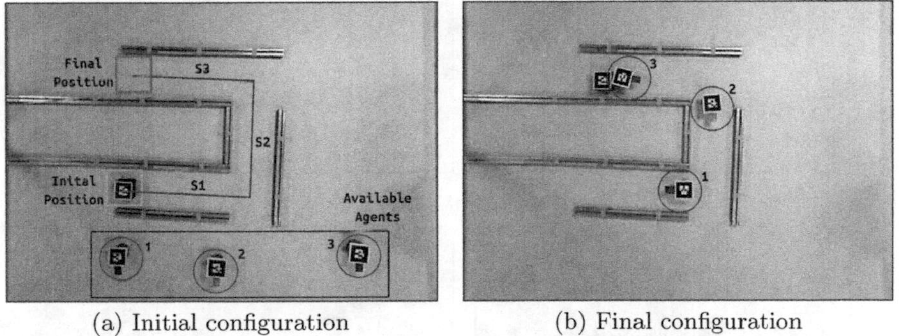

(a) Initial configuration (b) Final configuration

Fig. 8. Real experiment using three differential ground robots to transport one object. All robots contribute to object's transportation, cooperating to meet the system needs.

4.2 Real Experiment

Here we present a proof of concept experiment to demonstrate two main facets: (i) the real-time execution of our method, and (ii) the feasibility of the planned path in a simplified, real world conditions.

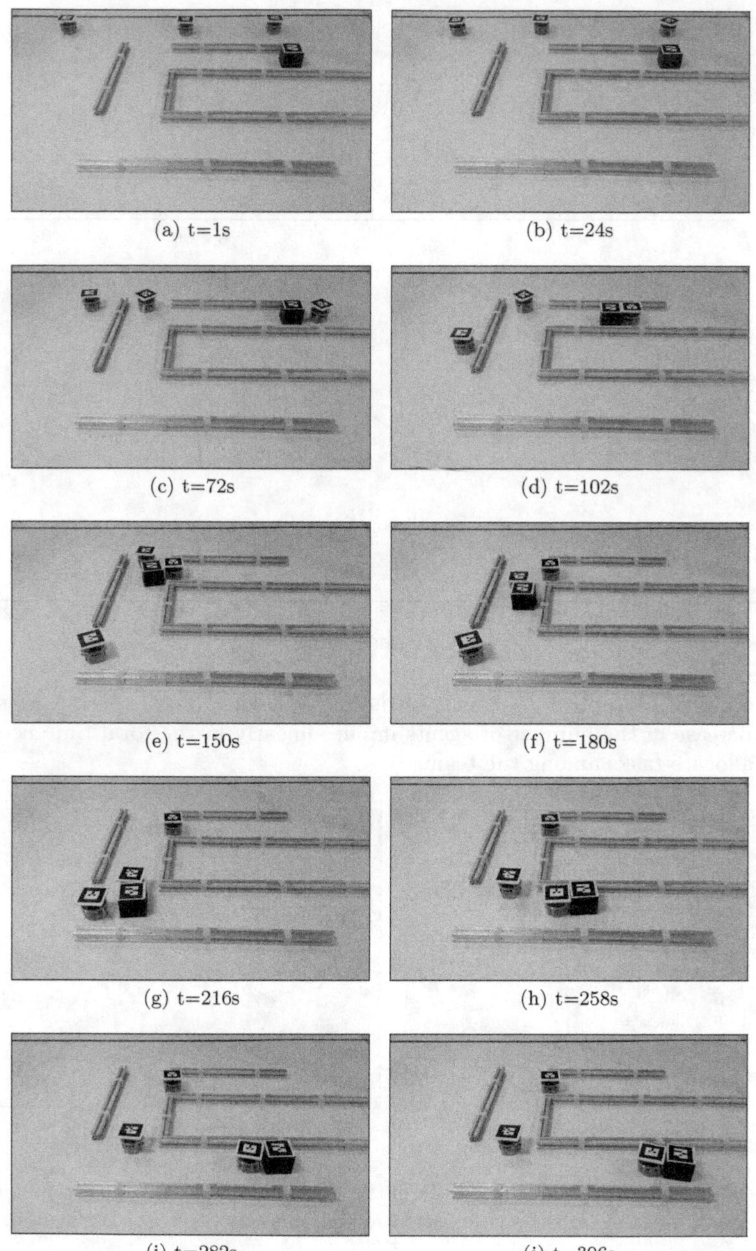

(a) t=1s (b) t=24s

(c) t=72s (d) t=102s

(e) t=150s (f) t=180s

(g) t=216s (h) t=258s

(i) t=282s (j) t=306s

Fig. 9. Transportation example of one object by a team of three real agents. After the object path planning, all agents evaluate which one is the best for each segment, resulting in the allocation of all robots to perform the total transportation of the object.

The experiment was performed in a table of approximate area of $2 \times 2\,\mathrm{m}$ which ensures a flat surface for the mobile robots. The localization process in this case was simplified using a tracking system that observes the marking codes in each item in the scene and informs to the system, creating a global positioning structure. The position of utilized obstacles were previously configured on the map of the experiment and informed to the agents.

The task consists in transporting an object with negligible weight using a team of three ground robots (e-Puck platform[2]), in a workspace with an arbitrary number of obstacles. Figure 8 presents the initial and final configurations of the robots and the object.

Figure 9 presents screenshots over time of the execution of the experiment. As cab be seen, all three agents were used during the transportation task, each one performing a part of the total path of the object. In the beginning, agent 1 is selected to start the transportation, while the other agents move and wait their turn to perform their actions. As soon as the object is moved along the first segment (S1 - Fig. 9(c)), agent 2 becomes the responsible and leaves the item to the end of the second segment (S2 - Fig. 9(e)), and passes to agent 3 the object, that will move it to its final position passing by the segment S3 (Fig. 9(g)). This implies that the system can use the available resources to better the result of the mission.

5 Conclusions and Future Work

In this work we presented a complete framework to tackle the object transportation problem using a group of heterogeneous robots. Our methodology considers the main phases involved in the transportation problem, such as path planning, task allocation and control.

The experiments showed the effectiveness and flexibility of the technique with respect to the task needs, using different types of robots, as well as creating different paths, according to the constants of the proposed utility function.

Future directions include the extension of the proposed methodology to consider a cooperative transport using more than one agent simultaneously, allowing the movement of larger objects. Another relevant improvement is the use of local information only to accomplish the task, requiring an exploration phase prior to all planning and coordination steps.

Acknowledgments. This work was developed with the support of the Conselho Nacional de Desenvolvimento Científico e Tecnológico, Coordenação de Aperfeiçoamento de Pessoal de Nível Superior and Fundação de Amparo à Pesquisa do Estado de Minas Gerais.

[2] http://www.e-puck.org/.

References

1. Ahmadabadi, M., Nakano, E.: A "constrain and move" approach to distributed object manipulation. IEEE Trans. Robot. Autom. **17**(2), 157–172 (2001)
2. Augugliaro, F., Lupashin, S., Hamer, M., Male, C., Hehn, M., Mueller, M.W., Willmann, J.S., Gramazio, F., Kohler, M., D'Andrea, R.: The flight assembled architecture installation: cooperative construction with flying machines. IEEE Contr. Syst. **34**(4), 46–64 (2014)
3. Bibuli, M., Bruzzone, G., Caccia, M.: Robot task allocation and path-planning systems in the minoas project framework. In: 2011 19th Mediterranean Conference on Control and Automation (MED), pp. 1194–1199. IEEE (2011)
4. Carvalho, F.F., Cavalcante, R.C., Vieira, M.A., Chaimowicz, L., Campos, M.F.: A multi-robot exploration approach based on distributed graph coloring. In: 2013 Latin American Robotics Symposium and Competition (LARS/LARC), pp. 142–147. IEEE (2013)
5. Fink, J., Ani Hsieh, M., Kumar, V.: Multi-robot manipulation via caging in environments with obstacles, pp. 1471–1476 (2008)
6. Gerkey, B., Matari'c, M.: Pusher-watcher: an approach to fault-tolerant tightly-coupled robot coordination, vol. 1, pp. 464–469 (2002)
7. Gerkey, B.P., Matarić, M.J.: Principled communication for dynamic multi-robot task allocation. In: Rus, D., Singh, S. (eds.) Exp. Robot. VII, pp. 353–362. Springer, Heidelberg (2001)
8. Guizzo, E.: Three engineers, hundreds of robots, one warehouse. IEEE Spectrum **7**(45), 26–34 (2008)
9. Harmo, P., Taipalus, T., Knuuttila, J., Vallet, J., Halme, A.: Needs and solutions-home automation and service robots for the elderly and disabled. In: 2005 IEEE/RSJ International Conference on Intelligent Robots and Systems, IROS 2005, pp. 3201–3206. IEEE (2005)
10. Lehman, A., Berg, K., Dumpert, J., Wood, N., Visty, A., Rentschler, M., Platt, S., Farritor, S., Oleynikov, D.: Surgery with cooperative robots. Comput. Aided Surg. **13**(2), 95–105 (2008)
11. Lindsey, Q., Mellinger, D., Kumar, V.: Construction with quadrotor teams. Autonom. Robots **33**(3), 323–336 (2012)
12. Lynch, K.M.: Nonprehensile Robotic Manipulation: Controllability and Planning. Carnegie Mellon University, Pittsburgh (1996)
13. Michael, N., Fink, J., Kumar, V.: Cooperative manipulation and transportation with aerial robots. Autonom. Robots **30**(1), 73–86 (2011)
14. Miyata, N., Ota, J., Arai, T., Asama, H.: Cooperative transport by multiple mobile robots in unknown static environments associated with real-time task assignment. IEEE Trans. Robot. Autom. **18**(5), 769–780 (2002)
15. Murphy, R., Kravitz, J., Stover, S., Shoureshi, R.: Mobile robots in mine rescue and recovery. IEEE Robot. Autom. Mag. **16**(2), 91–103 (2009)
16. Murray, R.M., Li, Z., Sastry, S.S.: A Mathematical Introduction to Robotic Manipulation. CRC Press, Baco Raton (1994)
17. Öztürk, S., Kuzucuoğlu, A.E.: Optimal bid valuation using path finding for multi-robot task allocation. J. Intell. Manufact. **25**(2), 1–14 (2014)
18. dos Santos, B.S.R., Givigi, S.N., Nascimento, C.L.: Autonomous construction of structures in a dynamic environment using reinforcement learning. In: 2013 IEEE International Systems Conference (SysCon), pp. 452–459. IEEE (2013)

19. dos Santos, B.S.R., Givigi, S.N., Nascimento, C.L.: Autonomous construction of multiple structures using learning automata: description and experimental validation (2014)
20. Song, P., Kumar, V.: A potential field based approach to multi-robot manipulation. In: IEEE International Conference on Robotics and Automation, Proceedings, ICRA 2002, vol. 2, pp. 1217–1222. IEEE (2002)
21. Sugar, T., Kumar, V.: Control of cooperating mobile manipulators. IEEE Trans. Robot. Autom. **18**(1), 94–103 (2002)
22. Tanner, H., Loizou, S., Kyriakopoulos, K.: Nonholonomic navigation and control of cooperating mobile manipulators. IEEE Trans. Robot. Autom. **19**(1), 53–64 (2003)
23. Tiganas, V., Kloetzer, M., Burlacu, A.: Multi-robot based implementation for a sample gathering problem. In: Proceedings of International Conference System Theory, Control and Computing (ICSTCC) (2013)
24. Wawerla, J., Vaughan, R.T.: A fast and frugal method for team-task allocation in a multi-robot transportation system. In: 2010 IEEE International Conference on Robotics and Automation (ICRA), pp. 1432–1437. IEEE (2010)
25. Willmann, J., Augugliaro, F., Cadalbert, T., D'Andrea, R., Gramazio, F., Kohler, M.: Aerial robotic construction towards a new field of architectural research. Int. J. Archit. Comput. **10**(3), 439–460 (2012)
26. Yamashita, A., Arai, T., Ota, J., Asama, H.: Motion planning of multiple mobile robots for cooperative manipulation and transportation. IEEE Trans. Robot. Autom. **19**(2), 223–237 (2003)

ND-NCD: Environmental Characteristics Recognition and Novelty Detection for Mobile Robots Control and Navigation

Antônio Soares, Valéria Santos, Cláudio Toledo, Fernando Osório(✉), and Alexandre Delbem

ICMC, University of São Paulo, São Carlos, SP, Brazil
ahelsonms@gmail.com, {valeria,claudio,fosorio,acbd}@icmc.usp.br
http://www.icmc.usp.br/

Abstract. Mobile robot applications usually perform a path planning and its execution considers a previous known map. On the other hand, some application must explore the environment, defining a path from a source to a destination point, without knowing the environment map. The environment exploration, path planning towards a goal and navigation control tasks should be done at the same time. This study proposes a new method for mobile robot control and navigation based on the environmental characteristics recognition and novelty detection, named ND-NCD (Novelty Detection with Normalized Compression Distance). This method can be used as a key component in environment exploration and topological mapping tasks. In a previous work, a Genetic Algorithm (GA) for exploratory path planning was implemented to create a topological map (graph) from the source to the destination point, generating a set of actions which the robot must perform to achieve the goal. Each action was associated to a different reactive behavior specifically designed for characteristic places of the environment, such as corridors, curves or intersections. The proposed method, ND-NCD is used to recognize such different environmental characteristics, allowing to activate/associate the adequate actions whenever the method recognizes a context change (new context). This allows us to integrate the GA based environment exploration method together with the robot control reactive behaviors, which can be properly selected and switched according to the environmental characteristics detected/discovered by the ND-NCD. The ND-NCD uses the robot perception (e.g. laser sensor) to detect novelty and to recognize already known characteristics, thus allowing an incremental representation of the environment structures. The experiments were performed in the Player/Stage simulator and in a real indoor environment. ND-NCD performance is compared with a Neural Network trained to recognize context changes in the same environment. The results indicate that ND-NCD is a promising approach to be used in exploration and navigation control for mobile robots with the advantage of detecting a context change just knowing an initial state (corridor) from the environment. The proposed method does not need to be trained previously in order to know all the states (supervised training), being able to incrementally discover the different environment configurations.

F. Santos Osório and R. Sales Gonçalves (Eds.): LARS 2015/SBR 2015, CCIS 619, pp. 192–209, 2016.
DOI: 10.1007/978-3-319-47247-8_12

Keywords: Novelty detection · Robot navigation control · NCD-Normalized Compression Distance · Genetic algorithm · Path planning

1 Introduction

The path planning task consists of defining a way for the robot to leave its initial position and reach its goal, it could be based on a previously provided map of the environment, or, it could be achieved by environment exploration, discovering the path to the destination. The navigation control task refers to the actions performed by the robot following a pre-defined trajectory or going towards some specific direction/position. Based on the previous work proposed by [1], this study presents a new hybrid approach to explore the environment, plan the robot trajectory, and control this robot through the navigation. A Genetic Algorithm is applied to find a sequence of actions that enable the robot reaches its goal. The actions are local reactive behaviors like turn left or go forward, according to the local context (environment characteristics). Then, to control the robot navigation, an algorithm based on the Normalized Compression Distance (NCD) is used to identify in which state/context the robot is and activate the correspondent action. The development of the ND-NCD based approach for navigation control of robots is the original contribution of this work, since it can also discover new configurations, and also, recognize previously known environmental characteristics. The proposed method uses the NCD to process the perception data obtained from the robot sensors, in order to recognize the different environmental characteristics.

Other approaches were proposed to control indoor robots during navigation. A system based on landmarks for wheelchairs was developed in [2]. The landmarks are composed by metallic paths and radio-frequency identification (RFID) tags. The path is generated using: a topological map, a metric map, the current localization and the destination asked by the user. According to the authors, the localization mechanism of the navigation system is robust for a partially-known static environment if there are enough installed landmarks, since the odometry error is incremental for wheelchair displacements. The disadvantage of this proposal is that the environment must have installed landmarks to work. Several alternative approaches, like this, rely on previously available environment knowledge and/or precise robot and destination localization, using maps or placing artificial landmarks, as proposed in the above cited work.

In [3], reactive and learning navigation algorithms for exploration robots that must avoid obstacles and reach goals in limited time and observations were evaluated. A rule-based navigation and neural-evolved navigation techniques for controlling the robot based on environment information obtained from sonar and inertial sensors were proposed. According to the authors, the results show that neuro-evolutionary algorithms obtained has better performance than rule-based algorithms in relatively-complex domains, in which the robot had to select paths that lead to goal while avoiding obstacles. The adoption of Evolutionary/Adaptive Algorithms allows to create more robust solutions that can adapt the robots perception and control, and also optimize/improve their behavior.

An approach to evolve behaviors in an unknown environment for navigation problems was tested in [4], in which complex behaviors were obtained from simple ones. Each behavior is supported by a controller based on an artificial neural network (ANN). Thus, a method for the generation of a hierarchy of neurocontrollers is developed and verified in simulated and real experiments. Several behavioral modules are initially evolved using neurocontrollers based on different ANN paradigms. The results show that is possible to obtain complex behaviors from simple behaviors coordination, benefiting the development of navigation systems for real-world scenarios.

In [5], it was developed a navigation subsystem of a mobile robot which operates in typical human environments to execute different tasks, such as transporting waste in hospitals or escorting people in exhibitions. This paper describes a hybrid approach that combines both offline and online navigation methods. The offline method generates a reference path based on a prior map (previously known environment). The online method adapts the generated path to avoid static and also dynamic (moving) obstacles. The authors discuss about the properties of the proposed method and the experimental results obtained from real-world experiments. Hybrid systems allow to explore the best properties of each system component in order to improve the individual components performance and limitations, then obtaining a better overall system performance.

A hybrid system using a topological navigation method was proposed in [6]. A topological map (graph) was used to represent the environment in which the nodes are characteristic points, like crosses and turns, and the edges represent the navigability paths among these points. An Artificial Neural Network (ANN) was trained to learn and recognize each characteristic point of the environment (state/context recognition). During the navigation, the ANN is used to recognize in what state/context the robot is and to activate an adequate reactive behavior for that situation. The ANN recognize states, and the sequence of states is represented by a Finite State Machine (FSM), which controls the current state, the current action being executed and associated to that state, and when to change to a next state, which is recognized by the ANN as a new state/context. This work was motivated the development of the present work, since it presents some limitations and possible improvements. More details about this system, known as Neuro-FSM, will be presented in the next section.

ANNs need samples of all possible states in the environment in order to train a navigation control system. In the new navigation system proposed based on NCD, only samples from a default initial state (data from *corridor* regions in this paper) are necessary in the offline step of the approach. The proposed algorithm is able to identify when the robot state changes from the default state and, then, it activates the suitable reactive behavior during the execution of the path.

This paper is organized as follows. Section 4.2 describes some related works. Section 2 presents the two previous works, Neuro-FSM and Hybrid GA-ANN Topological Navigation, that motivate and originate this work. Section 4 introduces Normalized Compression Distance Method (NCD) which was used to recognize/discover environmental characteristics. Section 3 describes the developed genetic algorithm for exploration and path planning. Section 5 proposes the new navigation system based on ND-NCD. Section 6 presents experiments and results. Finally, Section 7 concludes the paper.

2 From Neuro-FSM to GA-ANN and ND-NCD

Our initial motivation for this work was the topological mobile robot navigation system, Neuro-FSM, proposed by Sales and Osorio [6–9]. The Neuro-FSM (Neural Finite-State Automata) was developed in order to provide the control and navigation system for mobile robots based on previously known topological maps and previously defined paths, which are specified based in these topological maps.

In the Neuro-FSM proposed approach, the map, the path and also the robot actions were all pre-defined. The system FSM (Finite State Machine) is a sequencer that uses the current state to associate a specific action to the current context/situation, and, when the current state is changed (new context), a new action is then selected to be executed. For example, if the robot is following a path along a corridor, the present state is defined as corridor and the action is to move forward along this corridor (e.g. move forward, avoid collisions, follow the walls), and when the robot reaches the end of the corridor, or any other new situation representing a different context (e.g. left/right turns or corridor crossing), then the robot must switch the present action to another behavior that allows it to deal with this new situation. Sometimes, there are more than one possible action when a new situation is found, so, the FSM has coded into it the sequences of actions to follow each time we need to change from one current state to the next state. When the robot reaches the end of a corridor with a turn left, it recognizes this situation and switch to turn left behavior. Briefly, the FSM defines the sequence of states and actions the robot will execute, but what causes the decision of when to change from the current state and current action to a new state/action are the inputs that this FSM receives from the Neural module.

In this work an Artificial Neural Network (ANN) was adopted in order to detect the states (current state) and the state changes (context change), based on the robot sensors. The ANN is used to inform the FSM about context changes. The Artificial Neural Network (ANN - Neural Module) was used to recognize the current state and the state changes based on the robot sensor inputs. The complex pattern recognition and sensor inputs analysis is performed by an ANN, that perceives the environment and detects the current state. For example, the ANN is able to classify the input patterns (e.g. laser data) that define a corridor, and when the robot reaches a corridor intersection, the ANN will also be trained to identify this new pattern, and notify the FSM that the robot has reached an intersection. On the other hand, the FSM when is notified about a state change, defines the action to be selected and executed in this new configuration, for example, turn left into an intersection. The Fig. 1 presents an overview of the Neuro-FSM architecture.

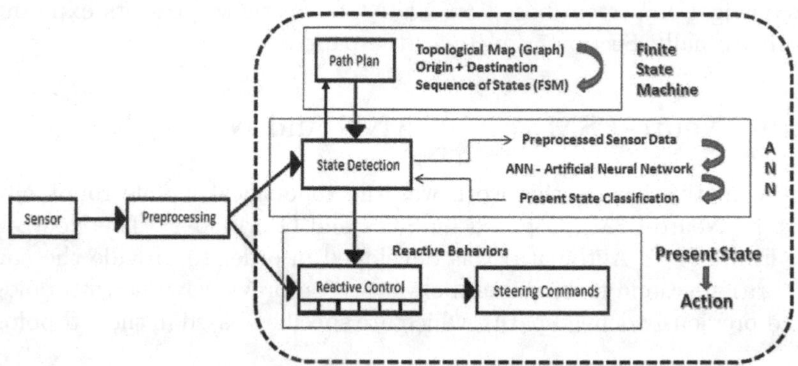

Fig. 1. Neuro-FSM system architecture overview [Adapted from Sales 2012].

The Neuro-FSM approach is interesting since it allows a mobile robot to navigate based only in a very simple previously provided topological map and set of actions. Although this approach has showed to be reliable, the user should first provide all the knowledge about the problem: map, states, actions/behaviors and path. Figure 2 shows an example of some states, perceptions, topological map and path (sequence of nodes in the graph). One of the most complex tasks is to train the ANN in order to recognize different states (situations), since we adopt a supervised learning algorithm. We need to train the ANN to recognize the different states (e.g. corridor, left turn, right turn, intersection) presenting a set of labeled examples. This can be hard to be done.

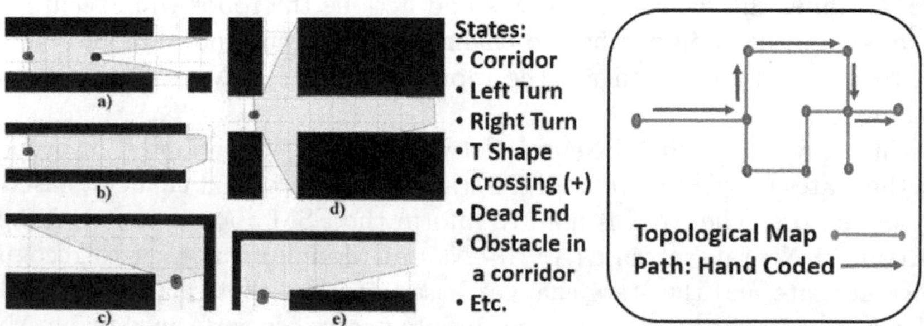

Fig. 2. Examples of robot states and perception (a)–(e) and the graph representing a topological map with a pre-defined path represented on it.

The topological map, path (FSM) and set of actions can be defined by the user, and usually the most difficult task is to find the path from the origin to the destination, when we are navigating into a more complex environment map. The set of actions usually are a set of simple and well known reactive/deliberative behaviors (e.g. move forward, follow the wall, turn 'n' degrees to the left/right, turn to the left/right up to have a free space in front of the robot). Concluding, the two most complex tasks are: learn and recognize/classify the different environment configurations (states) and discover the path from the origin to the destination (path planning).

In the following sections of the present work we will present solutions to deal with and tackle these two problems: the GA approach allows to discover the path considering the navigation based on a topological map, and the ND-NCD approach that allows to incrementally discover new states (situations). These two strategies allow us to envisage a more robust and less user dependent solution to robot navigation problems and applications, were the user do not need to provide pre-defined maps (environment exploration using GA), paths (path discovery and planning using GA) or describe all the states (environmental characteristics detected/discovered using ND-NCD). The robot control and navigation will still be done based on the FSM approach (sequencer) with states and actions, but considering the states and path can be automatically discovered.

3 Genetic Algorithm for Robot Navigation

This work extends the Genetic Algorithm (GA) proposed in [1]. The main goal is to explore the environment, with a squad of robots (GA population) in order to find the path from one initial point to a destination point, using only the approximate known position of the robot and the target destination (considered in the fitness function value estimation). The authors applied a GA to find a sequence of actions/movements of a mobile robot that must reach a goal position. The sequence of actions result in a trajectory when they are executed by the robot. Therefore, the individual is composed by a sequence of integers, each one representing a reactive behavior (action) that can be performed by the robot sequentially. The possible actions are listed next (Table 1):

Table 1. List of possible robot actions

Action number	Action description
0	Go forward in an intersection
1	Turn left
2	Turn right
3	Go forward in a right fork
4	Go straight
5	Go forward in a left fork
6	Turn around

A GA was developed to evolve this sequence of actions, which was encoded as an individual. A population of individuals is randomly initialized. The chromosome length and the population size were experimentally defined (Sect. 6). Each individual is evaluated executing a robot navigation using a graph representation of the environment (Sect. 6). A repair operator is executed when an action encoded in the gene of an individual leads to an unfeasible action. For instance, if an action decoded means turn left but there is no way at left, the repair operator replaces this action by another randomly selected, that is also feasible at this point of the graph.

The fitness function is based on the length of the path traveled by the robot. In Eq. 1, *sum* is the sum of Euclidean distances between each pair of nodes, from the set of nodes included in the path. If the action *turn around (6)* is chosen, the distance between the nodes is multiplied by a factor once the robot has entered in a dead-end corridor. If the goal is reached, the total sum is divided by a constant $\alpha > 1$. If the goal is not reached, the Euclidean distance between the current node and the goal node is added to the sum, multiplied by a constant $\beta > 1$. Equation 1 is the fitness function as proposed in [1], where α and β are constant values, v_a is the current node, v_g is the goal node and $d(v_a, v_g)$ is the euclidean distance between v_a and v_g nodes.

$$f = \begin{cases} sum/\alpha & \text{if the goal is reached} \\ sum + \beta d(v_a, v_g) & \text{if the goal is not reached} \end{cases} \tag{1}$$

The population at each generation t is formed by the best individual from population at generation $t-1$ and other individuals are determined by crossover and mutation. The tournament of two selects parents from population at generation $t-1$. The uniform crossover operator is applied over these parents generating a new individual. Next, mutation operator is always applied to the new individual. This operator randomly choices an amount of genes to be mutated, which means to select randomly another action to be performed. Finally, the fitness values of the new individuals are determined by Eq. 1.

The new individual is inserted into the population, except if its fitness value is equal to one of its parents. In this case, the reproduction process is repeated. The evolutionary process evolves until one of the stop criteria has been satisfied. Two stop criteria were adopted: if the maximum number of generations is reached or if the best individual found was not improved during 0.1 % of the maximum number of generations.

The proposed AG method allows to discover a sequence of actions describing the movements the robot should execute to reach the goal destination. This sequence of actions considers the adoption of the Neuro-FSM approach described in the previous section. The path is described by a FSM (Finite State Automata), being composed by a sequence of actions (local behaviors), each one of them being associated to the current state. When the current state changes, the next state is assumed and a new action is selected. The current state is recognized by a previously trained ANN (Artificial Neural Network) that classifies the present situation (sensors perception), indicating the current state and the state changes.

4 Normalized Compression Distance

This section presents NCD and describes some works illustrating relevant applications.

4.1 Definition

Normalized Compression Distance (NCD), from Information Theory [10], is a metric that can find relationships among data, determining similarity between variables by measuring the size of their data compressed. This approach doesn't require any specific knowledge of the application domain. According to [10], NCD is a universal and robust metric that has been successfully applied to areas such as genetics, literature, music and astronomy.

NCD is based on Normalized Information Distance (NID) [11], which considers the similarity between samples (variable, files, etc.) according to the dominant characteristic shared by them. The NID uses the concept of Kolmogorov complexity [12] to estimate distances, however, the corresponding calculation is computationally intractable. NCD performs an approximation of the NID based on the results of a compression algorithm. In practice, the distance between two samples (stored in files) calculated by NCD is a positive number ranging from $[0; 1+\varepsilon]$. This positive number represents how much two files are different, where ε is the upper bound to the error of the compressor used. The distance value between two files x and y is given by Eq. 2:

$$NCD(x,y) = \frac{C(xy) - min\{C(x), C(y)\}}{max\{C(x), C(y)\}} \tag{2}$$

where xy is the concatenated file from x and y, $C(xy)$ is the size of the file xy after the compression and $C(x)$ and $C(y)$ are the sizes of file x and y after their compression respectively.

The level of compressions affects the distance values determined by NCD. Thus, the choice of a compressor can be based on rate of compression of the files involved considering several of the compressors found in the literature [13]. The most commons used with NCD have been bzip, gzip, lzm and ppmz. Bzip compressor found the best rate of compressions for the data used in this work.

4.2 Related Works

Some previous papers have used NCD as a pre-processing technique since it can fast estimate dissimilarities involving several types of data. This section illustrates the usage of NCD by presenting some works from the literature.

The authors in [14] reported that NCD can determine similar software code, when this distance is combined with techniques Neighbor Joining and Fast Newman algorithm. The approach, named DAta MIning of COde REpositories (DAMICORE), works with different types of code representations, which usually hampers the performance of the most known data mining methods since it doesn't require previous knowledge about each code language used. NCD in DAMICORE basically generates a distance matrice from set of codes enabling DAMICORE to highlight useful software similarities from source code level.

In [15], NCD was applied to the problem of determining the fragment of files. NCD is combined with the k-nearest-neighbour, which is used as classification algorithm. The method was evaluated using a random selection of 3,000 file fragments with 512 bytes from 28 different types of files. The results showed around 70 % of accuracy.

In [16], a similarity measure has defined based on compression, named Fast Compression Distance (FCD), that combines the relative accuracy of NCD with the reduced complexity of Pattern Representation using Data Compression (PRDC). PRDC is a classification methodology based on compression with dictionaries extracted of data. PRDC is faster than NCD alone, however, with lower accuracy. FCD was tested for databases of images. In a first step, images were quantized in a color space and converted into strings, without losing texture information. Then, representative dictionaries were extracted from each object and similarities between individual images were calculated by comparing each pair of dictionaries. The experiments showed that FCD doesn't lose in accuracy compared to NCD and requires low computing time.

In [17], NCD was used together with Multidimensional Scaling (MDS [18]) in weaning, a process of discontinuing mechanical ventilation that replaces or assists in the breath of patients with respiratory problems. The proposed method considers heterogeneous data from a monitoring system for prediction of the weaning in three stages. In the first stage, NCD was applied to a number of temporal sequences labeled by failure or success, generating a distance matrix whose elements are a measure of the dissimilarities between pairs of sequences. In the second stage, the NCD matrix is projected into an N-dimensional space by the MDS technique, which maps a set of points into this space (each point is associated with a different sequence). Finally, the points are clustered by associating them with the labels of the original sequences in order to build a classifier that distinguishes between weaning success or failure. According to the authors, the results showed that the combination of NCD with MSD is able to differentiate heterogeneous temporal sequences.

In general, those papers show that NCD can be used in a relatively easy way to deal with complex data types without previous knowledge about an application.

5 Novelty Detection with Normalized Compression Distance

The proposed GA determines the actions to be executed by the robot to reach its goal. However, the navigation needs to be controlled such that the robot knows exactly when each action must be performed. This control should take into account the environment and its patterns (sensor data), so the robot must recognize the present context before executing an action.

The initial approach adopted in the Neuro-FSM requires to previously train an ANN in order to be able to recognize each different state (present context), represented by complex sensor data patterns. The ANN adopted in the Neuro-FSM approach was trained based on a supervised learning algorithm, that means, the user should previously prepare a set of training examples (labeled patterns) in order to train the ANN. In the present work, the adoption of the ND-NCD seeks to avoid the need of this previous step, training an ANN based on a supervised learning set. The ND-NCD is an unsupervised novelty detection and pattern recognition algorithm, so it can facilitate this task of states recognition, and also, it can be used to incrementally discover new states.

Note that the major part of the path usually is composed by corridors, where the robot can go on straight ahead. When the situation (state) changes, being different from a corridor, it should be detected, and a new action may be required. The present paper proposes a new method of novelty detection named Novelty Detection with NCD (ND-NCD) to recognize new states in a path traversal. The ND-NCD was previously known as NDN.

The method is divided into three distinct procedures:

1. Collecting data to build the knowledge-base: The environment is scanned during a traversal to compose the knowledge base. The scanning is performed by the robot on automatic control (e.g. an wall-following behavior). It is not necessary to scan all the corridors, just enough samples to represent the main aspects of corridors.
2. Calculating of Activation Sensitivity (AS): The calculation is performed using NCD to generate a vector U_k of distances for each sample vector V_k of the knowledge-base. The highest median $m_k = median_k\{U_k(i)\}$ defines a sensitivity value of the base, called Activation Sensitivity (AS). This means that the sample with median less than AS should be considered a corridor, as summarized in the diagram of Fig. 3.
3. Running: NCD calculates the distances between each sample (V_k) obtained by a laser sensor adapted to the robot and those samples from the knowledge-base. Thus, each detected sample results in a vector of distances U_K with a corresponding median value. If the median is greater than AS, the sample is not recognized as a corridor, as described in the diagram of Fig. 4.

To avoid mislabeling, a filter is applied to a distribution of medians for the calculation of AS. The filter used in ND-NCD is the average of the last 10 out of 50 medians. As a consequence, the classification of an external signal can only occur after computed a window of samples of size 10. Figure 5 shows the distribution of the averages with and without a filter. Finally, it is important to note that ND-NCD is disabled when the robot is performing an action.

6 Experiments and Results

This section presents the experiments and results obtained by GA for robot navigation using ND-NCD. All experiments have the same goal, go out from

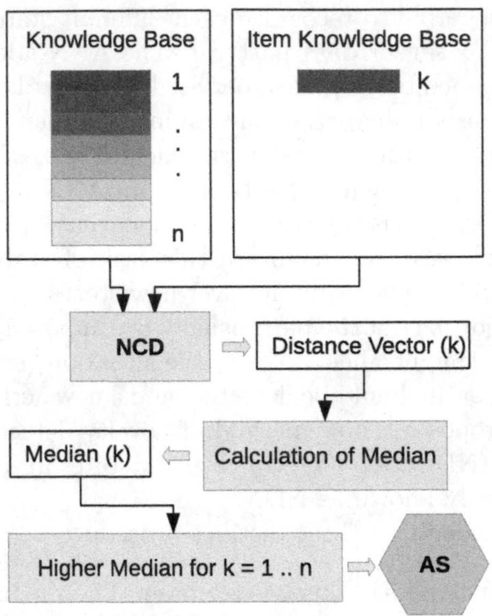

Fig. 3. Diagram of the calculation of the activation sensitivity.

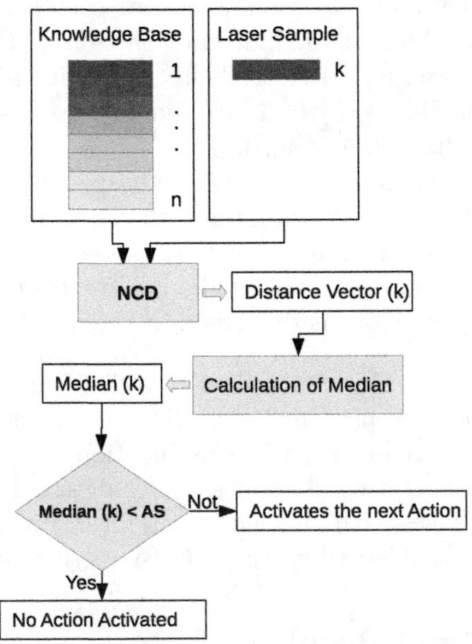

Fig. 4. Process of activation of actions.

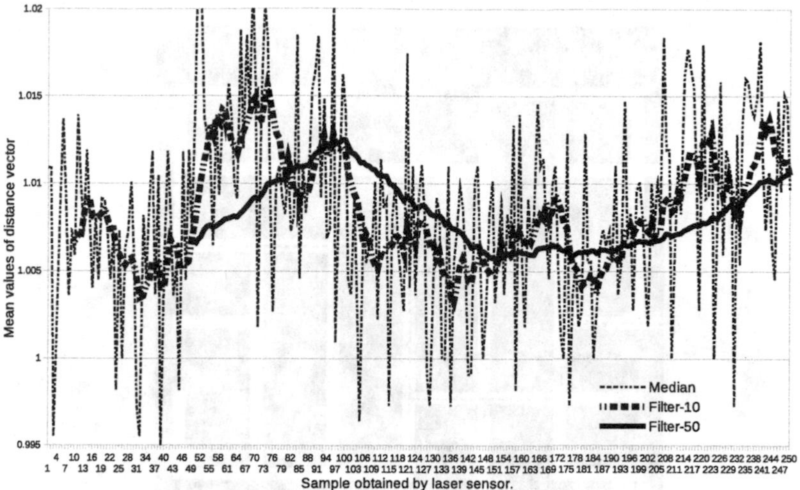

Fig. 5. Different filters for ND-NCD.

point "A" and reach point "B", following a sequence of actions returned by the GA.

6.1 Parameters of the GA Used

The GA for robot navigation was executed with a chromosome length of 30 (size of the sequence of actions) and population size of 100 individuals. It stops if the maximum number of 1000 generations is reached. A total of 10 trials of the GA was performed for each experiment using different seeds for the random number generator. The first experiment was performed in the map showed in Fig. 6.

The second experiment was done in the map showed in Fig. 7, which is larger than the first map. The maximum number of generations was set to 100,000 and the population size was 1000. However, the chromosome length remained the same used in Experiment 1, since it was enough to encode the necessary number of actions. The results achieved are presented in Table 2.

Table 2. Results from executions of the GA for robot navigation.

Map	Generations to converge	Best fitness
Map 1	11.7 ± 11.89	134.5
Map 2	25238.7 ± 22810.52	590.77 ± 20.66

Fig. 6. Map for simulation of Experiment 1.

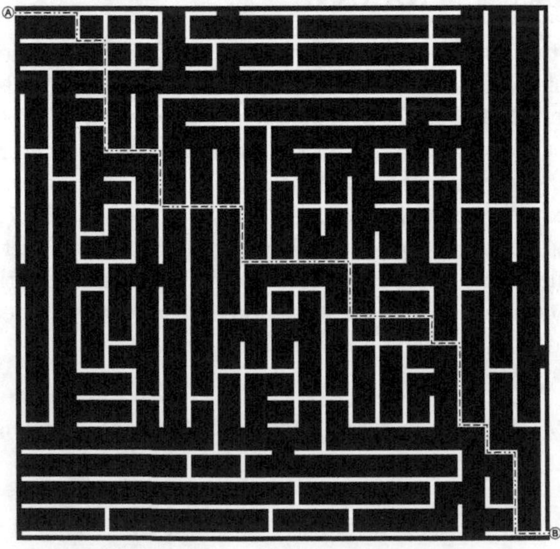

Fig. 7. Map for simulation of Experiment 2.

6.2 Parameters Used for ND-NCD

A total of four experiments were conducted to validate the performance of ND-NCD. Two of the experiments were simulations using Player/Stage and the other two over an indoor real environment. The robot Pioneer P3-AT with a 180° SICK Lidar sensor was used in this case.

ND-NCD recognizes if the robot is or is not in a corridor, through the AS value. A new action is performed by the robot whenever a change in environment is detected, which means the median value of the vector of distance is larger than AS value. When the robot follows its trajectory and the corresponding medians are values below AS, a corridor is being detected. This result disables the impact of the previous action performed and enables to run a new action, as soon as the AS value is exceeded again.

The knowledge-base for simulation is formed by 40 corridor samples collected from different parts of the map. The knowledge-base for indoor experiments is formed by 40 samples collected at beginning of the main corridor.

Experiment 1: Simulated in Player/Stage. ND-NCD has to recognize in the beginning of the path, a region different from corridor as shown in Fig. 6. For this map, the sequence of actions found by GA was {go forward in a left fork, turn right, turn left, go forward in an intersection, turn left, turn right, go forward in an intersection, go forward in a right fork, go forward in a left fork}. The complete presentation of Experiment 1 is available in https://goo.gl/5ZfqpU. Figure 8 shows how the medians of the distance vectors change as the robot walks.

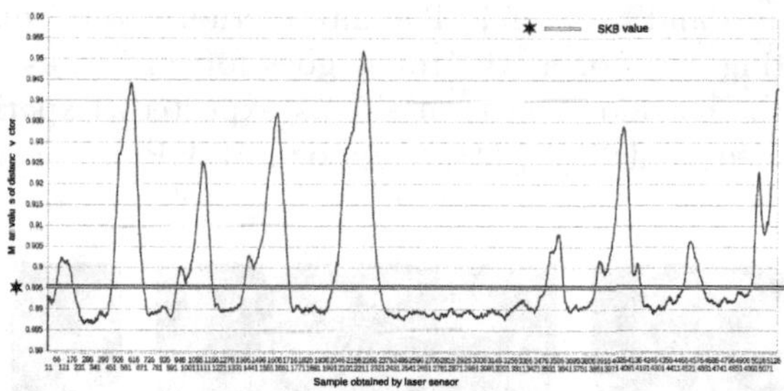

Fig. 8. Median value of the vector of distances along the path in Experiment 1.

Experiment 2: Simulated in Player/Stage. In this experiment the number of actions increases considerably, becoming harder the task of the GA for robot navigation, as shown in Fig. 7. The best individual found by GA was {turn right, turn left, turn right, go forward in an intersection, go forward in a right fork, go forward in a right fork, turn left, go forward in a left fork, turn right, turn left, go forward in an intersection, go forward in an intersection, turn right, turn left, go forward in a left fork, go forward in a left fork, turn right, turn left, go forward in an intersection, turn right, turn left, turn right, turn left, turn right, turn left, turn right, turn left, go forward in a left fork}. The complete execution of Experiment 2 is the video available in https://goo.gl/npyhfU.

Experiment 3: Indoor environment. Figure 9 shows the map of a real indoor environment where Experiment-3 was conducted. The simplicity of this map doesn't require that the GA runs for path planning. There is only one action to be performed {turn right}. However, the idea is to demonstrated that the robot control executed using ND-NCD is able to identify the intersection of routes and executed the action. Experiment 3 is in the video available in https://goo.gl/lveF8Q.

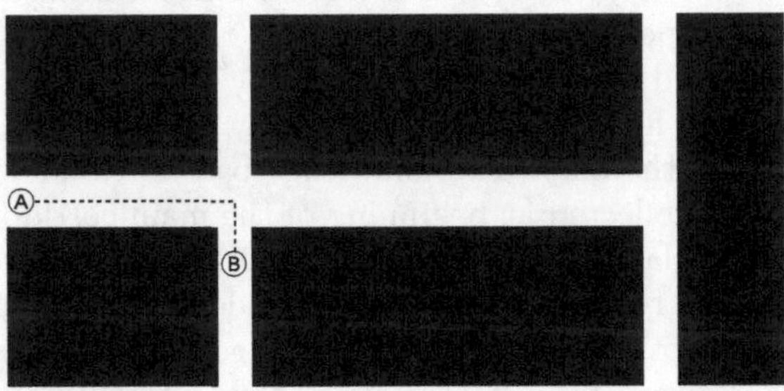

Fig. 9. Map of a indoor environment, Experiment 3.

Experiment 4: Indoor environment. Figure 10 shows the same map of Experiment 3, but the robot has now to perform the following actions: {go forward in an intersection of routes and turn left}. The intersection doesn't disturb ND-NCD capacity of detecting corridor as the robot goes forward, thus, it goes forward until the end of the corridor and turn left as expected. Experiment 4 can be followed by the video available in https://goo.gl/HyKPpi.

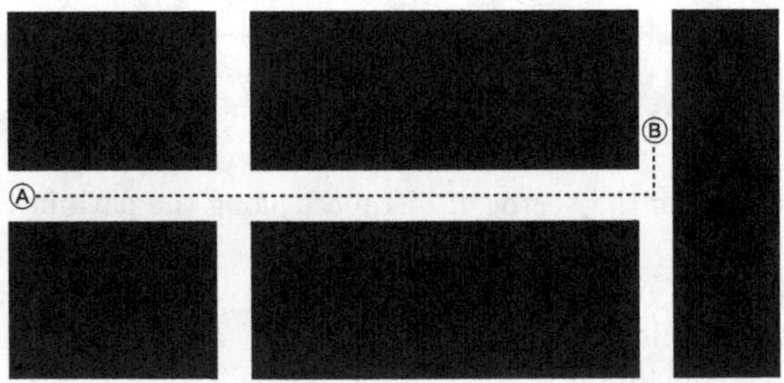

Fig. 10. Map for indoor Experiment 4.

As the work proposed in [6], an Artificial Neural Network (ANN) was also trained to recognize different types of regions of the environment (robot states). The ANN used was a multilayer perceptron that was trained with resilient propagation algorithm. First, in order to compare the ANN with ND-NCD, the database for training was composed only by the same knowledge-base of corridor regions used by ND-NCD. ANN failed in identify changing state. It always responses corridor during its validation using data from others environment regions (states).

Next, ANN was trained using a database with examples of each possible state: *corridor, crossing, left turn, right turn, right fork, left fork, T-shaped bifurcation and dead end.* As the goal was to distinguish the corridor state from the others, the data of corridor was labeled with 0 and all the others were labeled with 1. The samples of corridor state were the same used by ND-NCD. The data of the others states were obtained from new collected samples. The database contained 40 samples from the corridor state and 70 examples from all other states, a total of 10 samples for each state. 5-fold cross validation was used to evaluate the performance of ANN, witch 80 % of the database was used in the training and 20 % of them was used for test in each execution. In all executions, ANN reached a success rate of 100 %.

Thus, ANN was not able to distinguish the corridor state from others if only corridor samples are provided. The developed ANN requires data from others states to succeed. On the other hand, the experiments showed that ND-NCD is able to correctly recognize corridors using only 40 samples as knowledge-base.

7 Conclusions

The hybrid approach proposed in this paper combines successfully the path planning and a new control approach for a mobile robot walking in a topological map. This hybrid approach improves the previous proposed approach, the Neuro-FSM, which was dependent of several previously defined information (topological map, ANN states recognition, path planning). This new approach allows to automatically explore the environment, discovering a path to reach the goal destination, and also, it is able to recognize and discover new states (environment configurations), without the need of creating supervised data sets to train an ANN.

The approach autonomously determines a sequence of actions using a genetic algorithm and a new novelty detection method based on the normalized compression distance (NCD). This new method used to detect/discover environment configurations (states) was called Novelty Detection with NCD (ND-NCD). The proposal can detect environment regions different from corridors without knowing any other type of regions of the environment.

In this way, the proposed hybrid approach, GA with ND-NCD, can reach a goal from a sequence of actions to be performed in different regions from corridors (turns, intersections, etc.), without knowing the environment map, the precise localization of the robot or other information further than some samples illustrating what is a corridor of the environment.

The advantage of ND-NCD for the hybrid approach can be highlighted by comparing it with the use of an Artificial Neural Network (ANN) for the same purpose. An ANN needs databases for at least two classes, one for corridor samples and another for non-corridor samples in order to correctly recognize new regions of the environment (state changes). On the other hand, the proposed ND-NCD requires only one database containing exclusively samples from corridors. It is important to note that a robot using ND-NCD no longer needs to know the path to be followed, since it is only necessary to activate the next reactive behavior generated by the GA, when a non-corridor is detected by ND-NCD. Those aspects indicate the proposed hybrid approach using GA with ND-NCD is promising for mobile robot control.

Acknowledgments. The authors would like to thank FAPESP and CNPq for their financial support.

References

1. Santos, V.C., Sales, D.O., Toledo, C.F.M., Osório, F.S.: A hybrid GA-ANN approach for autonomous robots topological navigation. In: Proceedings of the 29th Symposium On Applied Computing. ACM (2014, to be published)
2. Cruz, C.D.L., Celeste, W.C., Bastos, T.F.: A robust navigation system for robotic wheelchairs. Control Eng. Pract. **19**(6), 575–590 (2011). 7th IFAC Symposium on Fault Detection, Supervision and Safety of Technical Processes (SAFEPROCESS), Special Section: Fault Diagnosis Systems, Barcelona, 30 June–3 July 2009. http://www.sciencedirect.com/science/article/pii/S0967066110002546
3. Knudson, M., Tumer, K.: Adaptive navigation for autonomous robots. Robot. Auton. Syst. **59**(6), 410–420 (2011). http://dx.doi.org/10.1016/j.robot.2011.02.004
4. Fernandez-Leon, J.A., Acosta, G.G., Mayosky, M.A.: Behavioral control through evolutionary neurocontrollers for autonomous mobile robot navigation. Robot. Auton. Syst. **57**(4), 411–419 (2009). http://www.sciencedirect.com/science/article/pii/S0921889008000936
5. Sgorbissa, A., Zaccaria, R.: Planning and obstacle avoidance in mobile robotics. Robot. Auton. Syst. **60**(4), 628–638 (2012)
6. Sales, D.O., Osório, F.S., Wolf, D.F.: Topological autonomous navigation for mobile robots in indoor environments using ANN and FSM. In: I CBSEC: Conferência Brasileira em Sistemas Embarcados Críticos, São Carlos, Brazil (2011)
7. Sales, D.O.: NeuroFSM: Aprendizado de Autmatos Finitos atravs do uso de Redes Neurais Artificiais aplicadas Robs Mveis e Veculos Autnomos. Master dissertation in Portuguese, CCMC, ICMC, Universidade de Sao Paulo, São Carlos, Brazil (2012)
8. Sales, D., Correa, D., Osório, F.S., Wolf, D.F.: 3D vision-based autonomous navigation system using ANN and kinect sensor. In: Jayne, C., Yue, S., Iliadis, L. (eds.) EANN 2012. CCIS, vol. 311, pp. 305–314. Springer, Heidelberg (2012)
9. Sales, D.O., Correa, D., Fernandes, L., Wolf, D.F., Osório, F.S.: Adaptive finite state machine based visual autonomous navigation system. Eng. Appl. Artif. Intell. **29**, 152–162 (2014)
10. Cilibrasi, R., Vitanyi, P.: Clustering by compression. IEEE Trans. Inf. Theor. **51**(4), 1523–1545 (2005)
11. Terwijn, S.A., Torenvliet, L., Vitnyi, P.M.: Nonapproximability of the normalized information distance. J. Comput. Syst. Sci. JCSS **77**(4), 738–742 (2011). AINA 2009. IEEE. http://www.sciencedirect.com/science/article/pii/S0022000010001029

12. Li, M., Vitnyi, P.M.: An Introduction to Kolmogorov Complexity and Its Applications, 2nd edn. Springer, New York (1997)
13. Ito, K., Zeugmann, T., Zhu, Y.: Clustering the normalized compression distance for influenza virus data. In: Elomaa, T., Mannila, H., Orponen, P. (eds.) Ukkonen Festschrift 2010. LNCS, vol. 6060, pp. 130–146. Springer, Heidelberg (2010)
14. Sanches, A., Cardoso, J., Delbem, A.: Identifying merge-beneficial software kernels for hardware implementation. In: International Conference on Reconfigurable Computing and FPGAs (ReConFig 2011), pp. 74–79 (2011)
15. Axelsson, S.: The normalised compression distance as a file fragment classifier. Digital Invest. **7**(Supplement), 24–31 (2010). The Proceedings of the Tenth Annual DFRWS Conference. http://www.sciencedirect.com/science/article/pii/S1742287610000319
16. Cerra, D., Datcu, M.: A fast compression-based similarity measure with applications to content-based image retrieval. J. Vis. Commun. Image Represent. **23**(2), 293–302 (2012). http://www.sciencedirect.com/science/article/pii/S1047320311001441
17. Lillo-Castellano, J., Mora-Jiménez, I., Santiago-Mozos, R., Rojo-Álvarez, J., Ramiro-Bargueño, J., Algora-Weber, A.: Weaning outcome prediction from heterogeneous time series using normalized compression distance and multidimensional scaling. Expert Sys. Appl. **40**(5), 1737–1747 (2013). http://www.sciencedirect.com/science/article/pii/S0957417412010810
18. Tipping, M.E., Bishop, C.M.: Probabilistic principal component analysis. J. Roy. Stat. Soc. Ser. B **61**, 611–622 (1999)

Implementing and Simulating an ALLIANCE-Based Multi-robot Task Allocation Architecture Using ROS

Wallace Pereira Neves dos Reis[1(✉)] and Guilherme Sousa Bastos[2]

[1] Federal Institute of Education, Science, and Technology of Rio de Janeiro, IFRJ, Campus Volta Redonda, Volta Redonda, RJ, Brazil
wallace.reis@ifrj.edu.br

[2] Institute of System Engineering and Information Technology - IESTI, Federal University of Itajubá - UNIFEI, Itajubá, MG, Brazil
sousa@unifei.edu.br

Abstract. In this chapter, we discuss the implementation and simulation results of a ALLIANCE-based architecture on Robot Operating System (ROS). In this approach, the system parameters were set empirically and we do not discuss system performance metrics. The focus is implementing the task allocation algorithm. After briefly review MRTA problem, we compare known architectures in some key aspects. Although only simulations validate the ALLIANCE-based approach, system flexibility and adaptivity is notable despite its runs variations.

Keywords: Multi-robot system · Task allocation · Robot operating system

1 Introduction

Multi-robot systems have been extensively studied and applied in many different areas in recent decades. These applications involve areas such as healthcare, safety, cleanliness, rescue, transportation, not to mention others. On Multi-Robot Systems (MRS), a critical issue is the proper task allocation successfully achieving the final goal from cooperation among robots. Complexity of the studied systems is rising and has two major sources: (a) larger robot teams sizes and (b) greater heterogeneity of robots and tasks [1]. Nevertheless, several reasons explain interests in these systems, but are not limited to: decreased design complexity and cost while increased efficiency [2], dealing task complexity, increased system reliability [3], and, in the future, cooperation between robot and human, which is a long-term goal for Gerkey and Matarić [1].

When related to multi-task problems, single-robot systems demand multi-resources robots. But the system performance might decrease depending on mission challenges. Moreover, the loss of such a robot during a mission is critical. For instance, when applied to space exploration, MRS can easily overcome a team member deficiency, but if there is only a monolithic robot in the mission,

F. Santos Osório and R. Sales Gonçalves (Eds.): LARS 2015/SBR 2015, CCIS 619, pp. 210–227, 2016.
DOI: 10.1007/978-3-319-47247-8_13

its failure means the loss of developing time, resources and a multi-million dollar budget [4]. For healthcare facilities applications, which is a dynamic environment and new emergency priority tasks might randomly appear, introducing robots means to reduce dependency on support staff, and according to Das, McGinnity, Coleman, and Behera, it is better to use a multi-robot system (MRS) consisting of heterogeneous robots than designing a single robot capable of doing all tasks [5]. However, an efficient task allocation algorithm is necessary to handle with all heterogeneity of robots, tasks, and the stochastic environment to attain the benefits of using a MRS [5]. In addition to previous statements, Cao *et al.* stresses that may also be derived from experiments with MRSs insight into social sciences (organization theory and economics), life science and cognitive science (psychology, learning, artificial intelligence) [2]. This work proposes joining the MRS advantages with the ROS advantages on implementing a Multi-Robot Task Allocation (MRTA) architecture. Thus, it examines the MRTA problem and demonstrates a behavior-based MRTA architecture implementation, based on well-known ALLIANCE [6,7], on a growing open-source framework for robotics development.

In this chapter, Sect. 2 presents some concepts and classifications comparing two MRTA systems, behavior-based and market-based approaches, emphasizing the architecture that underlies this work, ALLIANCE. Also, it shortly presents the Robot Operating System and its foundations. Section 3 defines the implemented architecture approach. Section 4 details the behavior sets implemented on robots discussing the undertaken simulations and the obtained results. Finally, Sect. 5 presents the conclusions.

2 Related Work

2.1 Multi-robot Task Allocation Architectures

MRSs differ from other distributed intelligent agents (DIA) system because of their *embodiment* and implicit *real world* environment, requiring more complex models than software domain, like databases or networks [2]. Among other classifications, we first highlight the decision making process [2], whether the system is centralized or decentralized, and type of cooperation, emergent or intentional [1,6]. A single control agent observing the whole environment and accessing full system information essentially characterize centralized systems. This agent makes decisions to maximize the global utility and, by having whole environment information, it is capable of making optimal decisions, with some environment size constraints. However, the central station needs a robust and permanent communication with each single robot in the team. And it is also a weak link, a bottleneck, meaning it can never fail for the successful robots mission completion.

Not to mention the system complexity increases as the number of robots in the team grows, a decentralized system lacks such single control agent and can be divided into two categories: distributed and hierarchical [2]. In hierarchical architectures, usually in larger robot teams applications, a group defines a team member as a local leader, centralizing on it the group decisions. Whereas, in a distributed architecture, all robots are equals in respect to control, even when they are heterogeneous. Another subcategory derives from distributed architectures: a fully distributed architecture, which is, no individual robot is responsible for others control [7]. The two last concepts differ, especially, in communication among robots and techniques used in task allocation. In a fully distributed architecture, robots communicate each other via broadcast messages, no two-ways communication is established. It means a robot does not depend on other robots acknowledgement to allocate a task although using the team members information. With regard to distributed architectures, robots need explicit or two-ways communication, since task allocation involves a negotiation [8]. As a result, a robot takes the control of task allocation process and becomes responsible for the messages concerning that task.

Decentralized architectures have several advantages in relation to centralized architectures. A decentralized system requires local communication among the robots and, if necessary, an intermittent communication with the central station. In addition to bringing fault tolerance, redundancy, reliability, and scalability to the system. In centralized schemes, reach the last two topics is a critical issue.

Another possible classification is regarding the type of cooperation among robots, as pointed out by Lynne Parker [6]. The emergent type of cooperation, likewise called swarm robotics, deals with a large number of homogeneous robots, each of which has deep capabilities constraints. This approach is well suited for applications which execution time is not a critical factor and that the tasks performed are repeated widely in relatively large areas, such as parking cleaning [6]. All robots in the Swarm-robotics system have the same control laws and usually are based on biological cooperative communities. Despite showing cooperative behavior, system individuals respond solely to stimulus from other individuals or the environment, there is no explicit negotiation or task allocation [1].

On the other hand, intentional cooperation deals with a team with a limited number of individuals, typically heterogeneous robots, performing several distinct tasks [6]. Robots know about teammates, their actions, and states [8]. The team cooperates intentionally to achieve an exact level of efficiency and complete the mission, according to Gerkey and Matarić [1], often through task-related communication and negotiation. However, Gerkey and Matarić [1] yet highlights MRS to show coordinated behavior must not be intentional. Using both methods is possible to demonstrate same task execution. The debate about contributions of each approach remains open.

In order to formalize MRTA problems analysis and treat it with a more theoretical view, Gerkey and Matarić defined a formal taxonomy [1]. On their definition robots, tasks, and assignments are the three axes that describe and represent MRTA problems. Single-task robots can execute a single task at a time (ST) and multi-task robots (MT) are those that can execute multiple tasks simultaneously. With respect to tasks, a single-robot task (SR) means that each task requests merely one robot. Whereas multi-robot task (MR) means a task can require multiple robots to be completed. Lastly, task assignment categories are the instantaneous assignment and the time-extended assignment. Instantaneous assignment (IA) allocation is bounded to instantaneous task assignments with respect to environment available data, the robots, and the tasks, with no further planning. By contrast, the time-extended assignment (TA) allocation means that more information is available, and tasks can be assigned over time. Concerning task assignment, Bastos [9] shows another perspective: in IA problems, the number of robots is greater than the number of tasks, and, in TA problems, the opposite occurs, i.e., the number of tasks is greater than the number of robots. Based on the above definitions, the analysis is extended to the combination between the robot type (either ST or MT), the task type (either SR or MR), and the type of task assignment (either IA or TA).

Even though Gerkey and Matarić formal taxonomy [1] describes a wide range of problems, there are task constraints, excluding a collection of MRTA problems. As their work assumes independent utilities, also are the tasks. For this reason, Korsah, Dias, and Stentz [10] expanded the standard taxonomy for more complex cases, calling it *iTax*, handling with interrelated utilities and constraints issues. The key concept of *iTax* is task decomposition, creating four dependencies classes, from elementary tasks to complex tasks, which involve many other tasks. And [10] adopts all previous concepts of robots, tasks, and assignment of [1]. As the implemented architecture, based on ALLIANCE and as well it, assumes independent tasks, from now on, only [1] taxonomy is considered, for simplicity.

To finish MRTA architectures classification in this chapter, intentional cooperation can be crudely divided into behavior-based and market-based approaches. There are other classifications available, the intent is not to cover it all. In behavior-based approaches, the task assignment commonly occurs without explicit robot team discussing [8]. As an example of fully distributed cooperation, robots use the external knowledge (such as feedback sensors, team members states and actions, team members' capabilities, mission status) and internal parameters to determine which robot should perform which task. Unlike behavior-based, market-based approaches involve an explicit communication among robots, since it needs to negotiate the required tasks [8]. Based on Economy market theory, robots typically try to optimize the overall utility based on individuals utilities that robots bid to perform a task. Commonly, task allocation occurs in a greedy fashion. The task negotiation needs a mediator. So a robot takes its responsibility, until the final task assignment. In market-based distributed architectures, any robot can play the mediator role. For this reason, robots could be responsible for different task assignments since all team members are equals.

As behavior-based architecture examples, there are ALLIANCE [6–8] and *Emergency Handling Task* [11], among others, in which robots have internal motivation that lead them to take a task. On ALLIANCE, *robot impatience* and *robot acquiescence* guide the motivation level, while on *Emergency Handling Task commitment* and *coordination* have this effect. No explicit negotiation occurs among robots, once a robot starts a task, the team must inhibit that task, based on previous broadcast messages sent. Then, that task will not be assign to other robot[1]. The internal motivation of a robot can be modified depending on how tasks are being executed, but not by another robot's behavior set motivation. Moreover, a robot broadcast a message with an already assigned task that was previously idle.

There are several market-based architectures in the literature, as MURDOCH [12], M+ [13], and *Dynamic role assignment* [14], besides many others based on the same premises. In these architectures, after the newly available task announcement, robots evaluate the proper metrics and calculate a bid for task execution. This bid is the task *fitness score* in [12], the task *cost* in [13], and the task *utility* in [14]. The bidders broadcast their messages and the mediator robot, called *auctioneer* in [12] or *attach leader* in [14], evaluates the bids to assign the task. In the specific case of MURDOCH, the auctioneer monitors task progress and if it is satisfactory, it renews the contract with the winner bidder, until the task accomplishment. If it is not, the task is back to actions. Despite broadcast communication in common to above architectures, MURDOCH presents an advantageous anonymous communication, often called *subject-based addressing*, based on resource centric, *publish/subscribe* model [8,12].

Gerkey and Matarić [15] suggest that one of the primary challenges facing MRS is how efficiently define the utility of a given action in course. Especially that most if not all coordination approaches rely on some form of utility, also with different designations as seen in previous paragraphs. In addition, according to Bastos [9] the greater majority of MRTA architectures use the non-variable utility in task allocation problem, but in real world tasks have priorities and time lifespan. Therefore, to better solutions, modeling of the problem should contemplate variable utility. ALLIANCE addresses this issue by using a type of variable utility, *robot impatience* and *robot acquiescence*, which are variables inherently valued over time [6,7,9], as better defined in Sect. 3. In contrast, MURDOCH *fitness score* bid depends on *metrics* forms, like simple or weighted sums of variables (such as feedback sensors and robot states), but not always taking into account the time.

With a view to formal taxonomy classification, ALLIANCE attempts to ST-SR-IA and ST-SR-TA problems [1,8]. As defined [1], in the simpler case of instantaneous assignment, the architecture iterates task assignment in a greedy fashion until completing the mission. In the time-extended assignment, the problem is a variant MRTA problem called *ALLIANCE efficiency problem* (AEP) [1], which consists of minimizing the time taken by a robot to accomplish its allocated tasks but given each robot a subset of tasks making up the mission, not a single task [16]. Still, MURDOCH is a variant of ST-SR-IA problem, defined as on-line assignment since the architecture solve MRTA problems in which tasks are randomly injected into the system, so when a new task is introduced, the fittest robot will execute it [1]. Finally, M+ also addresses ST-SR-IA and ST-SR-TA problems. In this case, as robots know a previous task execution ordering,

[1] But if there is a fault, e.g., a robot suddenly turns off; the team must be able to reassign tasks to complete the mission.

ST-SR-IA problems iterates negotiation until there is no task to be negotiated [1]. But, as robots can negotiate a task by anticipation, while executing a task, it also addresses ST-SR-TA problems [13]. In summary, ALLIANCE is, basically, a decentralized, fully distributed, behavior-based architecture addressing ST-SR-IA and ST-SR-TA problems with no explicit task communication among robot team members. Section 3 details it better.

2.2 Robot Operating System - ROS

Robot Operating System (ROS) is an open-source framework for robotic development that aim to meet a specific set of challenges when developing large-scale service robots [17]. According to the authors, the philosophies that guide the framework are unique and derived from specific previous designs to manage particular robotic systems. Among framework premises, one can highlight the multi-lingual and tools-based characteristics. The first allows the code reuse and coding in developer preference languages. The last allows the developer to decide what tools use for a specific application.

Nodes, messages, topics, and services are the fundamental concepts of the implementation [17]. Nodes are processes that perform computation. A ROS-system is commonly comprised of many nodes. And nodes communicate each other through messages, which are strictly typed data structure. Nodes publish messages in topics and other nodes should subscribe a topic to receive a message. ROS-topics follows the *subject-based addressing* concept and *publish/subscribe* model. Besides that, there are also services, analogous to web services, with a message of request and a message with response.

Arising from framework design philosophies, its main advantages are the reuse of code, the multi-lingual nature, and its adaptability since it is free and open-source. Kerr and Nickels describe in [18] a survey to determine the robotics framework that best fits an undergraduate lab. To this end, five criteria are established: easy to use, capable, adaptable, easy to install and maintain, and developmental stage. ROS scored good ratings on the criteria of adaptability and development stage. Nevertheless, it scored one of the lowest ratings on the easy to use criteria. This work shows that, despite all the advantages, ROS is not a user environment? And requires programing experience to developing. Even the authors of [17] acknowledge that the ROS is not the best framework for all robots. In addition, extend beyond, stating that they believe that this framework does not exist [17]. However, it should be held that the platform is in great development in recent years, adding increasingly features and robots.

3 Our ALLIANCE-Based Approach

Lynne E. Parker sets the premises for the ALLIANCE in [6]. Out of these, the ALLIANCE-based approach premises are:

- Robots of the team can detect the effect of their own actions, with probability greater than 0[2].
- Team members do not lie and they are not intentional enemies.
- The overall environmental knowledge is not available on a centralized storage[3].
- The means of communication is not guaranteed to be available and messages may be lost.
- The robots do no possess perfect sensors and actuators[4].
- If a robot fails, it does not necessarily communicate to others.

Based on architecture formal model, one defines a robot set $R = [robot_0, robot_1, robot_2, ..., robot_n]$ and a set of tasks $T = [task_1, task_2, task_3, ..., task_m]$. In addition, each robot has a behavior set which is in charge of allocating tasks through behavior sets activation. For instance, robot r_i has the *behavior sets* or *high-level task-achieving functions* $A_i = [a_{i1}, a_{i2}, ...]$.

The ALLIANCE-based approach uses the same input variables of original architecture in [6]. These variables are combined to calculate the motivation level for each behavior. The specific motivation level that activates behavior sets is the threshold of activation parameter, θ. Since behaviors impatience and acquiescence vary throughout the behavior sets, a single threshold of activation is sufficient. The main inputs are:

(1) *sensory_feedback:* This entry defines when a behavior is applicable (1) or not (0) from the information from the robot physical or *virtual* sensors at time t. Each behavior set has proper *sensory_feedback* function to determine when it is applicable. Some function inputs, are messages received from ROS topics, others are outputs from behavior sets.

(2) *comm_received:* this entry is responsible for the inter-robot communication and (1) determines its value. The robot messaging rate is called ρ_i and express the rate robot r_i publishes its current activity on the specific ROS topic: $/TaskAllocation$. There is also a time parameter which increases fault tolerance, τ_i, that determines how long robot r_i allows to pass without receiving message from another robot r_k before deciding that r_k has ceased its activity. So, inter-robot communication function is

$$comm\_received(i, k, j, t_1, t_2) = \begin{cases} 1, \text{if robot } r_i \text{ has received} \\ \quad \text{message from robot } r_k \\ \quad \text{concerning task } j \text{ in} \\ \quad \text{the time span } (t_1, t_2), \\ \quad \text{where } t_1 < t_2 \\ 0, \text{otherwise.} \end{cases} \tag{1}$$

[2] In this chapter, since all experiments are simulated, this probability is equal to 1. Even if sensors and robot models includes deviation errors, they never fail completely. But, the architecture covers the real robot case.

[3] For the simulated experiments, robots previously know map limits and each of which initial position.

[4] Sensors models in the simulator include a standard deviation error.

(3) *activity_suppression:* When a behavior set is activated, it simultaneously begins to inhibit the other robots behavior sets. On ALLIANCE-based approach, the behavior set *IDLE* does not inhibit any other, wherein the robot waits to activate a behavior set and execute a task. This entry receives value 1 for behavior set not inhibited and 0 for inhibited behavior sets.

(4) *impatience:* robot impatience basically involves two parameters, $\delta(t)$, and ϕ. The $\delta(t)$ parameter is the level of impatience which robot r_i can range if another robot r_k executes a task. There are two distinguished levels of impatience, $\delta_{fast}(t)$ and $\delta_{slow}(t)$. In the last case, the robot is more patient for a running task. The ϕ parameter is a value of time which determines how long robot's r_k communication influences the motivation of robot r_i. This brings, even more, fault tolerance. After ϕ time units, r_k's communication messages does not affect r_k's motivation. The parameter ϕ of impatience, unlike the original architecture, does not vary with time, although the final motivation varies.

(5) *impatience_reset:* this input resets the motivation of a certain behavior when a robot sends for the first time a broadcast message to the selected task. The limited influence aims to not let a slow robot delay system performance as a whole.

(6) *acquiescence:* acquiescence indicates the ability of the robot to realize that a certain behavior, shall read task, should be abandoned because it is not being carried out satisfactorily. There are two parameters related to this entry: ψ and λ. The ψ parameter indicates the time at which robot r_i maintain the behavior before allowing robot r_k to activate it. The λ parameter indicates how long r_i keep an active behavior before attempting another behavior, i.e., another task.

Finally, the motivational controller calculates each behavior set motivation by (2):

$$\begin{aligned} m_{ij}(0) &= 0, \\ m_{ij}(t) &= [m_{ij}(t-1) + impatience_{ij}(t)] \\ &\quad \times (sensory\_feedback_{ij}) \\ &\quad \times (activity\_suppression_{ij}) \\ &\quad \times (impatience\_reset_{ij}) \\ &\quad \times (acquiescence_{ij}). \end{aligned} \tag{2}$$

In the original task selection algorithm, a robot first separates tasks into two categories: (a) those that he knows he can perform better than the other robots and that no other is carrying out, and (b) such other tasks, as it can perform. After splitting into two categories, firstly, robot selects at first task category until there is no more idle tasks and then searches tasks in the second category. It runs until the *sensory_feedback* warn there is no more tasks. ALLIANCE-based implementation in ROS, the priority of a given task was made from different parameters, *impatience* and *acquiescence* values for each task and each robot. Thus, the highest priority task has a higher level of *impatience*, causing the motivation for this task reaching the threshold of activation before others. Still on the original work, robots have previous experience with its team, learning the ability of each other in carrying out a task. The author of [6] and [7] shows the L-ALLIANCE learning architecture [16] that is a learning tool in MRTA problems. Proving it simplicity, the parameters of ALLIANCE-based architecture, such as impatience and acquiescence levels, waiting time, and communication time were empirically set.

The architecture verification used the MobileSim simulator for robots that use an open-source ARIA library [19]. This simulator was chosen to contain the real robots to be used in the next steps of the research. In addition, using real robots or this simulator along with ROS, it is necessary to use an interface between ROS and ARIA, the RosAria package [20]. This package provides an interface between ROS and Adept MobileRobots and ActivMedia robot bases. The robots used in the simulation were Amigobots by Adept Mobile Robotics. These robots are available in the robotics laboratory of the Federal University of Itajubá. Amigobot, Fig. 1, is a small differential-drive robot assembled with an array of eight sonars, wheel encoders, buzzer, bumpers and other sensors.

Fig. 1. Amigobot: differential-drive robot by Adept MobileRobots used in this work.

Additionally, Amigobots carry wireless access point for control and communication. In their case, an off-board machine processes all data. To address data exchanging and robot control, the motivational controller subscribe only the needed sensors RosAria topics and publishes robots velocity commands in the specific topic, besides using the same wireless mean to publish/subscribe messages of task allocation.

There is an important comparison to make about the communication between robots in the original work [6,7] and implementation on ROS. In the original work, communication between robots takes place through an integrated radio-based positioning system. This system consists of a transceiver coupled to each robot and two sonar base stations for use in triangulating the robot positions, reporting its position to itself and the other team members. For applications restricted to small environments without physical interference with communication, such as walls, this system works well. Howbeit, the larger the area to be covered by the robots, more critical it is communication, and more complex will be using an external system. Thusly, ROS brings benefits to architecture, since communication among robots takes place through the TCP/IP protocol, i.e., over wireless means. Robots can cooperate just kilometers away over the internet. Indoors like office corridors, the already common use of routers solves the problem. Furthermore, for more complex applications such as outdoors, the location held by only one type of sensor can contain very large errors. While on location, ROS has packages and various implementations of Simultaneous Localization and Mapping (SLAM), as [21,22], which can improve the tracking system of the robot team.

All in all, Fig. 2 shows the system communication scheme. Robots publishes their current tasks with an identification, called *robot_id*, at the */TaskAllocation* topic. Similarly, all robots are subscribers of this topic updating their motivation levels according to *comm_received* input. The following topics *[...]/cmd_vel*, *[...]/sonar*, and *[...]/odom* are the interface between ROS nodes and ARIA nodes, made by RosAria package. The ROS topics aforementioned are responsible for the robot speed commands, range measurement and odometry measurement, respectively. Each robot node, responsible for ALLIANCE-based architecture processing, processes the data of subscribed topics and publishes control messages according the active behavior set.

For complex behaviors, RosAria also supports other topics that have not been treated in this study because they are not used for the sensors. In addition, ROS allows other devices are also used together with the robots, such as on-board cameras and lasers scans.

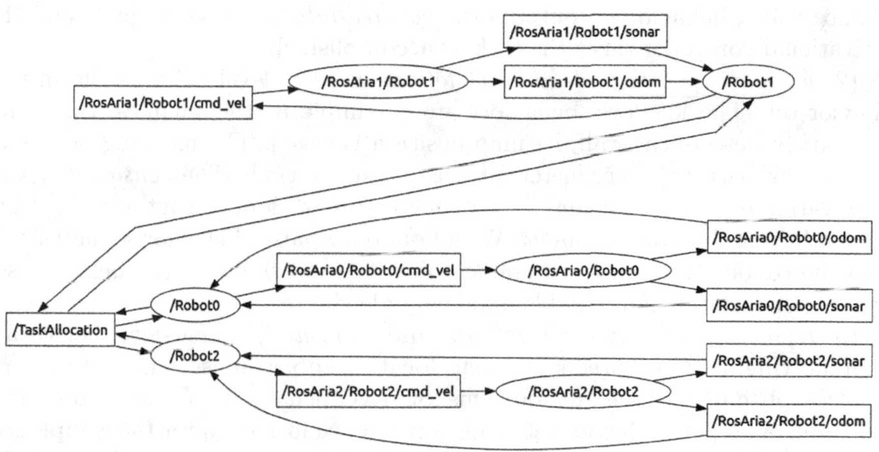

Fig. 2. Graphical representation of communication among nodes (ellipses) and topics (rectangles) of the implemented ALLIANCE-based architecture on ROS for a three robots team. */Robot0*, */Robot1*, and */Robot2* are the ALLIANCE-based nodes.

4 Behavior Sets Details and Simulations Results

We have simulated ALLIANCE-based approach numerous times under different conditions. In this section, we describe the developed behaviors sets programmed on robots nodes and detail three experiments and their results. We present the results in simulation snapshots and robots motivation values during each simulation run. In addition to these preliminary definitions, the experiments map has always the same length and the boxes do not change places. Tasks are geographically distributed around the map, so to speak. It will be defined next.

4.1 The Behavior Sets

In ALLIANCE-based architecture, the tasks are previously programmed on robots as behavior sets. By activating a behavior set, a robot executes a task. The following behavior sets were implemented:

(1) *wander-right-side* and *wander-left-side*: two low-level behaviors composed these behavior sets: *wander* and *side-keep*. The first is responsible for making the robot wander the environment by an obstacle avoidance algorithm, exploring the environment. The second keeps the robot in the map correct side, either left or right, from the initial position and odometry information. The behavior set *sensory_feedback* input variables are *[...]/odom* and *covered-total-distance*. Likewise, the behavior set output variables are *[..]/cmd_vel*, *covered-distance*, and *covered-total-distance*. The *covered-distance* variable outputs the distance robot has already covered with the behavior set activated. When it reaches a predefined distance value, behavior set outputs the *covered-total-distance* variable and the motivational controller takes the task as accomplished.

(2) *boundary patrol*: *find-wall* and *follow-wall* low-level behaviors form this behavior set. The low-level behaviors are as simple as their names. But, once robot safely detects the wall, its map position is saved. The meaning of it is to robot surely patrol the perimeter, at least, a single time. The *sensory_feedback* input variables are *[...]/odom* and *perimeter-round*, and output variables are *[...]/cmd_vel* and *perimeter-round*. With boundary patrol behavior set activated, when the robot gets back to the same point it found the wall, the behavior set outputs perimeter-round variable and conclude the task.

(3) *report*: encloses *go-to-report-area* and *publish-report* low-level behaviors. The reporting area is a predefined map location robots must achieve to report the accomplishment of the mission. This low-level behavior is a simple go-to-goal algorithm. As report behavior set is appropriate exclusively after the completion of other tasks, *sensory_feedback* input variables are *covered-total-distance*, for both sides of the map, *perimeter-round*, and *[...]/odom*. And the output variables are *[...]/cmd_vel* and *report-published*. When the behavior set outputs the last one, the mission is successfully over.

4.2 Simulations Results

The conducted experiments aim to demonstrate some of the architecture key features supported by a patrolling mission. Therefore, the mission requires robots

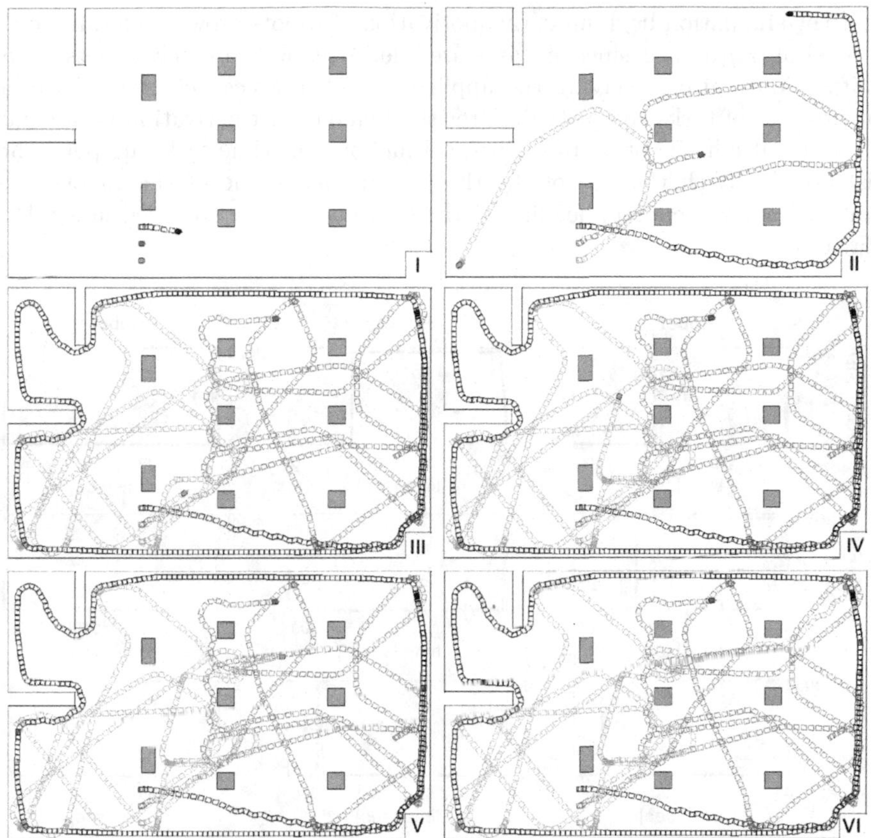

Fig. 3. Experiment 1: task allocation without robot failure or task reassignment. Robots are enumerated r_0 (black trails), r_1 (gray trails), and r_2 (light gray trails), from top first position to bottom. In frame I, r_0 has activated boundary patrol behavior set and executes its task. In the second frame, we observe that all robots have activated behavior sets. In frame III, r_0 and r_1 have already finished their tasks, however, *report* behavior set is still inhibited by *sensory_feedback*. Additionally, in the same frame, r_2 still executes first allocated task. In frame IV, r_2 complete its task and, for now on, *report* behavior set motivation starts to grow over time. In frame V, r_2 activates the last behavior set available, *report*, and accomplishes the mission, in frame VI.

to patrol a room dividing the tasks into patroling the room border, patrolling the left and the right sides of the room, and report tasks completion. So that, the set of tasks for the specified mission request a single task of each behavior set. The first experiment serves as a default for a normal system run, without any robot failure or communication problems. The robots activities evolution while executing the tasks is shown in Fig. 3, and robots motivation over time in Fig. 4.

At the simulation beginning, the motivation of robots grows according to the fast level of *impatience*, since all tasks are idle. As shown in behavior sets, *report* manifests only after receiving the applicable *sensory_feedback* from the other behavior sets of tasks completion. After the behavior set activation with regard each robot, it inhibits other behaviors, throughout the time tasks are performed satisfactorily. With the end of the three main tasks, one of the robots must report the mission ending, heading to the location already specified in low-level behaviors.

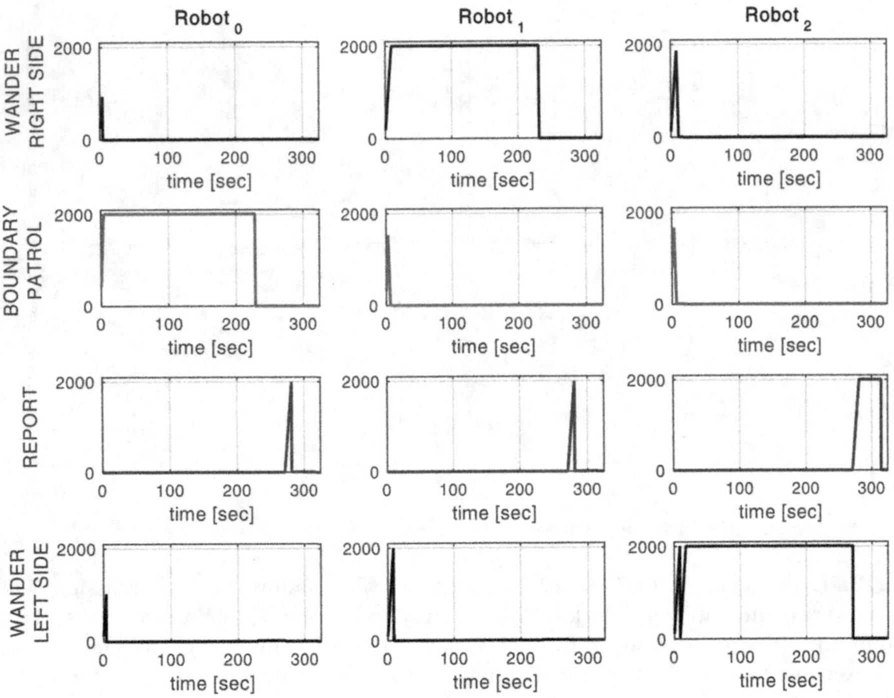

Fig. 4. Evolution of motivation values over time for the first experiment. As *sensory_feedback* for *report* behavior set, of all robots, inhibits itself, its motivation only grows when other tasks are complete. The threshold value for all behavior sets is 2000. Each behavior evolves to a particular level, because of different levels of *impatience.*

The second experiment examines architecture robustness of the point of view of failure, namely the lack of progress and noisy communication, as Figs. 5 and 6 show. Here, we consider a lack of progress a non-critical robot failure, i.e., failures during the task execution, such as delays, that do not represent a total collapse of a robot.

During the simulation, r_0 and r_2 could not avoid a collision and end up stuck for a while. This lack of progress causes a delay in completing the task, which

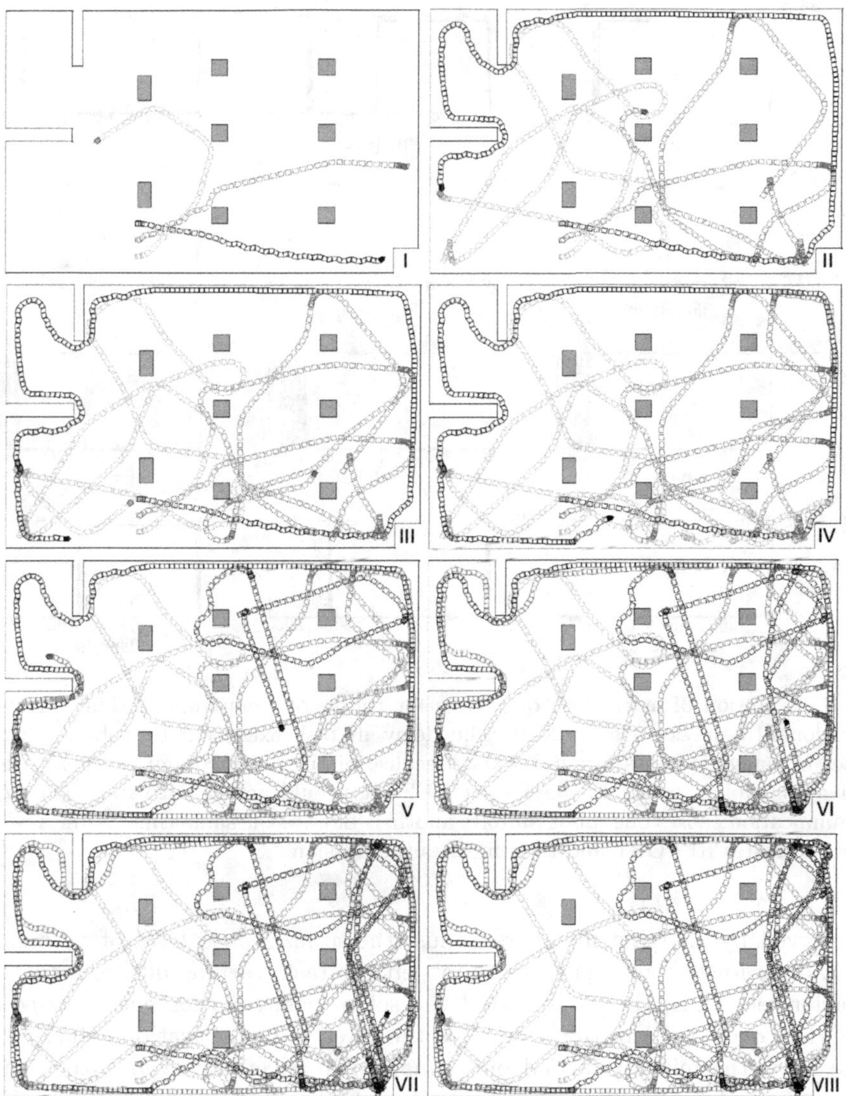

Fig. 5. Experiment 2: lack of progress and noisy communication. Robots are enumerated as the previous simulation. In frame I and II, robots activate behavior sets and start tasks execution. Still in the second frame, r_0 gets stuck in r_2, what harmed their task execution time. In frame III, robots still executing the same tasks. But in frame IV, r_0 has activated *wander-right-side* and r_1 has activated *boundary-patrol* behavior sets. Frame V shows tasks evolution, after the task reassignment. In frame VI, robots have finished their primary tasks, but due to noisy communication, *report* behavior set could not be activated. Finally, r_0 activates *report* behavior set in frame VII, and in frame VIII, it finishes the mission.

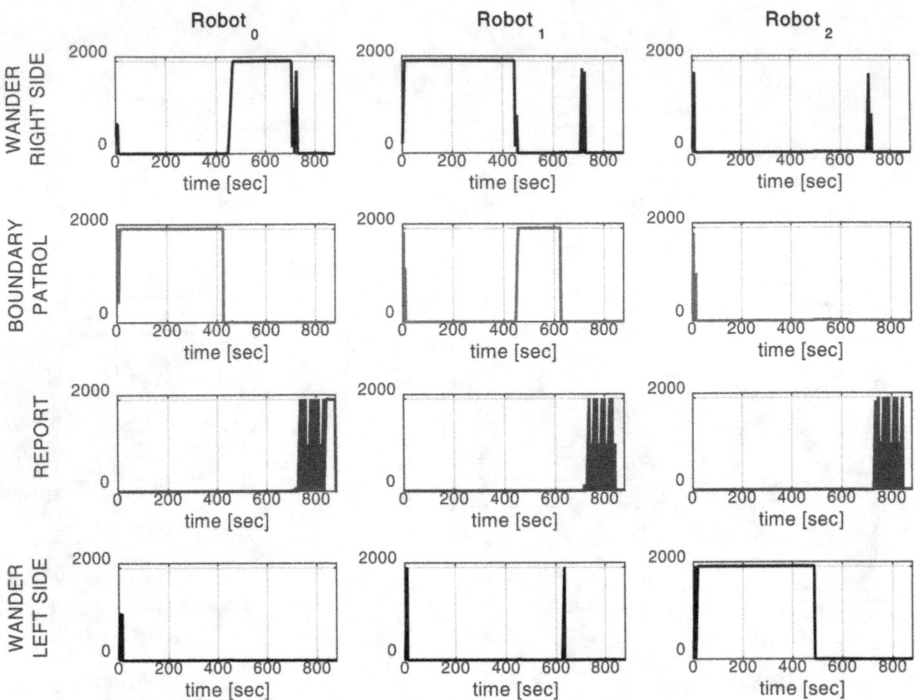

Fig. 6. Evolution of motivation over time for the second experiment. The threshold value for all behavior sets is 2000. The delay in task execution by robots "stuck" causes exchange in tasks, around 400 s. Because of the noisy communication, robots motivation evolves but the behavior set REPORT could not be activated. The noisy communication causes one robot inhibits another, but ate the same time. After a while, $robot_0$ activates REPORT and then finishes the mission.

is observed by all team members. Thus, the *acquiescence level* of r_0 regarding *boundary-patrol* grew and the robot left the task, turning off the respective behavior set. Thus, the r_1 activated the set of behaviors for *boundary-patrol*, since their degree of *impatience* reached the threshold of activation value. Along these lines, the mission carried out remained in execution even with a tasks reassingment. Continuing with the experiment, the effects of noisy communication appear in the report activation since this behavior set requires the outputs of the other sets. Thus, lost messages inhibit the activation of behavior sets. After a period of conflict, a robot actives behavior set and ends the mission.

With a new team member the third experiment aims to verify architecture robustness as regards to robot total failure, however with redundant resources. Noisy communication still affects robots. With the resource redundancy, we observe the task allocation of the failed robot is virtually instantaneous, not causing harm to the mission, as we can see by robots motivation in Fig. 7. During the simulation, Fig. 8, robot r_1 fails and immediately stops communicating with other team members. We recognize that task reassignment in such case

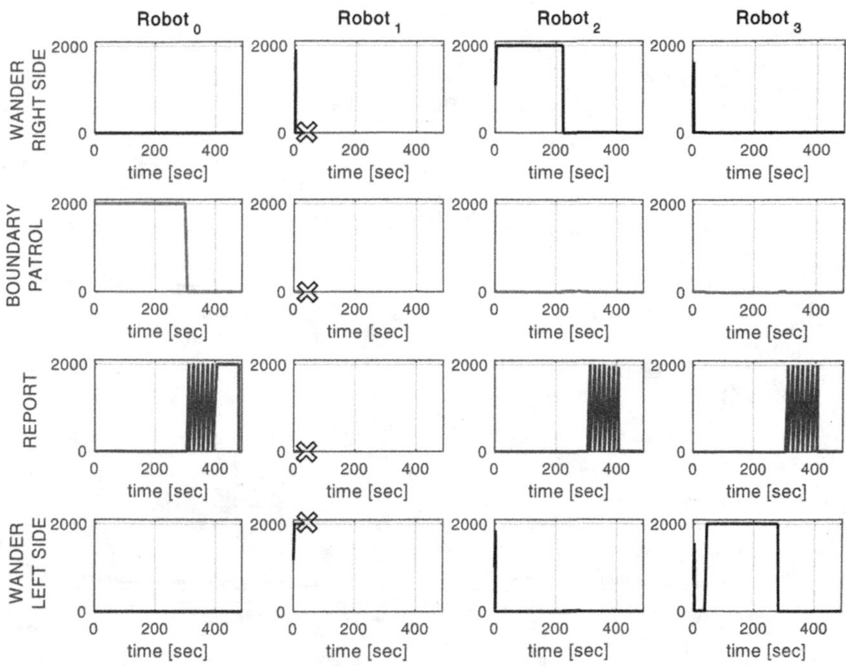

Fig. 7. Evolution of motivation over time for the third experiment. The threshold value for all behavior sets is 2000. The crosses in graphs show the moment that $robot_1$ fails.

occurs in a very quickly manner than when failed robot can not perceive its failure for a while. Thus, acquiescence parameter so important, to prevent a mission to stops because of a team member.

5 Conclusions

This paper describes the implementation and simulation of a fully distributed, behavior- based, decentralized, and fault-tolerant multi-robot task allocation based on architecture proposed in [6] using ROS as development framework.

One of the advantages of the approach is the ease of implementation of communication between robots that already have developed interface packages. As ROS works with simple messages submission, another important advantage is the possibility of increasing the number of robots with little change in the main code. Despite having handicaps as the initial difficulty programming for beginners, the volume of packages and applications development in ROS demonstrates how the framework has gained ground among developers.

For the proposed experiments, the system worked properly, particularly regarding the dynamically allocation and reallocation of tasks, adaptability, and fault tolerance. Although only simulations validate the ALLIANCE-based approach, unexpected failure can occur, such as robots crashing. Even so, it is possible to perceive system flexibility and adaptivity despite its runs variations.

Fig. 8. Experiment 3: Introduction of a new team member and occurrence of a robot total failure. With a new member, robots are enumerated r_0 (black trails), r_1 (gray trails), r_2 (light gray trails), and r_3 (gray trails), from top first position to bottom. In frame I, r_0, r_1, and r_2 execute the available tasks. Frame II shows r_2 failure. The new team member, r_3 activates the behaviors set in frame III. In frame IV, both wander-side tasks are complete. So, r_0 finishes its task in frame V. Also r_0 activates REPORT and finishes the mission, as showed in frame VI.

References

1. Gerkey, B.P., Matarić, M.J.: A formal analysis and taxonomy of task allocation in multi-robot systems. Int. J. Robot. Res. **23**(9), 939–954 (2004)
2. Cao, Y.U., Fukunaga, A.S., Kahng, A.B., Meng, F.: Cooperative mobile robotics: antecedents and directions. In: Proceedings of 1995 IEEE/RSJ International Conference on Intelligent Robots and Systems 1995, Human Robot Interaction and Cooperative Robots, vol. 1, pp. 226–234. IEEE (1995)

3. Badreldin, M., Hussein, A., Khamis, A.: A comparative study between optimization and market-based approaches to multi-robot task allocation. Adv. Artif. Intell. **2013**, 12 (2013)
4. Yliniemi, L., Agogino, A.K., Tumer, K.: Multirobot coordination for space exploration. AI Mag. - Am. Assoc. Artif. Intell. **35**(4), 61–74 (2014)
5. Das, G.P., McGinnity, T.M., Coleman, S.A., Behera, L.: A distributed task allocation algorithm for a multi-robot system in healthcare facilities. J. Intell. Robotic Syst. **80**(1), 33–58 (2015)
6. Parker, L.E.: ALLIANCE: an architecture for fault tolerant multirobot cooperation. IEEE Trans. Robot. Autom. **14**(2), 220–240 (1998)
7. Parker, L.E.: On the design of behavior-based multi-robot teams. J. Adv. Robot. **10**(6), 547–578 (1995)
8. Parker, L.E.: Multiple mobile robot systems. In: Siciliano, B., Khatib, O. (eds.) Handbook of Robotics, pp. 921–941. Springer, Heidelberg (2008)
9. Bastos, G.S., Ribeiro, C.H.C., de Souza, L.E.: Variable utility in multi-robot task allocation systems. In: IEEE Latin American Robotic Symposium, LARS 2008, pp. 179–183. IEEE (2008)
10. Korsah, G.A., Stentz, A., Dias, M.B.: A comprehensive taxonomy for multi-robot task allocation. Int. J. Robot. Res. **32**(12), 1495–1512 (2013)
11. Østergård, E.H., Matarić, M.J., Sukhatme, G.S.: Distributed multi-robot task allocation for emergency handling. In: IEEE/RSJ International Conference on Proceedings of Intelligent Robots and Systems, 2001, vol. 2, pp. 821–826. IEEE (2001)
12. Gerkey, B.P., Matarić, M.J.: Sold!: Auction methods for multirobot coordination. Int. J. Robot. Autom. **18**(5), 758–768 (2002)
13. Botelho, S.C., Alami, R.: M+: a scheme for multi-robot cooperation through negotiated task allocation and achievement. In: Proceedings of 1999 IEEE International Conference on Robotics and Automation, vol. 2, pp. 1234–1239. IEEE (1999)
14. Chaimowicz, L., Campos, M.F., Kumar, V.: Dynamic role assignment for cooperative robots. In: Proceedings of IEEE International Conference on Robotics and Automation, ICRA 2002, vol. 1, pp. 293–298. IEEE (2002)
15. Gerkey, B., Matarić, M.J.: Are (explicit) multi-robot coordination and multi-agent coordination really so different. In: Proceedings of the AAAI Spring Symposium on Bridging the Multi-agent and Multi-robotic Research Gap, pp. 1–3 (2004)
16. Parker, L.E.: L-ALLIANCE: task-oriented multi-robot learning in behavior-based systems. J. Adv. Robot. **11**(4), 305–322 (1996)
17. Quigley, M., Conley, K., Gerkey, B., Faust, J., Foote, T., Leibs, J., Wheeler, R., Ng, A.Y.: ROS: an open-source robot operating system. In: ICRA Workshop on Open Source Software, vol. 3 (2009)
18. Kerr, J., Nickels, K.: Robot operating systems: bridging the gap between human and robot. In: 44th Southeastern Symposium on System Theory (SSST 2012), pp. 99–104. IEEE (2012)
19. Adept Mobile Robots: Mobilesim simulator. Accessed 25 June 2015
20. ROS Wiki: Rosaria package summary. Accessed 15 July 2015
21. Machado Santos, J., Portugal, D., Rocha, R.P.: An evaluation of 2D SLAM techniques available in robot operating system. In: IEEE International Symposium on Safety, Security, and Rescue Robotics (SSRR 2013), pp. 1–6. IEEE (2013)
22. Zaman, S., Slany, W., Steinbauer, G.: ROS-based mapping, localization and autonomous navigation using a pioneer 3-dx robot and their relevant issues. In: Saudi International Electronics, Communications and Photonics Conference (SIECPC 2011), pp. 1–5. IEEE (2011)

Humanoid Robot Gait on Sloping Floors Using Reinforcement Learning

Isaac J. Silva[1(✉)], Danilo H. Perico[1], Thiago P.D. Homem[1,3], Claudio O. Vilão Jr.[1], Flavio Tonidandel[2], and Reinaldo A.C. Bianchi[1]

[1] Electrical Engineering Department, Centro Universitário FEI, São Bernardo do Campo, São Paulo, Brazil
{isaacjesus,dperico,thiagohomem,cvilao,rbianchi}@fei.edu.br

[2] Computer Science Department, Centro Universitário FEI, São Bernardo do Campo, São Paulo, Brazil
flaviot@fei.edu.br

[3] Computer Science Department, Instituto Federal de São Paulo, Boituva, São Paulo, Brazil
http://www.fei.edu.br, http://www.ifsp.edu.br

Abstract. Climbing ramps is an important ability for humanoid robots: ramps exist everywhere in the world, such as in accessibility ramps and building entrances. This works proposes the use of Reinforcement Learning to learn the action policy that will make a robot walk in an upright position, in a lightly sloped terrain. The proposed architecture of our system is a two-layer combination of the traditional gait generation control loop with a reinforcement learning component. This allows the use of an accelerometer to generate a correction for the gait, when the slope of the floor where the robot is walking changes. Experiments performed on a real robot showed that the proposed architecture is a good solution for the stability problem.

Keywords: Humanoid robots · Gait pattern stabilization · Reinforcement learning

1 Introduction

Analyzing the research related to the development of humanoid robot locomotion, it can be noted that the vast majority of researchers are looking for the stability of humanoid robots during walking on flat floors. However, in a real environment, there is also the need to empower the walking of these robots on inclined floors. What motivates the research in humanoid robots is the fact that locomotion with legs is the best form of locomotion in environments with discontinuities in the floor [1].

Climbing ramps is an important ability for humanoid robots: ramps exist everywhere in the world, such as in accessibility ramps and building entrances. Also, in the RoboCup Domain, ramps are used in the Rescue Robot League

F. Santos Osório and R. Sales Gonçalves (Eds.): LARS 2015/SBR 2015, CCIS 619, pp. 228–246, 2016.
DOI: 10.1007/978-3-319-47247-8_14

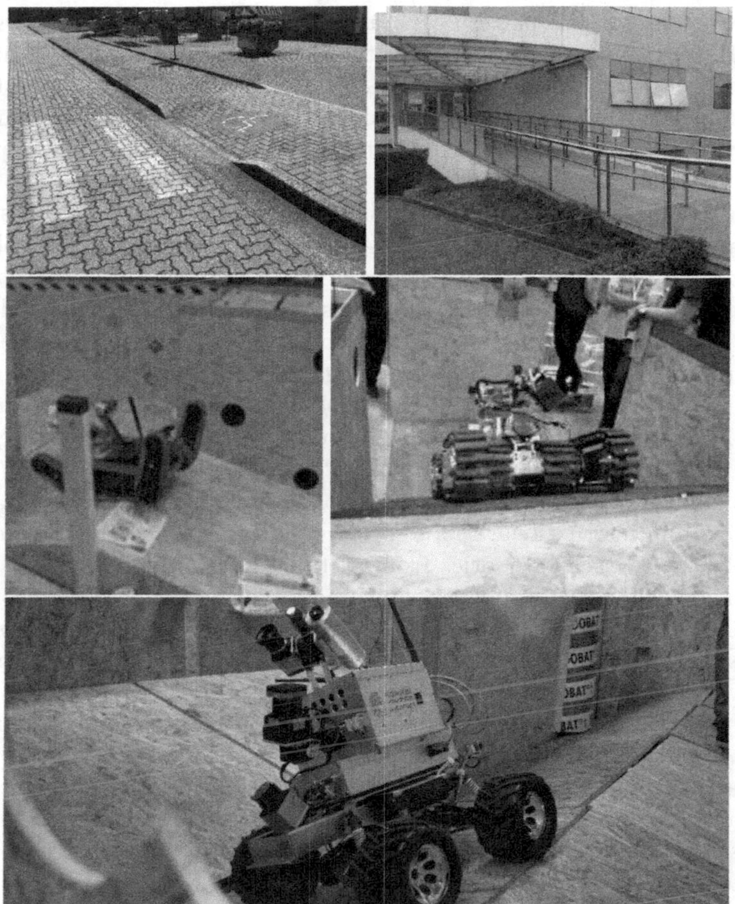

Fig. 1. Ramps in the world: accessibility ramps, building entrance ramps, RoboCup Rescue Robots in their arenas.

arenas, which have 15° ramps to challenge localization and mapping implementations and to increase complexity. On other leagues, such as the Humanoid Leagues, stability on slightly sloped floors is important, because the arenas are not completely leveled. Some examples of ramps are presented in Fig. 1.

This work proposes the use of Reinforcement Learning [2] to learn the action policy that will make a robot walk in an upright position, in a lightly sloped terrain. We assume that the terrain has only variation perpendicular to the frontal plane. In order to achieve that, we propose a system with 2 layers: one that generate the gait of the robot, making use of a gyroscope in a closed feedback loop to correct the robot's gait; and a second layer – the RL component – that makes use of an accelerometer to generate a correction for the gait, when the slope of the floor where the robot is walking changes.

Our robot uses a gait pattern generation based in coupled oscillators developed by Ha, Tamura and Asama [3] for the Darwin-OP robot [4], and they also developed a closed loop control able to retrieve the robot from a possible fall, which is done by reading the gyroscope data. The oscillators perform sinusoidal trajectory of movement synchronized in time, so the robot is capable to perform dynamically balanced gait.

The first humanoid robot developed by Centro Universitário FEI was named Milton, and it was composed of 22 RX-28 Dynamixel motors. This robot has evolved to Newton robot [5], in which the major difference was the use of an Intel computer, named NUC, and the Cross Architecture software model that, together, allowed the removal of the microcontroller. After having created the Newton robot, our group evolved, once again, the original project and built more four robots, named B1, B2, B3 and B4. These four robots were a mixture between Newton robot's software and unit of processing (that is an Intel NUC) with the mechanical concepts of Darwin-OP robot [4].

Some humanoid robots have been developed in Brazil, most of them aiming to participate in international robot soccer competitions such as the RoboCup and the Latin American Robotics Competition (LARC). Among these robots, we can highlight the Eva Robot [6], which has 20 degrees of freedom, composed of MX-106R and EX-106R motors; and the Dimitri Robot [7], that uses elastic actuators in its joints, being composed of 28 MX-106 and 3 MX-64R Dynamixel motors with 1241.79 mm of height.

This work was motivated by changes in the rules of the humanoid league for 2015: from this year onwards, the fields in which the robots will play will be made of artificial grass, with the intent to make the game closer to the real soccer environment. Artificial grass fields are not perfectly plain, and variations in the turf cause small slopes, causing the robots to fall. The system proposed here intends to address this problems, however, the experiments performed in this article demonstrates the model efficiency to solving the more general problem of walking on sloped floors.

The remaining of this paper is organized as follows: Sect. 2 presents theoretical background, including Reinforcement Learning and Gait Generation techniques. Section 3 reviews related work. Section 4 presents details of the proposed system, and Sect. 5 shows the experiments and results obtained. Finally, Sect. 6 concludes this work.

2 Research Background

This section presents a brief description of the two main topics approached in this paper: Reinforcement Learning and Robot Humanoid Walking. The reinforcement learning algorithm that will be presented is the Q-Learning [8]. The subsection Robot Humanoid Walking describes about the gait pattern generate developed by Ha, Tamura and Asama [3] to the Darwin-OP humanoid robot and also used in our humanoid robot.

2.1 Reinforcement Learning

Reinforcement Learning algorithm, broadly studied by Sutton and Barto in [2], uses a scalar reinforcement signal $r(s, a)$ that is received by the agent as a reward or penalty in each state s update. Each action a taken by the agent is defined in order to maximize the total reward it receives, using the knowledge obtained until the present moment.

RL can be formalized by the Markov Decision Process (MDP), being the learning modeled by a 4-tuple $<S, A, T, R>$, S: is a finite set of states; A: is a finite set of possible actions; R: $S \times A \rightarrow \Re$ is a Reinforcement function; T: $S \times A \rightarrow \Pi(S)$ is a state transition function, where $\Pi(S)$ is the state transition function mapping probability. As the goal of the agent, in RL, is to learn an optimal policy of actions that maximizes a function, MDP enables the mathematical modeling of a RL agent.

One of the most popular RL algorithms is the Q-Learning [8]. Q-Learning aims to maximize the function Q for each state, in which Q represents the expected return for taking action a when visiting state s and following policy π. The Q-Learning rule is:

$$Q(s_t, a_t) = Q(s_t, a_t) + \alpha[r(s_t, a_t) + \gamma \max_{a_{t+1}} Q(s_{t+1}, a_{t+1}) - Q(s_t, a_t)] \tag{1}$$

Where s_t is the current state; a_t is the action performed in s_t; $r(s_t, a_t)$ is the reward received after taken a_t at s_t; s_{t+1} is the new state; a_{t+1} is the action performed in s_{t+1}; γ is the discount factor ($0 \leq \gamma < 1$); and α is the learning rate ($0 < \alpha < 1$).

A common strategy used to chose the actions during the exploration process is known as *ε–Greedy*. This strategy pursues the action that results in the higher value to Q with probability $1 - \varepsilon$. The *ε–Greedy* action choice rule, can be written as:

$$\pi(s_t) = \begin{cases} a_{random} & \text{if q} \leqslant \varepsilon, \\ \arg\max\limits_{a_t} \; Q(s_t, a_t) & \text{otherwise.} \end{cases} \tag{2}$$

Where q and ε are parameters that define the exploration/exploitation trade-off; and a_{random} is an action randomly chosen among those available in state s_t. The complete Q-Learning is shown in Algorithm 1.

2.2 Robot Humanoid Walking

The humanoids robots development research consists of diverse interests, ranging from the desire to replace humans in dangerous activities (mining, nuclear energy, military interventions, dangerous activities in disaster environments), as to assistance in daily activities (attendants, assist in domestic and repetitive activities) [1]. Another advantage of humanoid robots is because of the locomotion with legs be the best form of locomotion in environments with discontinuities in the floor, such as steps and stones [1].

Algorithm 1. Q-Learning

Ensure: $Q(s, a)$ arbitrarily
repeat
Visit state s_t
Select an action a_t according to the action choice rule.
Execute the action a_t
Receive the reward $r(s_t, a_t)$ and observe the next state s_{t+1}
Update the values of $Q(s_t, a_t)$ according to the Q-Learning rule:
$Q(s_t, a_t) \leftarrow Q(s_t, a_t) + \alpha[r(s_t, a_t) + \gamma \max_{a_{t+1}} Q(s_{t+1}, a_{t+1}) - Q(s_t, a_t)]$
Update the state $s \leftarrow s_{t+1}$
until s is terminal

Where s_{t+1} is the next state and a_{t+1} is the action performed in s_{t+1}

The work of Marder [9] presents studies showing that for the locomotion control in vertebrates and invertebrates animals, there are neuronal circuits responsible for producing rhythmic motor patterns such as walking, breathing, flying, and swimming. These biological neural networks are called Central Pattern Generators (CPG). Based on CPG, some researchers developed walking patterns generators for humanoid robots.

Currently one of the most difficult problems to be solved in humanoid robots is the walking ability. Many techniques have been developed over the years [3,10–12], however the current humanoid robots still show instability in the dynamic floor compared with the human walk.

The walking control of humanoid robots can be divided in two categories: Passive Dynamics Walking (PDW), where the gravity executes the movement [10], and the walk with active control (fully actuated biped robots), where the idea is the walking control with energized joints, usually motors. And there are two categories of control algorithms: time-dependent and time-invariant algorithms [1].

The passive walk was motivated by energy saving and by the human walking, where, in some periods, low level of muscle activities are found [10]. Although it is important to investigate bipedal robots based on passive-dynamic walking properties, in practice, any bipedal robot needs to have their joints energized, even in a semi-passive mode.

Using the same concept PDW, there are the Virtual Slope walking [13], where the robot walks as whether down a virtual incline ramp, using gravity to shortening leg at each step, thus it is able to walk. However, the leg of the robot can not be shortened infinitely. Then after an half oscillation phase (hip swing phase), it is necessary that the shortened leg must be extended with an amount of energy added to the system.

During the walk, two different situations happen in sequence: one phase of double support, where the robot supports their feet in the floor and it is statically stable (center of mass is inside the support polygon); and the phase where the robot supports only one foot in the floor [1]. We also know that during a race, one can be often with no foot in contact with the ground. These characteristics should be considered to develop advanced algorithms for humanoid robots control with dynamically balanced locomotion.

One of most used criteria to solve the humanoid robot stability is the Zero Moment Point (ZMP) [14]. ZMP is defined as the point on the ground where the sum of all the moments of the active forces are equal to zero [11]. If the robot keeps the ZMP over the convex hull of the foot polygon (support polygon), the reaction force in the floor compensates the active forces caused by the robot movement, keeping a dynamically balanced gait [11]. Considering a robot in a slow walking, the ZMP coincide with the center of mass of the robot.

Some researchers have developed other Gait Generator models like the 3D Linear Inverted Pendulum Model (3D-LIPM) developed by KAJITA et al. [12], in order to reduces the complexity of dynamic calculations.

One other way to generate the gait pattern is the one used by the DARwIn-OP (Dynamic Anthropomorphic Robot with Intelligence - Open Platform) robot, developed by Ha, Tamura and Asama [3]. It is based on coupled oscillators, and also developed as a control loop – using a gyroscope as sensor – capable of stabilizing the robot during the walk.

In this gait generator, one for each foot OSC_{move} (movement oscillator) and one for the center of mass OSC_{bal} (balance oscillator). Each oscillator develops a synchronized sinusoidal trajectory, allowing the robot to have a dynamically stable gait. The gait with oscillators does not perform the real time ZMP calculus, however, our results show that this method perform a walk based on a dynamic model of ZMP, with an advantage: low computational cost. We can express the balance oscillator as in 3 and the movement oscillator as in 4:

$$OSC_{bal} = \rho_{bal} sin(\omega_{bal} t + \delta_{bal}) + \mu_{bal} \tag{3}$$

$$OSC_{move} = \begin{cases} \rho_{move} & \left[0, \frac{rT}{4}\right) \\ \rho_{move} sin(\omega_{move} t + \delta_{move}) & \left[\frac{rT}{4}, \frac{T}{2} - \frac{rT}{4}\right) \\ -\rho_{move} & \left[\frac{T}{2} - \frac{rT}{4}, \frac{T}{2} + \frac{rT}{4}\right) \\ \rho_{move} sin\left(\omega_{move}\left(t - \frac{rT}{2}\right) + \delta_{move}\right) & \left[\frac{T}{2} + \frac{rT}{4}, T - \frac{rT}{4}\right) \\ \rho_{move} & \left[T - \frac{rT}{4}, T\right) \end{cases} \tag{4}$$

Where:

- ρ is the sinusoidal amplitude;
- ω is angular velocity;
- δ is a phase shift;
- μ is an offset;
- r is the Double Support Phase (DSP) ratio;
- T is the walking period.

Ha, Tamura and Asama [3] assumed the dynamic model of the robot as a inverted pendulum to develop the feedback model. The computed compensation torque is provided to the balance oscillator parameter μ_{bal}. According to Ha, Tamura and Asama, μ_{bal} was assumed to operate as a torque generator, so OSC_{bal} can be written according to Eq. 5:

$$OSC_{bal} = \rho_{bal} sin(\omega_{bal} t + \delta_{bal}) + k_p \theta_r + k_d \dot{\theta_r} \quad (5)$$

Where:

- $\dot{\theta_r}$ is the gyroscope sensor data;
- θ_r is the joint angle sensor data;
- θ_r is the joint angle sensor data;
- k_p and k_d are found by the experimental method.

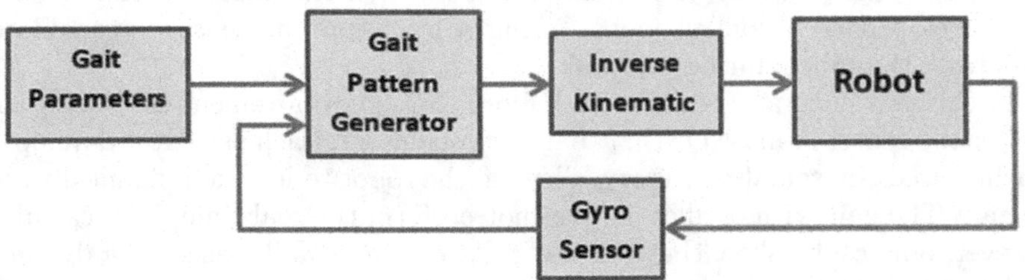

Fig. 2. Gait generator model.

The gait pattern generator can be configured by parameters. In Darwin-OP the values of these parameters can be changed by the file *config.ini*. Our humanoid robot is based on DARwIn-OP robot. The gait parameters are read from config.ini file when initializing the control process. Then the gait generator calculates the positions of coupled oscillators, and these values are passed to inverse kinematics module, that calculates the position values of each of the 18 servo motors, and the gyro sensor reading is performed during the execution of the gait, and if there is any fluctuation in the gyro sensor that can cause the robot fall, the gait generator will perform a correction to try avoid falling, can be seen in Fig. 2.

3 Related Work

As far as we know, the first work that specifically addressed the walk of an humanoid, biped, robot walking on sloped terrain was published by Zheng and Shen [15], who proposed a scheme to enable a biped robot to climb sloping surfaces, by using position sensors on the joints and force sensors underneath the heel and toe to detect the transition of the supporting terrain from a flat floor to a sloping surface. Their algorithm "for the biped robot control system evaluate the inclination of the supporting foot and the unknown gradient, and a compliant motion scheme is then used to enable the robot to transfer from level walking to climbing the slope".

While Zheng and Shen may be the first paper to address the problem, Chew, Pratt and Pratt [16] is currently the most cited one: they proposed a control strategies for walking dynamically and steadily over sloped terrain with unknown slope gradients and transition locations. The proposed algorithm is based on geometric considerations, being very simple, detecting the ground blindly using foot contact switches. Whit this algorithm, they were able to make a simulated 7-link (six degree-of-freedom, the robot has two legs and a body) planar biped to walk up and down slopes.

More recent work on this subject includes: Zhow et al. [17], that proposed a dynamically stable gait planning algorithm that use zero-moment point (ZMP) constraints for climbing a sloping surface; Hong et al. [18], that proposed a command state (CS)-based modifiable walking pattern generator for modifiable walking on an inclined plane in both pitch and roll directions; Huang et al. [19], that computes the future ZMP locations taking into account a known slope gradient. Literature about biped walking on uneven terrain – such as grass, sands and rocks – is much larger, and unfortunately cannot be included in this work.

The system proposed in this paper differs from all the published research because, as far as we know, we are the first combining Reinforcement Learning techniques and Gait generation, in a two-layer architecture, to achieve the goal of climbing up and down slopes.

4 Proposed System's Architecture

The proposed architecture of our system, presented in Fig. 3, is a two-layer combination of the traditional gait generation control loop with a reinforcement learning component.

The low level layer implements Ha, Tamura and Asama [3] gait generation system, where the control generates the sinusoidal commands for the servomotors, and implements a closed feedback control loop using a gyroscope to correct the robot's gait.

The Reinforcement Learning layer uses an accelerometer to generate a correction for the gait, when the slope of the floor that the robot is walking changes. To be able to correct the posture of the robot, this layer has to learn the optimal action policy needed to stabilize the robot in the stand-up position.

The problem that the RL layers is trying to learn can be described as a finite state MDP, where the states are the accelerometer values, discretized in 61 states, using the accelerometer values (the discretization created 30 states where the robot tilts forward, 30 where it tilts backward, and one state that is the central position, which means an upright robot). In this layer, the robot can only perform two actions: increments or decrements the position of the ankle joint. The transition function is not deterministic, because one action will make the robot move the ankle joint to the nearby servo positon angle, but that does not mean that the action resulted in a transition to the next state. Finally the reward given is a positive reinforcement when the robot reaches the upright position and a negative one for all other states. We use only 61 states because as learning was performed in a real robot, in this case there is the need to reduce as much as possible the state space, thus reducing training time.

The algorithm in the RL layer acquires the measures of the accelerometer in X and Z axis. The X axis indicates whether the robot is leaning forward or backward in a qualitative way and the Z axis indicates quantitatively how much the robot is inclined forward or backward. The accelerometer is positioned near to the center of mass of the robot.

To combine the 2 layers of control, the RL layer is able to change the parameter $pitchOffset$ of the gait generation ($pitchOffset$ is the angle offset around the axis Y in the ankle joint). In this way, when the floor inclination changes, the accelerometer perceives this change, and the RL layers defines a new $pitchOffset$ value to make the robot get in the upright position again. In this case the RL must learn the accelerometer angle mapping to orientate the correct position of the joints' motor by the parameter value $pitchOffset$.

In the model presented in the Fig. 3, the layer that feeds the $pitchOffset$ could be used another technique as in the work Baltes, Iverach-Brereton, Anderson [20], where they used fuzzy to make a system for enabling a humanoid robot to balance on a bongo board (a simple apparatus consisting of a deck resting on a free-rolling wheel) or using proportional-derivative (PD) controller as the work of Iverach-Brereton et al. [21]. However, the advantage of using reinforcement learning compared with the PID, the PID equations need not be calculated in RL, because the reinforcement learning is seeking extracting a solution to the problem, a second advantage is that learning by strengthening even after that there was a convergence of policy π, the RL still seeking the optimal policy π^*. In the next Section we present the results of using this system on a real humanoid robot.

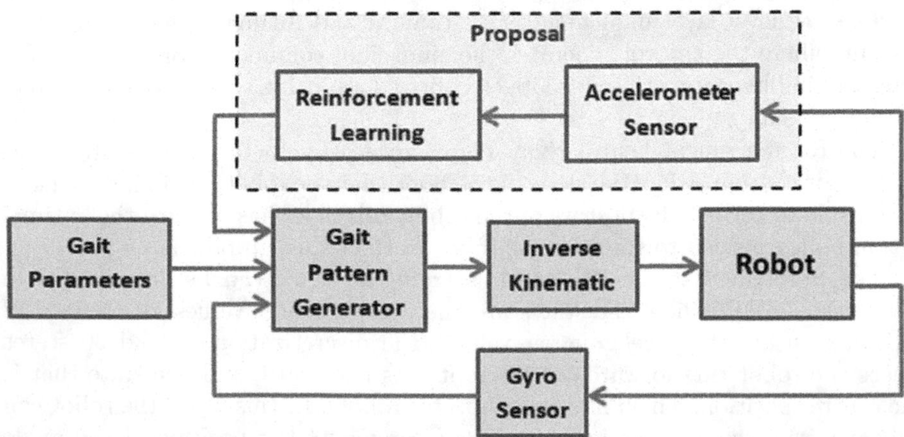

Fig. 3. The two-layer architecture proposed.

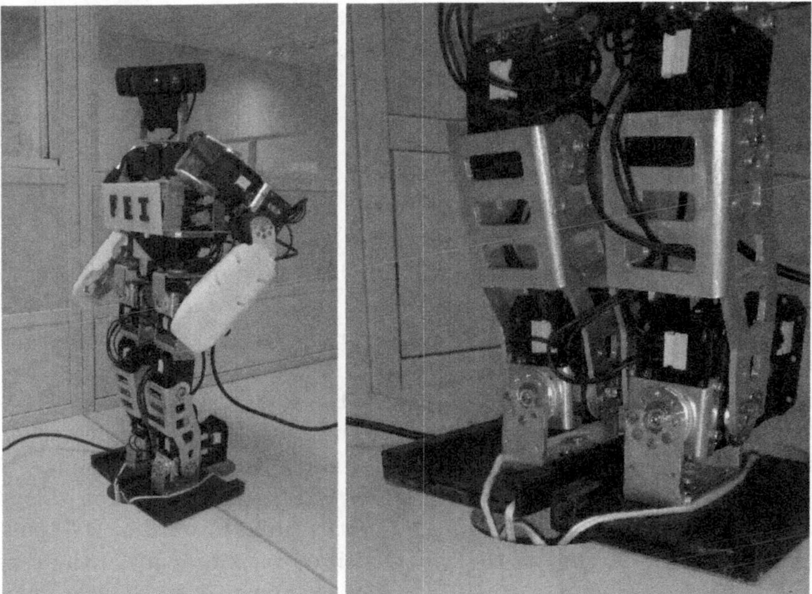

Fig. 4. The robot used to perform the experiments (left), with its' feet tied to the ground in order to prevent falls during the learning step (right).

5 Experiments

The experiment made to validate the proposed system is divided in three parts: in the first one, the Q-Learning RL algorithm is used to learn how to make a robot stay in the upright position; In the second one, the learned behavior was used in a robot standing up, positioned over a board, which is inclined to test if the robot could maintain its stability; Finally, the third part is to apply the learned behavior as feedback in the loop control during walking in a sloped terrain. All experiments were performed on a real robot.

The robot used in the experiments is based on DARwIn-OP [4] that consists on a robot with open source hardware and software and it is widely used by the teams in the RoboCup Humanoid KidSize League. The DARwIn-OP robot was developed by three American universities: Virginia Tech (RoMeLa - Robots and Mechanisms Laboratory), Pennsylvania University (GRASP - General Robotics, Automation, Sensing and Perception) and Purdue University, and ROBOTIS.

The main difference between the robot used in this experiment and the DARwIn-OP is that our robot is using the Intel NUC i5 processing board, instead of FitPC (present in the original DARwIn-OP project) and a new electronics and software architecture, that is presented in [22]. The robot, that have been developed by RoboFEI-HT team to participate in the RoboCup Humanoid KidSize League, can be seen in Fig. 4.

The robot has 20 DOF, being 6 per leg, 3 per arm, 2 on the head, height 490 mm, weight 3.0 Kg, walking speed 10 cm/s. Has an inertial measurement unit (IMU) comprised of 3-axis gyroscope, 3-axis accelerometer and 3 axis-magnetometer; a Logitech camera Full HD Pro C920 that is an USB-based camera placed on the head of the robot; and the servo motors are RX-28, and it has a running degree of 0° to 300°. The IMU used on the robot is the UM6 ultra-miniature produced by CH Robotics.

The values that are extracted from the IMU is approximately $-1 \leq x \leq 1$, $-1 \leq y \leq 1$ and $-1 \leq z \leq 1$. When the robot is standing perpendicular to the ground z is -1, and z is 1 if the robot stand upside down. When the robot is standing (axis $z \approx 1$), the axes x and y are parallel to the ground presenting the values of $x \approx 0$ and $y \approx 0$.

Considering the IMU ranges ($-1 \leq x \leq 1$, $-1 \leq y \leq 1$ and $-1 \leq z \leq 1$.) and that during the walking we limited the inclination to a maximum of 20°, therefore, the useful IMU range required to use in this experiment on the z axis was -0.77 to -1, so the discretization states was performed within this range -0.77 to -1 divided into 30 parts (if the robot is tilted forward at 20° the value of z axis is $z \approx -0.77$, since if the robot is tilted backward at 20 the value of z axis is the same value of $z \approx -0.77$). And for the x-axis used the following rule: if $x \leq 0$, then the robot is tilted forward; if $x < 0$, then the robot is tilted backwards.

If the robot tilt the body to 20°, the robot will be in a position where its Center of Mass is not inside in the support polygon, and it will take the fall, so values above 20° are not interesting in this experiment.

5.1 Learning the Robot's Upright Position Using Q-Learning

During the first step, the robot's feet were tied to the ground to prevent falls, as depicted in Fig. 4, and the robot would start with its inclination at an random selected angle. The Q-Learning algorithm would then learn how to make the robot stand up, using the accelerometer information.

In this step only the hips' servomotors were used, so the robot could swing like an inverted pendulum, tilting the chest forward and backward. The experiment was conducted in a real robot, and due to the time that the algorithm needed for the robot to finish each action, the duration of the complete learning was around few hours.

In the beginning of learning the values of $Q(s, a)$ were initialized with random values between 0 and 1, the exploration rate ϵ was equal to 0.1 (10 % of exploration) and the learning rate was 0.2. The reward given to the robot was 100 when it reached the goal state (with the robot standing up), and -1 in all other states. A trial ends when the robot reach the goal state, and is restarted with the robot at a random inclination. Figure 5 shows the evolution of this learning phase, In this graph (Fig. 5) we can see that the number of steps to reach the goal was decreasing every episode.

The graph of Fig. 6 shows that in the early episodes the standard deviation is high, it means that there is great variation from the average of the values of the

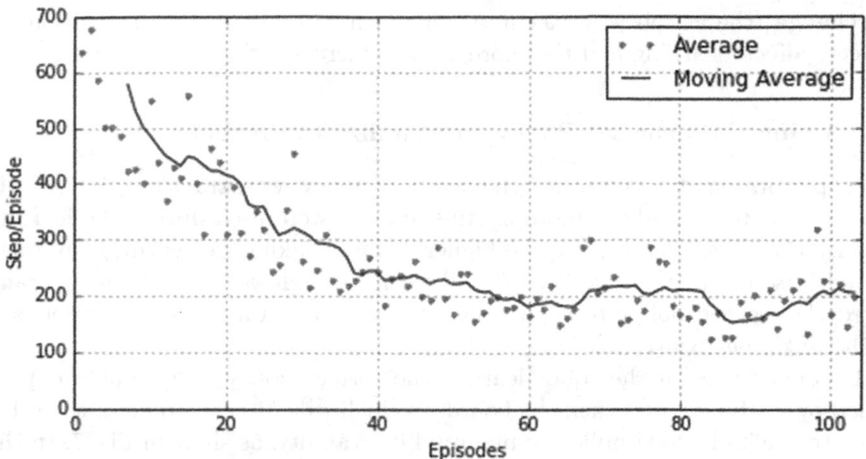

Fig. 5. Average steps to reach the goal state in function of the number of episodes, during the learning phase (with moving average).

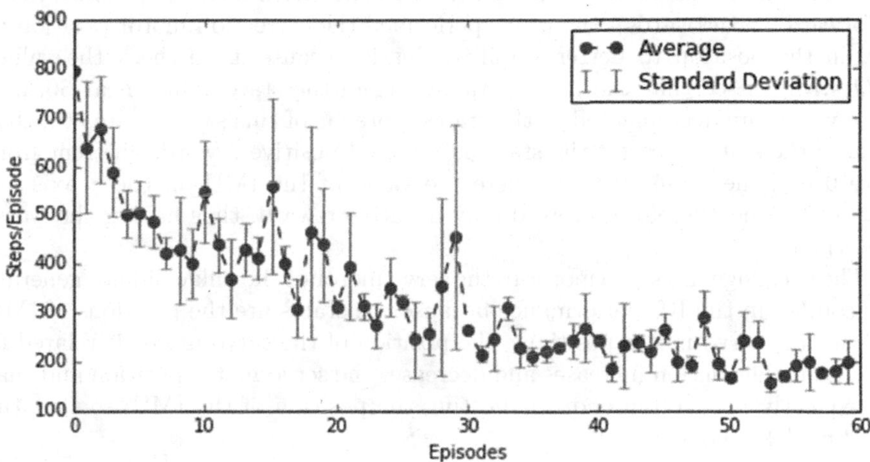

Fig. 6. Average steps to reach the goal state in function of the number of episodes, during the learning phase (with mean and standard deviation).

number of steps up to the episode, as we will running more episodes the standard deviation decreases, thus the $Q(s, a)$ values are approaching to the convergence, so the tendency is that the policy becomes increasingly stable.

In this experiment we noticed that the discretization size also affects learning. The model of this MDP is not deterministic, so when the interval between a state (the state is the axis position x and z IMU) and another is a high value (the value of z between certain states by the applied discretization) the execution of the action will increment or decrement the servant position but the state does not change (the action was executed but did not result in state transition), it directly affects learning and therefore greatly increases the learning time.

5.2 Using the Learned Policy to Stabilize the Robot

After performing the learning, the learned policy was transferred from the hips' motors to the ankles' motors. Hips' motors were used during the learning because the ankles' motors require higher torque, making the learning impractical, because the motors overheated and turned off. However, after being transferred the learned policy to the ankles' motors, let run a few more episodes to fit the ankle behavior.

In order to test if the robot learned the correct policy, the robot was positioned over a board, and then the board was inclined. At this moment, the robot needs to use its learned policy to maintain its stability, as show in Fig. 7. In this experiment the robot's feet were not tied to the board, so, if the inclination was larger than 20°, the robot would fall. We assume that the board has only variation perpendicular to the frontal plane.

As shown in Fig. 7, independent of the slope direction board (the board tilted to forward or backwards), the robot performed the correction motor position to stay in the position to better stability, this is because if we check the values of $Q(s, a)$, we see that the highest values are in the state values near 30, and these values are decremented in the states more far of this state (state 30), that because the state's near of the state 30 received positive reward, different from state 0 and the state 60 that where the value of the IMU on the z axis are $z \approx -0.77$, and therefore received only negative reward, thus having the worst $Q(s, a)$ value.

The exchange hip servo motor to the servo motor of the ankle did not generate any conflict in the RL performance because the states are the positions of IMU and not the servo motor positions, the position of the servo motor is related to the action. The action increases and decreases the servo motor position and that consequently most often end up affecting the position of the IMU, so that the MDP model is nondeterministic.

5.3 Combining the Gait Generator with the Learned Policy

This experiment was to apply the learned behavior as feedback in the loop control during walking in a sloped terrain, this makes the robot to be able to walk in a lightly sloped terrain and keeping a dynamically balanced gait. How in the experiments we assume that the terrain has only variation perpendicular to the frontal plane, the robot walked only forward (the robot is able to walk forward and backward on slopes).

As already described, in order to use the policy learned by the Q-Learning algorithm, the RL layer of or system adjusts the variable $pitchOffset$. This variable modifies the pitch angle of the ankle, tilting the body of the robot forward or backward. For stability control the Darwin-OP already uses the gyroscope

Fig. 7. Robot using the policy learned during the first phase to maintain stability on an inclined board.

sensor as feedback in the closed loop, maintaining the walking stability of the robot, as described in [3].

The $pitchOffset$ was only adjusted by the RL when there was a high level of certainty about a change on the ground slope. Therefore, to change the $pitchOffset$, a moving average about 15 readings of accelerometer was calculated. The data were read every 10 ms (5 ms is the time that the accelerometer spend to update the values), so a higher level of confidence about the ground slope was achieved. The variable $pitchOffset$ controls the angle in degrees. Every time an action was selected in the RL, an increase or a decrease was performed in $pitchOffset$. The angle was also limited in $-20° \leq pitchOffset \leq 20°$, because outside this range the robot's feet slid.

Our humanoid robot performe an omnidirectional walk, *setXAmplitude*, *setYAmplitude* e *setAAmplitude*.

- *setXAmplitude* is the forward footstep length, the maximum distance between the two legs at the instant one leg is opposite to another during the walk (used for walk forwards or walk backwards).
- *setYAmplitude* is the side length of the footstep (used to sidle).
- *setAAmplitude* is the angle of the gait that allows the robot to rotate during the walk.

In this experiment the values of the variables were: $setXAmplitude = 10$, $setYAmplitude = 0$ and $setAAmplitude = 0$, with these parameters the robot performs a walk only forward.

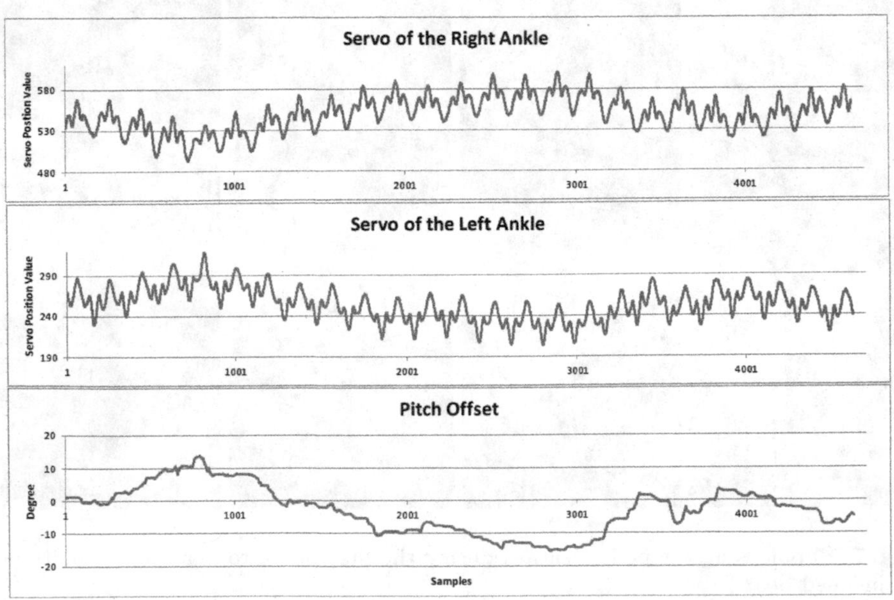

Fig. 8. The ankles' servomotors being adjusted according to the value of $pitchOffset$.

The graph of Fig. 8 shows the ankles' servo motors acting during the walking, the axis x is the time in milliseconds and the axis y are the values of the servo motor position (the servo motor ranges is 0–1024 related to the angle 0° to 300°). We can see the servo motors perform a movement of sinusoids composition.

The graph depicted in Fig. 8 shows the ankles' servomotors being adjusted according to the value of $pitchOffset$. It can be seen in this figure that there is a sinusoid movement superimposed in a lower frequency one. The higher frequency wave is the consequence of the gait movements, while the lower frequency is the result of changing the inclination of the board.

Fig. 9. Robot climbing an accessibility ramp.

To capture these graph values, the robot was positioned on a board, while performing the gait movement in the same position. Then the board was inclined and the robot sought the position of best stability by adjusting the $pitchOffset$. When there is not inclination in the board (parallel to the ground), the value of $pitchOffset$ was zero. In this graph of the Fig. 8, 5000 samples were collected. In the robot, the rate of writing in the servomotors is 8 ms, so was collected 125 samples per second.

In order to conclude this experiments, we finally took our robot outside the laboratory and made it go up and down several ramps. Figure 9 shows the robot climbing an accessibility ramp located in one street of our campus. It can be seen that the robot is able to climb the ramp. This was no small feat, as the robot is very small in relation to the size of the ramp, and the day was windy, making the robot to lose its balance. A video showing this same movement can be seen at https://youtu.be/JMT6tTcCaFQ.

As a matter of comparison, we tried to make the robot go up the ramp using only Ha, Tamura and Asama [3] gait generation system: the result was that the robot would fall instantaneously, every time.

Finally, we tried to detail our system in order to enable the repeatability of this work. But sometimes details are still missing. In order to facilitate the use of this system by other researchers, all the software is available for download at http://fei.edu.br/~rbianchi/software.html.

6 Conclusions

With the introduction of artificial grass fields in the RoboCup Humanoid League, the study of new equilibrium methods for robot locomotion became a necessity. The use of a gyroscope is not sufficient to maintain the stability of the robot in this type of ground. The use of other sensors – such as an accelerometer – combined with Reinforcement Learning techniques to help the robot to stabilize itself during the walk seems very promising.

The proposed architecture of our system, a two-layer combination of the traditional gait generation control loop with a reinforcement learning component that adjusts the robot's body inclination, improved the walking of the robot, enabling it to walk in different situations of sloping floors (ascending and descending floors).

All the presented experiments were performed in a real robot, this forces us to make the MDP state of space has to be smaller compared to an experiment performed in a simulator to keep the same time spent on the experiment, the difference is that the simulator we can speed up the simulation.

Using the same architecture described in the article, other models of MDP will be implemented in a future research, using a 3D simulator as was done in some research applying reinforcement learning on the simulator [23,24]. The advantage of using a simulator is that a large number of trials may be executed. On the other hand, the learned behavior to the simulator is difficult to be transferred to a real robot, as explained by Stone [23].

One drawback of this system is that, during the walk process, the robot needs to decrease the step size as the degree inclination increases. So, as a future works, we propose to adapt the step height according the degree inclination to maintain the same step size and use Reinforcement Learning to update other variables, improving the stability of the robot during the walk.

Acknowledgment. The authors would like to acknowledge the Centro Universitário FEI and the Robotics and Artificial Intelligence Laboratory for supporting this project. The authors would also like to thank the scholarships provided by CAPES and CNPq.

References

1. Westervelt, E.R., Grizzle, J.W., Chevallereau, C., Choi, J.H., Morris, B.: Feedback Control of Dynamic Bipedal Robot Locomotion, vol. 28. CRC Press, Boca Raton (2007)
2. Sutton, R.S., Barto, A.G.: Reinforcement Learning: An Introduction, vol. 1. MIT Press, Cambridge (1998)
3. Ha, I., Tamura, Y., Asama, H.: Gait pattern generation and stabilization for humanoid robot based on coupled oscillators. In: IEEE/RSJ International Conference on Intelligent Robots and Systems (IROS), pp. 3207–3212 (2011)
4. Ha, I., Tamura, Y., Asama, H., Han, J., Hong, D.W.: Development of open humanoid platform DARwIn-OP. In: SICE Annual Conference (SICE), pp. 2178–2181 (2011)
5. Perico, D.H., Silva, I.J., Vilão Jr., C.O., Homem, T.P., Destro, R.C., Tonidandel, F., Bianchi, R.A.: Newton: a high level control humanoid robot for the RoboCup Soccer KidSize League. In: Osório, F.S., Wolf, D.F., Branco, K.C., Grassi, V., Becker, M., Romero, R.A.F. (eds.) Robotics. CCIS, vol. 507, pp. 53–73. Springer, Heidelberg (2015). doi:10.1007/978-3-662-48134-9_4
6. Eva Robot (2016). http://fei.edu.br/brahur2016/artigos/Artigo%202%20-%20UFU.pdf
7. WF Wolves and Taura Bots - Team Description Paper (2016). http://www.robocup2016.org/media/symposium/Team-Description-Papers/Humanoid/RoboCup_2016_Humanoid_TeenSize_TDP_WF_Wolves_Taura_Bots.pdf
8. Watkins, C.: Learning from delayed rewards. Doctoral dissertation, University of Cambridge (1989)
9. Marder, E., Bucher, D.: Central pattern generators and the control of rhythmic movements. Curr. Biol. **11**, R986–R996 (2001)
10. Mcgeer, T.: Passive dynamic walking. Int. J. Robot. Res. **9**, 62–82 (1990)
11. Vukobratović, M., Borovac, B.: Zero-moment point thirty five years of its life. Int. J. Humanoid Rob. **1**, 157–173 (2004). World Scientific
12. Kajita, S., Kanehiro, F., Kaneko, K., Yokoi, K., Hirukawa, H.: The 3D linear inverted pendulum mode: a simple modeling for a biped walking pattern generation. In: IEEE/RSJ International Conference on Intelligent Robots and Systems, vol. 1, pp. 239–246 (2001)
13. Zhao, M., Dong, H., Zhang, N.: The instantaneous leg extension model of virtual slope walking. In: IEEE/RSJ International Conference on Intelligent Robots and Systems, IROS, pp. 3220–3225 (2009)
14. Vukobratović, M., Juricic, D.: Contribution to the synthesis of biped gait. IEEE Trans. Biomed. Eng. **16**(2), 1–6 (1969)

15. Zheng, Y.F., Shen, J.: Gait synthesis for the SD-2 biped robot to climb sloping surface. IEEE Trans. Robot. Autom. **6**, 86–96 (1990)
16. Chew, C.M., Pratt, J., Pratt, G.: Blind walking of a planar bipedal robot on sloped terrain. In: IEEE International Conference on Robotics and Automation, vol. 1, pp. 381–386 (1999)
17. Zhou, C., Yue, P.K., Ni, J., Chan, S.B.: Dynamically stable gait planning for a humanoid robot to climb sloping surface. In: IEEE Conference on Robotics, Automation and Mechatronics, vol. 1, pp. 341–346 (2004)
18. Hong, Y.D., Lee, B.J., Kim, J.H.: Command state-based modifiable walking pattern generation on an inclined plane in pitch and roll directions for humanoid robots. IEEE/ASME Trans. Mechatron. **16**, 783–789 (2011)
19. Huang, W., Chew, C.M., Zheng, Y., Hong, G.S.: Pattern generation for bipedal walking on slopes and stairs. In: 8th IEEE-RAS International Conference on Humanoid Robots, pp. 205–210 (2008)
20. Iverach-Brereton, C., Baltes, J., Postnikoff, B., Carrier, D., Anderson, J.: Fuzzy logic control of a humanoid robot on unstable terrain. In: Almeida, L., Ji, J., Steinbauer, G., Luke, S. (eds.) RoboCup 2015. LNCS, vol. 9513, pp. 202–213. Springer, Heidelberg (2015). doi:10.1007/978-3-319-29339-4_17
21. Baltes, J., Iverach-Brereton, C., Anderson, J.: Human inspired control of a small humanoid robot in highly dynamic environments or Jimmy Darwin rocks the Bongo Board. In: Bianchi, R.A.C., Akin, H.L., Ramamoorthy, S., Sugiura, K. (eds.) RoboCup 2014. LNCS, vol. 8992, pp. 466–477. Springer, Heidelberg (2015). doi:10.1007/978-3-319-18615-3_38
22. Perico, D.H., Silva, I.J., Vilao, C.O., Homem, T.P., Destro, R.C., Tonidandel, F., Bianchi, R.: Hardware and software aspects of the design and assembly of a new humanoid robot for RoboCup Soccer. In: Robotics: SBR-LARS Robotics Symposium and Robocontrol (SBR LARS Robocontrol), pp. 73–78 (2014)
23. Farchy, A., Barrett, S., MacAlpine, P., Stone, P.: Humanoid robots learning to walk faster: from the real world to simulation and back. In: Proceedings of the 2013 International Conference on Autonomous Agents and Multi-agent Systems, pp. 39–46 (2013)
24. Hester, T., Quinlan, M., Stone, P.: Generalized model learning for reinforcement learning on a humanoid robot. In: IEEE International Conference on Robotics and Automation (ICRA), pp. 2369–2374 (2010)

Structure-Control Optimal Design of 6-DOF Fully Parallel Robot

Fabian Andres Lara-Molina[1(✉)], Didier Dumur[2], and Edson Hideki Koroishi[1]

[1] Department of Mechanical Engineering, Federal University of Technology - Paraná, Cornélio Procópio, PR, Brazil
{fabianmolina,edsonh}@utfpr.edu.br

[2] Laboratoire des Signaux et Systèmes, CentraleSupélec-CNRS-Université Paris Sud, Université Paris Saclay, Control Department, 91192 Gif sur Yvette cedex, France
Didier.Dumur@centralesupelec.fr

Abstract. This contribution aims at introducing an optimal design methodology for the Stewart Platform robot that considers structure and control design variables simultaneously. This methodology intends to maximize the positioning accuracy in order to optimize the overall performance of the robot for a specific task. The structure design variables of the mechanism combined with the gains of the controller are the structure-control design variables, this global set is considered simultaneously in the optimal design methodology. A position control scheme, based on a PD controller, and the complete dynamics of the robot are considered to compute the overall tracking position as function of the structure-control design variables. A sensitivity analysis is performed to evaluate the effect of the structure-control design variables on the tracking position accuracy of the robot. The associated optimization problem is solved by using metaheuristic optimization methods. Simulation results demonstrate that the proposed design procedure is effective to increase the positioning accuracy, as well as to improve the closed loop dynamics performance of the robot.

Keywords: Optimal design · Optimization · Stewart-Gough platform · Position control · Tracking accuracy

1 Introduction

The optimal design of parallel robots aims at determining a set of design variables to satisfy an optimal performance criterion. Several studies have encompassed the optimal design of parallel robots by using different approaches that depend on the characteristics to be optimized. In those works, the optimal design has considered several optimum criteria.

The growing number of new applications of parallel robots demands to increase the positioning accuracy. In order to reach this objective, the researchers have worked in two main directions: optimal structural design and control system design. The optimal structural design determines design variables of the

F. Santos Osório and R. Sales Gonçalves (Eds.): LARS 2015/SBR 2015, CCIS 619, pp. 247–266, 2016.
DOI: 10.1007/978-3-319-47247-8_15

mechanism to reach an optimal kinematic criterion, such design variables are: the lengths of the links or kinematic chains. In addition, the control system design applies advanced control techniques to motion control of the robot, and thus to guarantee an adequate dynamic performance to execute a task. However, the optimal structure design and the design of the control system have been addressed separately. In most cases, the complete design of the robot has been executed sequentially: first, the structure design, and finally, the control system design.

The structure design of the mechanism has been figured up to obtain high kinematic performance, some works have addressed this problem by: maximizing kinematic accuracy [1], maximizing the stiffness [2], minimizing position error in the movable platform [14], and maximizing the required maximum workspace [15]. Those works only consider kinematic aspects in the design procedure, nevertheless, aspects related to the dynamics and control are not taken into account.

The design of the control system intends to improve the dynamic behavior to perform a specific task regarding the tracking accuracy performance of the robot. Thus, some works have applied advanced model-based controllers to enhance the positioning accuracy of parallel robots: predictive control [5,6], adaptive control [7], and iterative learning control [8], among others. However, additional analyses are necessary in order to evaluate how the kinematic properties influence the dynamic response of the controlled parallel robot, and thus, establishing additional criteria to select the structure design variables of the mechanism.

The mechatronic design has been used as an alternative to optimal design of parallel robots, this methodology takes into account structural-control design variables of the system simultaneously during the design procedure. Mechatronic design has been applied to parallel robots [9–11]. Villareal-Cervantes et al. (2009) [9] proposed structure-control mechatronic design of the planar parallel robot. Silva et al. (2013) [10] figured up the optimal design of a pick-and-place robotic system by using mechatronic design concepts. Villareal-Cervantes et al. (2013) [11] developed a robust formulation for the mechatronic design of planar parallel robot. All the aforementioned works are concerned with planar parallel robots. Nevertheless, some applications require high-performance positioning systems with spatial parallel mechanism.

In this direction, this paper proposes an optimization of structure-control design variables simultaneously to maximize position tracking accuracy of the Stewart-Gough robot to enhance the overall performance. Based on the complete model and control approach, the structure-control design variables are defined, this global set consists of the design variables of the mechanism combined with the gains of the controller. A PD controller is considered as position control scheme and the tracking position error is evaluated as function of the structure variables. A sensitivity analysis, based on variance decomposition method, is performed to quantify the effect of structure and control variables on position tracking accuracy. The optimization problem is solved by using Genetic Algorithms. Simulation results show that the optimal design method aids to select the structure-control that increases the positioning accuracy.

The remain of this chapter is organized in several sections. Section 2 presents the robot modeling to introduce the structure-control design variables and the objective function. In Sect. 3, a sensitivity analysis is carried out to quantify the effect of the design variables on the objective function. In Sect. 4, the optimization problem to solve the optimal design is presented. Section 5 gives the simulation results. Finally, Sect. 6 enumerates some conclusions.

2 Robot Modeling

The 6-*UPS* (Universal Prismatic Spherical) Stewart-Gough manipulator, shown in Fig. 1 (a), has six identical legs (see Fig. 1 (b)) connecting the fixed base to the movable platform by universal joints denoted by U at points B_i and spherical joints denoted by S at points P_i (for $i = 1, \ldots, 6$), respectively. Both the universal and the spherical joints are passive. Each leg has an upper and a lower member connected by an active prismatic joint denoted by P that extends and retracts the leg. The movable platform has six degrees of freedom, three translational and three rotational motions.

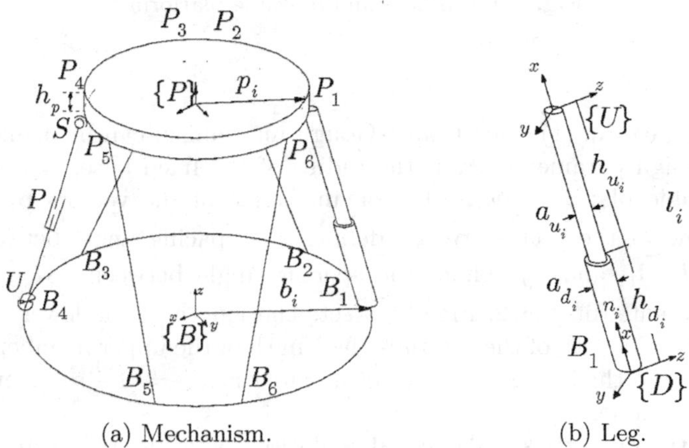

(a) Mechanism. (b) Leg.

Fig. 1. Stewart-Gough robot.

2.1 Structural Modeling

Two coordinate frames $\{P\}$ and $\{B\}$ are attached to the movable and fixed base respectively. The vector $\mathbf{b}_i = \left[b_{ix}\, b_{iy}\, b_{iz} \right]^T$ describes the position of the reference point B_i with respect to the frame $\{B\}$; in the same way, the vector $\mathbf{p}_i = \left[p_{ix}\, p_{iy}\, p_{iz} \right]^T$ describes the position of the reference point P_i with respect to the reference frame $\{P\}$ (see Fig. 2).

$$\begin{aligned}\mathbf{b}_i &= \left[r_b \cos(\psi_i)\; r_b \sin(\psi_i)\; 0\right]^T = \left[b_{ix}\; b_{iy}\; b_{iz}\right]^T \\ \mathbf{p}_i &= \left[r_p \cos(\boldsymbol{\Psi}_i)\; r_p \sin(\boldsymbol{\Psi}_i)\; 0\right]^T = \left[p_{ix}\; p_{iy}\; p_{iz}\right]^T\end{aligned} \tag{1}$$

where

$$\begin{aligned}\psi_i &= \tfrac{i\pi}{3} - \tfrac{\phi_b}{2} & \boldsymbol{\Psi}_i &= \tfrac{i\pi}{3} - \tfrac{\phi_p}{2} & i &= 1,3,5 \\ \psi_i &= \psi_{i-1} + \phi_b & \boldsymbol{\Psi}_i &= \boldsymbol{\Psi}_{i-1} + \phi_p & i &= 2,4,6\end{aligned} \tag{2}$$

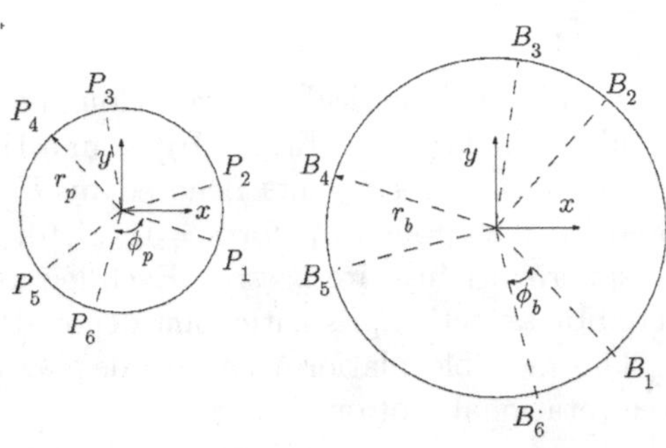

Fig. 2. Fixed base and movable platform.

According to Eq. (1), the Stewart-Gough mechanism can be defined by four structure design parameters: r_b is the radius of the fixed base, r_p is the radius of the movable platform, ϕ_b is the spacing angle of the vectors $\mathbf{b}_i$, ϕ_p is the spacing angle of the vectors $\mathbf{p}_i$. ϕ_b defines the spacing angle between $\widehat{B_2B_3}$, $\widehat{B_4B_5}$ and $\widehat{B_6B_1}$, and ϕ_p defines the spacing angle between $\widehat{P_2P_3}$, $\widehat{P_4P_5}$ and $\widehat{P_6P_1}$, as presented in Fig. 2. Finally, s sets the length of the lower member as function of total length of the leg, thus the length of the upper member is defined as $h_{d_i} = s.l_i$ and the length of the lower member is $h_{u_i} = (1-s)l_i$, with l_i being the length of the leg.

The inertial properties of the movable platform and the six legs are defined as function of the structure parameters: r_p, ϕ_p, r_b, ϕ_b and s. The geometric shape of the rigid bodies of the Stewart-Gough mechanism is defined as cylinders (see Figs. 1 (a) and (b)).

The inertia matrix of the lower and upper members of the legs is defined with respect to the coordinate frames $\{U\}$ and $\{D\}$ (Fig. 1(b)) and the centers of mass are $\mathbf{r}_u = \left[-\frac{1}{2}h_u\; 0\; 0\right]^T$ and $\mathbf{r}_d = \left[\frac{1}{2}h_d\; 0\; 0\right]^T$, respectively. The following parameters should be imposed in order to describe the inertia of all rigid bodies as function of the structural parameters: thickness of the movable platform h_p, radius of the upper member a_u and lower member a_d of the legs, density of the material of the movable platform ρ_p and legs ρ_d. The inertia tensor of the upper and lower members of the legs are $\mathbf{M}_u$ and $\mathbf{M}_d$, respectively.

$$\mathbf{I}_u = \begin{bmatrix} \frac{1}{2}m_u.a_u^2 & 0 & 0 \\ 0 & \frac{1}{12}m_u.(3a_u^2 + 4h_u^2) & 0 \\ 0 & 0 & \frac{1}{12}m_u.(3a_u^2 + 4h_u^2) \end{bmatrix} \tag{3}$$

$$\mathbf{I}_d = \begin{bmatrix} \frac{1}{2}m_d.a_d^2 & 0 & 0 \\ 0 & \frac{1}{12}m_d.(3a_d^2 + 4h_d^2) & 0 \\ 0 & 0 & \frac{1}{12}m_d.(3a_d^2 + 4h_d^2) \end{bmatrix} \tag{4}$$

where $m_u = \rho_u h_u a_u^2 \pi$ and $m_d = \rho_d h_d a_d^2 \pi$.

The center of mass of the movable platform is attached to the coordinate frame $\{P\}$. The inertia tensor of the movable platform $\mathbf{M}_p$ is:

$$\mathbf{I}_p = \begin{bmatrix} \frac{1}{12}m_p.(3r_p^2 + h_p^2) & 0 & 0 \\ 0 & \frac{1}{12}m_p.(3r_p^2 + h_p^2) & 0 \\ 0 & 0 & \frac{1}{2}m_p.r_p^2 \end{bmatrix} \tag{5}$$

where $m_p = \rho_p \pi r_p^2 h_p$.

2.2 Dynamic Model

The dynamic equation of the 6-UPS Stewart-Gough manipulator was derived in closed form through the Newton-Euler approach by [3], written below in the joint-space:

$$\mathbf{f} = \mathbf{J}^{-1}\mathbf{M}(\mathbf{q})\mathbf{J}^{-T}\ddot{\mathbf{q}} + \mathbf{J}^{-1}[\eta(\mathbf{q},\dot{\mathbf{q}}) - \mathbf{M}(\mathbf{q})\mathbf{J}^{-T}\mathbf{u}] \tag{6}$$

where

- $\mathbf{f} = \begin{bmatrix} f_1 \dots f_6 \end{bmatrix}^T \in \mathbb{R}^{6\times 1}$ is the actuator force vector.
- $\mathbf{J} = \begin{bmatrix} \mathbf{n}_1 & \dots & \mathbf{n}_6 \\ \mathbf{p}_1 \times \mathbf{n}_1 & \dots & \mathbf{p}_6 \times \mathbf{n}_6 \end{bmatrix} \in \mathbb{R}^{6\times 6}$ is the Jacobian matrix, where $\times$ denotes cross product of vectors, and $\mathbf{n}_i$ is the unit vector along each leg, for $i = 1,\dots,6$ (see Fig. 1(b)).
- $\mathbf{q} = \begin{bmatrix} l_1 \dots l_6 \end{bmatrix}^T \in \mathbb{R}^{6\times 1}$ is the leg length vector.
- $\dot{\mathbf{q}} = \begin{bmatrix} \dot{l}_1 \dots \dot{l}_6 \end{bmatrix}^T \in \mathbb{R}^{6\times 1}$ is the leg velocity vector.
- $\ddot{\mathbf{q}} = \begin{bmatrix} \ddot{l}_1 \dots \ddot{l}_6 \end{bmatrix}^T \in \mathbb{R}^{6\times 1}$ is the leg acceleration vector.
- $\mathbf{M} = \mathbf{M}_p + \sum_{i=1}^{6} \mathbf{M}_{li} \in \mathbb{R}^{6\times 6}$ is the total inertia matrix which considers the inertia of the six legs, $\mathbf{M}_{li}$, and the inertia of the movable platform, $\mathbf{M}_p$. Both terms, $\mathbf{M}_{li}$ and $\mathbf{M}_p$, are defined with respect to fixed frame $\{B\}$. It is worth

to mention that the inertia matrix of movable platform depends on $\mathbf{I}_p$ defined in Eq. (5). Additionally, the inertia matrix of legs, $\mathbf{M}_{li}$, depends on the inertia of upper member, $\mathbf{I}_u$, and lower member, $\mathbf{I}_d$, of the leg previously defined in Eqs. (3) and (4).

- $\eta = \eta_{plat} + \sum_{i=1}^{6} \eta_i \in \mathbb{R}^{6\times 1}$ is the Coriolis, gravitation, centrifuge force vector of the movable platform and each leg, and viscous friction forces at the joints for the 6-UPS.
- $\eta_i = \begin{bmatrix} \mathbf{v}_i \\ \mathbf{o}_i \times \mathbf{v}_i - \mathbf{f}_i \end{bmatrix}^T$, where $\mathbf{f}_i = c_s(\mathbf{w}_i - \omega)$, $\mathbf{o}_i = \mathbf{R}\mathbf{p}_i$ and $\mathbf{v}_i = (m_u \mathbf{n}_i.\mathbf{u}'_{4i} + c_p \dot{l}_i - m_u \mathbf{n}_i.\mathbf{g})\mathbf{n}_i - \frac{1}{l_i}\mathbf{n}_i \times \mathbf{u}'_{5i}$.
- $\mathbf{w}_i$ is the angular velocity of the leg, ω is the angular velocity of the movable platform, both vectors are expressed in cartesian coordinates.
- $\mathbf{R}$ is the orientation matrix of the movable platform and $\mathbf{g}$ is the acceleration vector due the gravity.
- c_u, c_p, c_s are the coefficients of viscous friction in the universal, prismatic and spherical joints, respectively.
- $\mathbf{u}'_{4i}$, $\mathbf{u}'_{5i}$ and $\mathbf{u} \in \mathbb{R}^{6\times 1}$ is an expression related to the acceleration of the legs.

Additional details of the formulation of the dynamic equation can be obtained in [3]. Equation (6) can be written in a simplified way:

$$\mathbf{A}(\mathbf{q})\ddot{\mathbf{q}} + \mathbf{h}(\mathbf{q}, \dot{\mathbf{q}}) = \mathbf{f} \tag{7}$$

where $\mathbf{A}(\mathbf{q}) = \mathbf{J}^{-1}\mathbf{M}(\mathbf{q})\mathbf{J}^{-T}$ and $\mathbf{h}(\mathbf{q}, \dot{\mathbf{q}}) = \mathbf{J}^{-1}[\boldsymbol{\eta}(\mathbf{q}, \dot{\mathbf{q}}) - M(\mathbf{q})\mathbf{J}^{-T}\mathbf{u}]$.

The geometric, inertial and dynamic properties of the dynamic model of the parallel robot are completely defined by the vector of parameters $\boldsymbol{\lambda}_s \in \mathbb{R}^{5\times 1}$ assuming that all parameters required to define the geometry, inertia and friction are known and fixed, thus:

$$\boldsymbol{\lambda}_s = \begin{bmatrix} r_p \ \phi_p \ r_b \ \phi_b \ s \end{bmatrix}^T \tag{8}$$

2.3 Tracking Position Control

As mentioned before, several advanced control techniques have been applied for motion control of parallel robots. However, in this contribution, a PD joint-space controller is considered, because it is straightforward to introduce the gains of this controller as control design variables.

Six independent joint-space PD controllers are used to track a desired trajectory as Fig. 3 shows.

Assuming that the desired trajectory for each actuator is specified with the desired joint-space position $\mathbf{q}^d$ and velocity $\dot{\mathbf{q}}^d$ meaning that this trajectory is differentiable and smooth enough, the control law is:

$$\mathbf{f}_c = \mathbf{K}_P(\mathbf{q}^d - \mathbf{q}) + \mathbf{K}_D(\dot{\mathbf{q}}^d - \dot{\mathbf{q}}) \tag{9}$$

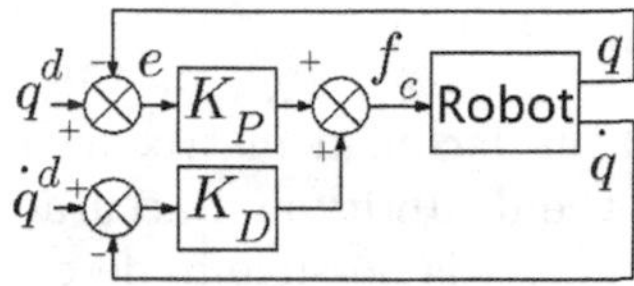

Fig. 3. Joint space PD controller of Stewart-Gough robot.

where $\mathbf{K}_P = diag(k_{p_1}, \ldots, k_{p_6})$, $\mathbf{K}_D = diag(k_{d_1}, ..., k_{d_6})$. It is worth to mention that the joint space position error is defined by $\mathbf{e} = [e_1 \ldots e_6]^T \in \mathbb{R}^{6\times 1}$. According to [4], the characteristic polynomial of one joint PD controller is $\Delta(s) = s^2 + k_d s/m_i + k_p/m_i$, where s is the Laplace variable, and m_i is the highest mass seen by the linear actuator of each leg which depends on $\mathbf{A}(\mathbf{q})$ of Eq. (7). Thus, the highest mass is presented when $m_i = m_p + m_u$, with m_p being the total mass of the movable platform, and m_u being the mass of upper member of the leg. In view of the fact that a parallel robot is being analyzed, m_i is equal for the six joints, consequently, one can assume that the proportional and derivative gains of the six PD controllers are equal. Additionally, the gains of the controller are defined as:

$$\begin{aligned} k_p &= m_i \omega_n^2 \\ k_d &= m_i 2\xi\omega_n \end{aligned} \tag{10}$$

Where ω_n is the natural frequency, and ξ is the critical damping. The tracking position controller can be parametrized by the vector $\boldsymbol{\lambda}_c \in \mathbb{R}^{2\times 1}$.

$$\boldsymbol{\lambda}_c = \left[k_p \; k_d\right]^T \tag{11}$$

Several methods to tune the PD controllers of robot manipulators have concluded that positive gains stabilize robot [12]. In addition, the closed-loop dynamics of the robot with the PD controller is formulated by using the state space formalism.

$$\dot{\mathbf{x}} = f(\mathbf{x}, \mathbf{x}^d, \boldsymbol{\lambda}, t) = \begin{bmatrix} \dot{\mathbf{q}} \\ -\mathbf{A}^{-1}(\mathbf{q})\mathbf{h}(\mathbf{q}, \dot{\mathbf{q}}) \end{bmatrix} + \begin{bmatrix} 0 \\ -\mathbf{A}^{-1}(\mathbf{q}) \end{bmatrix} \mathbf{f}_c \tag{12}$$

with $\dot{\mathbf{x}} = \left[\dot{\mathbf{q}}^T \; \ddot{\mathbf{q}}^T\right]^T \in \mathbb{R}^{12\times 1}$, $\mathbf{x}^d = \left[\mathbf{q}^{dT} \; \dot{\mathbf{q}}^{dT}\right]^T \in \mathbb{R}^{12\times 1}$. And, $\boldsymbol{\lambda} \in \mathbb{R}^{7\times 1}$ is the vector of the structure-control design variables formed by the parameters described in Eqs. (8) and (11).

$$\boldsymbol{\lambda} = \left[\boldsymbol{\lambda}_s^T \; \boldsymbol{\lambda}_c^T\right]^T = \left[r_p \; \phi_p \; r_b \; \phi_b \; s \; k_p \; k_d\right]^T \tag{13}$$

As seen in Eq. (12), the closed loop dynamics of the Stewart-Gough robot depend on the structure-control design variables. Furthermore, the tracking position error will also depends on the structure-control parameters, $\boldsymbol{\lambda}$.

For the optimal design procedure of the robot, an objective function that considers the structure-control design variables should be stablished as an optimal design criteria in order to minimize the tracking position of the robot over a desired trajectory.

2.4 Objective Function

Performance criteria based on Jacobian matrix have been widely used in the optimal design to improve the dexterity and accuracy of serial robots [13]. By using Jacobian matrix analysis, it is possible to determine the singularity loci of Stewart-Gough platform. The 6-UPS Stewart-Gough platform is a spatial mechanism. The conventional Jacobian matrix expresses a coupled relation of both translational and rotational motions. The elements of the conventional Jacobian matrix have nonhomogenous physical units. Therefore, the use of performance indices such as the condition number of the Jacobian matrix may lead to a lack of physical meaning [14].

Kinetostatic performance indices indicate when the parallel robot is closed to a singular configuration as an alternative to indices based on the Jacobian matrix [16]. The singularity zones in the workspace can be characterized with the aid of kinetostatic performance indices. When the parallel manipulator is close to the singularity zone, it loses its stiffness and its quality of motion transmission, this affects the position accuracy of the robot.

Furthermore, it has been demonstrated that the motion through such singularity loci is feasible and the singularities can also be examined based on the dynamics of the robot. Nevertheless, at this specific condition during motion the position accuracy decreases significantly [17]. Consequently, performance criterion based on the closed-loop dynamics are suitable for the optimal design of the Stewart-Gough robot. The tracking position error of PD position controller in Eq. (9) is selected since the objective of this contribution is to improve the position tracking accuracy of the robot.

Integrating Eq. (12) over an imposed trajectory $\mathbf{x}^d$ leads to the actual closed loop position of the robot. Thus, the position tracking error of the six legs is $\mathbf{e}(\boldsymbol{\lambda}) = (\mathbf{q}^d - \mathbf{q}) \in \mathbb{R}^{6\times1}$. The objective function to be minimized J is the total tracking error evaluated by means of the Root Mean Square Error ($RMSE$) of the six legs:

$$J = RMSE(\mathbf{e}(\boldsymbol{\lambda})) = \frac{1}{6}\sum_{i=1}^{6}\sqrt{e_i^T e_i} \tag{14}$$

3 Sensitivity Analysis

The previous sections presented the robot modeling, the parametrization of the structure-control design variables and the objective function for the optimal design. It would be interesting to evaluate the effect of each variable independently on the variation of the position accuracy of the robot. Additionally, the sensitivity analysis allows to understand the effect of each design variable within the search space to meet the optimum criterion.

The sensitivity analysis aims at determining the influence of each structure-control variable of Eq. (13) on the dynamic response. Consequently, this analysis allows to indicate the degree of influence of each variable on the variation of the dynamic response, specifically on the position accuracy of the robot.

Among the various methods used to analyze the sensitivity, the variance-based sensitivity analysis decomposes the variance of the output of the model into fractions which are associated with the variation of each variable [18]. This method allows to quantify the effect of the variation of an individual variable on the dynamic response of the robot by means of a probabilistic framework based on the Monte Carlo Simulation method. Additionally, this method copes with nonlinear models, which is suitable to quantify the sensitivity of the robot.

Considering the model under the form $y = f(\boldsymbol{\lambda})$, where y is a scalar output and $\boldsymbol{\lambda} = \left[\lambda_1 \ldots \lambda_7\right]^T \in \mathbb{R}^{k \times 1}$ is the vector of the design variables. These variables are considered as independently and uniformly distributed within the unit hypercube, i.e., $\lambda_i \in [0, 1]$ for $i = 1, \ldots, 7$. $f(\boldsymbol{\lambda})$ is decomposed as:

$$y = f(\boldsymbol{\lambda}) = f_0 + \sum_{i=1}^{7} f_i(\lambda_i) + \sum_{i<j}^{7} f_{ij}(\lambda_i, \lambda_j) + \cdots + f_{12\ldots,7} \tag{15}$$

The decomposition of the variance expression is [20]:

$$V(y) = \sum_{i=1}^{7} V_i + \sum_{i<j}^{7} V_{ij} + \cdots + V_{12\ldots 7} \tag{16}$$

where $V_i = V_{\lambda_i}(E_{\boldsymbol{\lambda}_{\sim i}}(y|\lambda_i))$, $V_{ij} = V_{\lambda_{ij}}(E_{\boldsymbol{\lambda}_{\sim ij}}(y|\lambda_{ij}))$, and so on; with λ_{ij} being a generic value for factor λ_i taken from row j of λ_i. A variance based first order effect for a generic design variable λ_i is:

$$V_{\lambda_i}(E_{\boldsymbol{\lambda}_{\sim i}}(y|\lambda_i)) \tag{17}$$

where λ_i is the i-th variable and $\boldsymbol{\lambda}_{\sim i}$ denotes the matrix of all variable except λ_i. The meaning of the inner expectation operation is that the mean of y is taken over all possible values $\boldsymbol{\lambda}_{\sim i}$ while keeping λ_i fixed. The associated sensitivity measure denominated first-order sensitivity index is defined as:

$$s_i = \frac{V_{\lambda_i}(E_{\boldsymbol{\lambda}_{\sim i}}(y|\lambda_i))}{V(y)} \tag{18}$$

s_i states the effect of the variation of λ_i only, however divided by the variation in other variables. Nevertheless, the total effect-index s_{Ti} measures the contribution to the output variance of λ_i, including all the effects of its interactions with any other input variable.

$$s_{Ti} = \frac{E_{\boldsymbol{\lambda}_{\sim i}}(V_{\lambda_i}(y|\boldsymbol{\lambda}_{\sim i}))}{V(y)} = 1 - \frac{V_{\boldsymbol{\lambda}_{\sim i}}(E_{\lambda_i}(y|\boldsymbol{\lambda}_{\sim i}))}{V(y)} \tag{19}$$

s_{Ti} measures the total effect that takes into account the interactions of λ_i. Thus, considering $V_{\boldsymbol{\lambda}_{\sim i}}(E_{\lambda_i}(y|\boldsymbol{\lambda}_{\sim i}))$ the first order effect of $\boldsymbol{\lambda}_{\sim i}$, consequently, $V(y)$ minus $V_{\boldsymbol{\lambda}_{\sim i}}(E_{\lambda_i}(y|\boldsymbol{\lambda}_{\sim i}))$ expresses the contribution of all terms in the variance decomposition which contain λ_i.

The Monte Carlo Simulation combined with the Latin Hypercube sampling [21] is used to calculate the total-effect indices. The Monte Carlo Simulation demands producing a sequences of samples of $\boldsymbol{\lambda}$ contained into the unite Hypercube, these sequence of random distributed samples are applied in the expressions presented previously to compute the factors necessary to determine the sensitivity indices. The total number of model evaluation necessary to compute the total-sensitivity index is:

$$N = n_s(k+1) \tag{20}$$

where $k = 7$ for this contribution, and n_s is the number of the Monte Carlo samples [18].

4 Optimization Problem

In this contribution, the optimal design aims at selecting the optimal structure-control design variables according to dynamic and geometric constraints. The optimization problem is solved to minimize position tracking error over a required workspace trajectory. For practical purposes, a required workspace trajectory is defined as a circular path that involves the motion of the movable platform in the xyz axes. The related joint-space position reference trajectory $\mathbf{q}^d$ is obtained by means of the inverse kinematic model. The joint-space velocity reference $\dot{\mathbf{q}}^d$ is proportional to the workspace velocity as stated by [3,19]. Moreover, based on the definition of the joint-space PD controller gain of Eq. (10), an inequality constraint for the derivative gain k_d is defined to guarantee a well damped closed-loop response. The optimization problem to select the structure-control design variables $\boldsymbol{\lambda}$ of the parallel robot is given by:

$$\begin{aligned}
&\min_{\boldsymbol{\lambda}}\{J = RMSE(e(\boldsymbol{\lambda}))\}\\
&\text{subject to}\\
&r_p, r_b \in [r_{min}, r_{max}]\\
&\phi_p \in [\phi_{min_p}, \phi_{max_p}], \phi_b \in [\phi_{min_b}, \phi_{max_b}]\\
&s \in [s_{min}, s_{max}]\\
&k_p \in [k_{p_{min}}, k_{p_{max}}], k_d \in [k_{d_{min}}, k_{d_{max}}]\\
&m_i 2\xi_{min}\sqrt{k_p} < k_d < m_i 2\xi_{max}\sqrt{k_p}\\
&\mathbf{f}_c \in [\mathbf{f}_{min_c}, \mathbf{f}_{max_c}]\\
&\forall \mathbf{q} \in \mathbf{q}^d, \forall \dot{\mathbf{q}} \in \dot{\mathbf{q}}^d
\end{aligned} \tag{21}$$

Metaheuristic algorithms for optimization have been successfully applied to nonlinear and constraint problems in order to find the global minima. Thus, this optimization problem is solved by using Genetic Algorithm [22].

4.1 Genetic Algorithm

Genetic Algorithms (GAs) are heuristic search algorithms based on the mechanism of natural selection and natural genetics initially proposed by [22]. GAs are high performance and robust optimization methods to solve engineering problems.

In general, a genetic algorithm has four basic characteristics: (i) A genetic representation of solutions to the problem; (ii) A way to create an initial population of solutions; (iii) Selection of the population for next generation, an evaluation function rating solutions in terms of their fitness; (iv) Genetic operators that alter the genetic ascendants during reproduction. The flowchart of GA is shown in Fig. 4.

GAs start with an initial set of random solutions, this set of solutions are called the population. Each individual of the population, which is a chromosome, represents a potential solution to the problem. The encoding is a genetic representation of the chromosome. In the evaluation, a measure of fitness is assigned to each individual. Individuals called parents are selected, the parents contribute to the population at the next generation. Some individuals of the population suffer genetic operations to create new individuals through stochastic transformations. There are two types of genetic operations: crossover and mutation. Crossover creates new individuals by combination of the parts of two parents; and mutation creates new individuals by randomly altering chromosome characteristics to guarantee genetic diversity in the population. New individuals of the population are called offspring. A new population is formed by selecting the more fit individuals from the present population and the offspring population. After successive iterations called generations, the algorithm converges to the best individual, which hopefully represents an optimal solution to the problem.

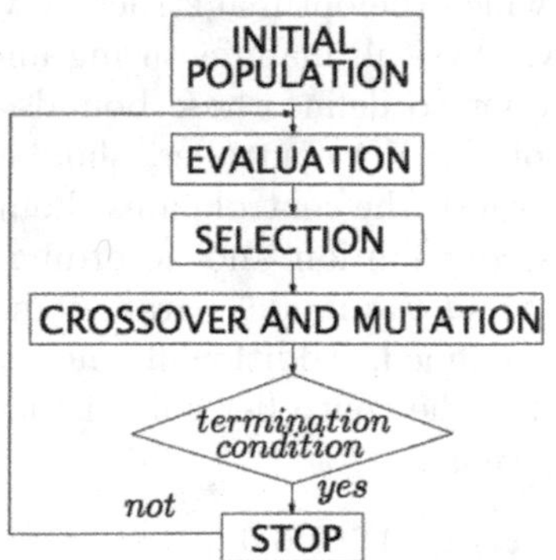

Fig. 4. Flowchart of Genetic Algorithm.

5 Simulation Results

This section presents the results of the sensitivity analysis and the optimal design. For the proposed optimization problem, the sensitivity analysis helps to evaluate the effect of the design variables within the search space on the objective function.

Table 1. Model parameters.

ρ_p	ρ_l	a_u	a_d	h_p	c_u	c_p	c_s
7874 kg/m^3	2697 kg/m^3	0.03 m	0.03 m	0.01 m	1×10^{-4} Ns/m	0.001 Ns/m	2×10^{-4} Ns/m

As presented in Sect. 2.2 the model parameters of Table 1 should be imposed to define completely the parameters of dynamic equation of the robot as function of the structure-control design variables $\boldsymbol{\lambda}$. The simulations were implemented using MATLAB.

5.1 Sensitivity Analysis

The sensitivity analysis is performed based on the model of the controlled Stewart-Gough robot presented in Sect. 2. This analysis is performed over an imposed circular workspace trajectory (see Fig. 6), nevertheless any other trajectory could be considered. The total effect-indices of the structure-control design variables $\boldsymbol{\lambda}$ of Eq. (13) are computed by using the variance-based sensitivity analysis presented in Sect. 3.

The sensitivity analysis is performed within a specific bound of the structure-control design variables $\boldsymbol{\lambda}$. The bounds of the structure variables are defined in order to cover a range in which the optimal structure variables are supposed to be contained, additionally, physical, manufacturing and assembling limitations should be taken into account to define these bounds. In the other hand, the control design variables, of the PD controller, should be positive in order to stabilize the robot [12]; however, the control gains should be bounded to restrict the control effort. In this contribution and according to the modeling of the robot, presented in Sect. 2 and the model parameters of Table 1, the limits of the structure variables are defined. Additionally, according to the definition of Eq. (10), the upper bound of the controller gains is defined for $\omega_n = 100$ rad/s, $\xi = 1.2$, and $m_i = 15.2$ kg; thus:

$$\begin{aligned} & r_p, r_b \in [0.2\text{m}, 0.6\text{m}] \\ & \phi_p \in [60^o, 120^o], \phi_b \in [0^o, 60^o] \\ & s \in [0.3, 0.7] \\ & k_p \in [0, 152000], k_d \in [0, 3648] \end{aligned} \tag{22}$$

In order to perform the sensitivity analysis, each design variable of $\boldsymbol{\lambda}$ is modeled as a normal distributed random variable. The mean, $\overline{\lambda}_i$, and standard deviation, σ_i, of each normal random variable presented in Table 2, $\mathcal{N}(\overline{\lambda}_i, \sigma_i^2)$, are selected in order to generate random values of the variables contained into the limits defined by the bounds of Eq. (22).

Table 2. Parameters of normal random variables.

	r_p[m]	$\phi_p[^o]$	r_b[m]	$\phi_b[^o]$	s	k_p	k_d
$\overline{\lambda}_i$	0.4	90	0.4	30	0.5	75000	1824
σ_i	0.0667	10	0.0667	10	0.0667	50666	1216

The number of computation of the Monte Carlo samples required to perform the sensitivity analysis was fixed at $n_s = 100$ to ensure an accurate solution as stated by the convergence analysis carried out previously, in which was verified that increasing the number of samples n_s, it was not obtained a numerical improvement in the solution [21]. Considering $k = 7$ design variables, the total number of model evaluations is $N = 800$ according to Eq. (20).

The total effect-indices of the design variables $\boldsymbol{\lambda}$ for the circular trajectory are shown in Fig. 5(a). As seen, the position accuracy is more sensitive to the proportional gain k_p of PD position controller than the other variables. This is expected since the position error is inverse proportional to this gain. However, among the structure variables the radius of the movable platform r_p exhibits a significant sensitivity (see Fig. 5(b)).

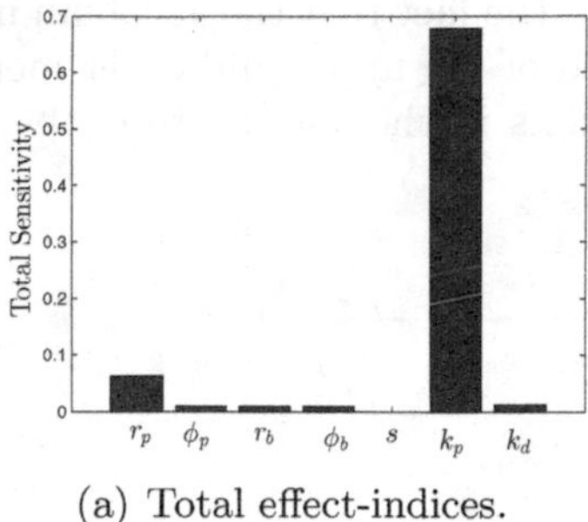

(a) Total effect-indices.

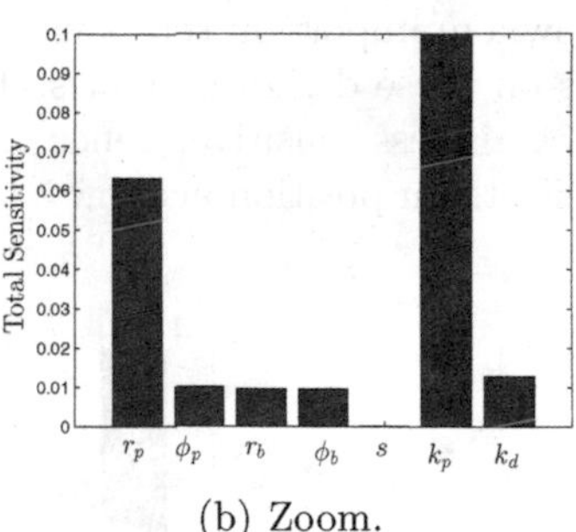

(b) Zoom.

Fig. 5. Total effect-indices over the reference trajectory.

In order to determine the sensitivity of the structure variables for the same circular trajectory of the previous analysis, the control variables are considered as constant. The constant values of control are their means $\overline{k}_p$ and $\overline{k}_d$ of Table 2. Figure 6 shows the circular reference trajectory and the trajectories obtained by using the structure variables of the Monte Carlo samples. This indicates that, even with constant controller gains, the position accuracy is very sensitive to the structural design variables.

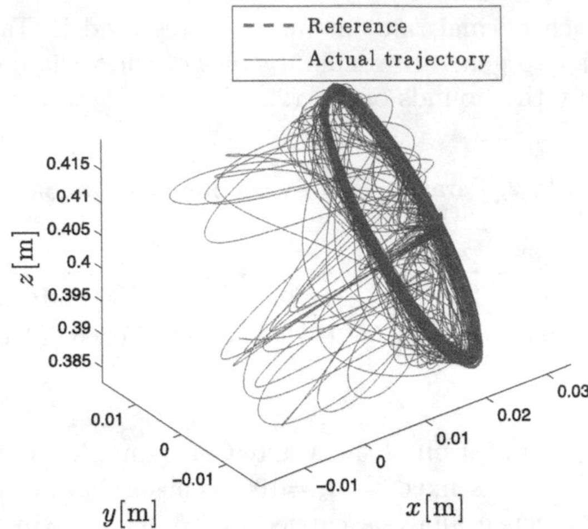

Fig. 6. Sensitivity analysis with k_p and k_d fixed to their mean value: reference and actual trajectory.

The total effect-indices of structure design variables for the circular trajectory are shown in Fig. 7 while k_p and k_d are fixed. This demonstrates that the position accuracy is highly sensitive to the radius of the movable platform, r_p, as seen in the sensitivity analysis of Fig. 5 (b), because r_p directly increases the positioning error of movable platform more than the other structure design variables. The radius of the fixed base, the spacing angles of the movable platform and fixed base show a unimportant sensitivity in view of the fact that the Jacobian matrix depends on these design variables. The length of the upper and lower members of the legs is less sensitive, hence the variations in the length of member have minor effects on position accuracy.

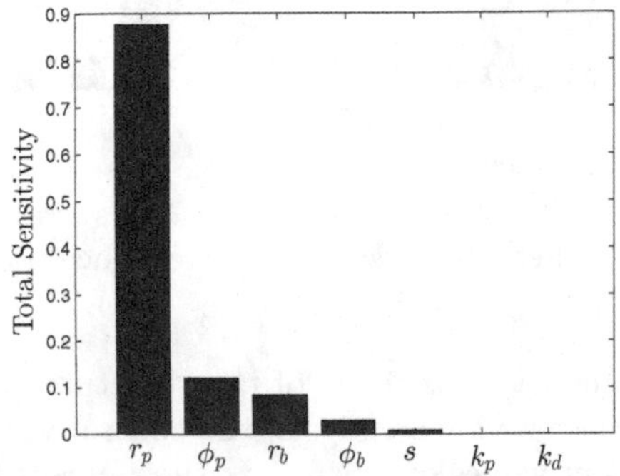

Fig. 7. Total effect-indices with fixed controller gains.

5.2 Optimization

The structure-control design variables of vector $\boldsymbol{\lambda}$ of Eq. (13) are optimized for obtaining the minimum RMS error over an imposed workspace trajectory (see Fig. 6), i.e. to maximize the position accuracy. The constraints of the design variables are limited by manufacturing tolerances and the maximum power of the actuators. The optimization problem is formulated as follows:

$$\begin{aligned}
&\min_{\boldsymbol{\lambda}}\{J = RMSE(e(\boldsymbol{\lambda}))\}\\
&\text{subject to}\\
&r_p, r_b \in [0.2\text{m}, 0.6\text{m}]\\
&\phi_p \in [60^o, 120^o], \phi_b \in [0^o, 60^o]\\
&s \in [0.3, 0.7]\\
&k_p \in [0, 152000], k_d \in [0, 3648]\\
&24.32\sqrt{k_p} < k_d < 36.48\sqrt{k_p}\\
&\mathbf{f}_c \in [-50\text{N}, 50\text{N}]\\
&\forall \mathbf{q} \in \mathbf{q}^d, \forall \dot{\mathbf{q}} \in \dot{\mathbf{q}}^d
\end{aligned} \tag{23}$$

The constraints of $\boldsymbol{\lambda}$ for this optimization problem are equal to the bounds defined in Eq. (22). Additionally, $\xi_{min} = 0.8$ and $\xi_{max} = 1.2$ is defined for the inequality constraint of k_d. The parameters of GA optimization algorithm were derived from previous contributions [1]. These parameters are presented in Table 3.

Table 3. Parameters used in the GA algorithms.

Parameter	GA
Max. Generation number	100
Population size	70
Crossover probability	0.5
Mutation rate	0.08

In order to evaluate the solution of the optimization problem of Eq. (23), the evolution of the objective function along the generations using the GA optimization algorithm is presented in Fig. 8. When GA optimization method is used, the objective function converges after 55 generations and its value is $RMSE(\mathbf{e}) = 0.2436 \times 10^{-3}$ m, the objective function was computed 3920 times among which 394 exhibited a singularity condition.

In the results of Table 4, one can observe that the gains of the PD controller, specially k_p does not reach the upper bound value due to the constraint imposed to the control action $\mathbf{f}_c$. The structure design variables over the circular reference

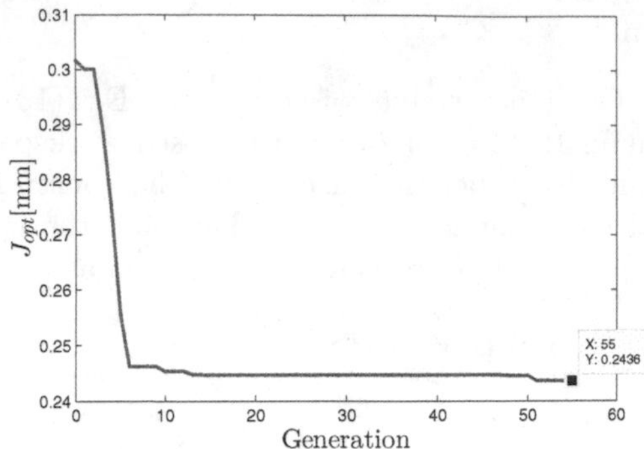

Fig. 8. Objective function evolution using GA.

Table 4. Optimization results: optimal structure-control design variables

	r_p[m]	ϕ_p[^{o}]	r_b[m]	ϕ_b[^{o}]	s	k_p	k_d
GA	0.2188	90.1067	0.2642	1.4077	0.6560	99864	2800.3

trajectory were selected by the optimization algorithm to avoid configurations closed to singularity zones taking into account the closed-loop dynamics of the robot. Consequently, the optimized structure variables minimize the degradation of position accuracy during the motion.

Based on the sensitivity analysis, one can consider the most sensitive design variables in the vector $\boldsymbol{\Lambda} = \left[r_p\ k_p\ k_d\right]^T$ with the purpose of performing the optimization only with the most sensitive design variables. It is considered that the other design variables assume their mean value shown in Table 2, thus, $\phi_p = 90^o$, $r_b = 0.4$m, $\phi_b = 30^o$ and $s = 0.5$. In this way, the optimization problem is simplified to:

$$\begin{aligned}
&\min_{\boldsymbol{\Lambda}}\{J = RMSE(e(\boldsymbol{\Lambda}))\}\\
&\text{subject to}\\
&r_p \in [0.2\text{m}, 0.6\text{m}]\\
&k_p \in [0, 152000], k_d \in [0, 3648]\\
&24.32\sqrt{k_p} < k_d < 36.48\sqrt{k_p}\\
&\mathbf{f}_c \in [-50\text{N}, 50\text{N}]\\
&\forall \mathbf{q} \in \mathbf{q}^d, \forall \dot{\mathbf{q}} \in \dot{\mathbf{q}}^d
\end{aligned} \tag{24}$$

The optimization problem of Eq. (24) is solved by using GA. The objective function converges after 51 generations and its value is $RMSE(\mathbf{e}) = 0.3010 \times 10^{-3}$m, the objective function was computed 1560 times among which 40 exhibited a singularity condition. The resulting optimal design variables are given by $\boldsymbol{\Lambda}_{opt} = \left[0.2710m\ 98388\ 2888\right]^T$. It is observed that the suboptimal solution, $\boldsymbol{\Lambda}_{opt}$, is acceptable when compared with the optimal solution of Table 4 that consider all the design parameters; although, the suboptimal $RMSE(\boldsymbol{\Lambda}_{opt})$ is greater than $RMSE(\boldsymbol{\lambda}_{opt})$, the objective function was computed fewer times during the optimization solution, i.e., the solution of the optimization problem with the most sensitive design variables reduces the computation intensity.

Additional simulations are considered to evaluate the performance of the robot with the initial $\boldsymbol{\lambda}_0$, the optimized $\boldsymbol{\lambda}_{opt}$ structure-control variables that were obtained with GA (see Table 4), and the suboptimal solution $\boldsymbol{\Lambda}_{opt}$. In these simulations, the circular reference trajectory of Fig. 6 is considered to assess: the workspace error, joint-space error and the force in the legs.

As expected, Fig. 9 shows that the joint-space error is minimized by using the optimal structure-control design variables, $\boldsymbol{\lambda}_{opt}$, obtained by GA optimization when compared with the initial design variables, $\boldsymbol{\lambda}_0$. This result shows that the optimization problem of Eq. (23), is properly solved. Moreover, the optimal $\boldsymbol{\lambda}_{opt}$ and suboptimal $\boldsymbol{\Lambda}_{opt}$ designs variables exhibits a similar performance.

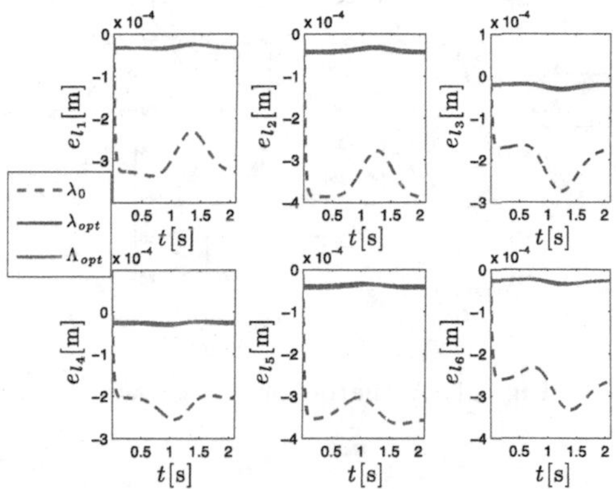

Fig. 9. Joint-space error with $\boldsymbol{\lambda}_{opt}$.

Figure 10 shows that the workspace error over the circular trajectory is minimized by using the optimal structure-control design variables $\boldsymbol{\lambda}_{opt}$. As seen in Fig. 10 the reduction in the workspace error is predominant in the orientation of the movable platform. However, the minimization of workspace error is smaller than the joint-space error (see Fig. 9), since the workspace reference trajectory is close to singular configurations.

Finally, the actuator force is also evaluated with the obtained optimal control-structure design variables, $\boldsymbol{\lambda}_{opt}$. The actuator force is the control of the robot. Figure 11 shows that the amplitude of the force of each actuator is also reduced over the motion. In addition, it is verified that the force of each actuator is contained within the constrains imposed in the optimization of problem of Eq. (23).

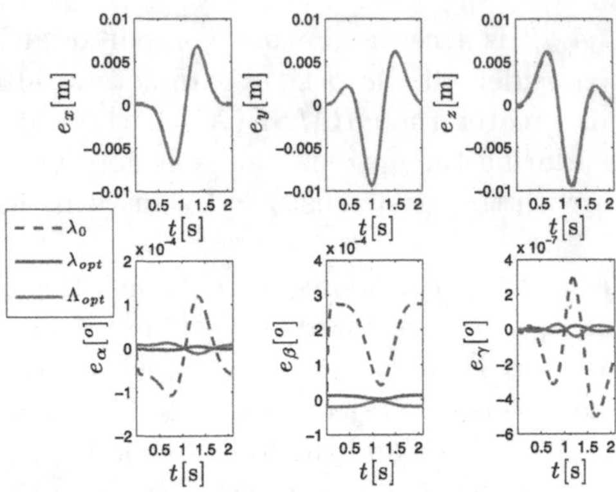

Fig. 10. Workspace error with $\boldsymbol{\lambda}_{opt}$.

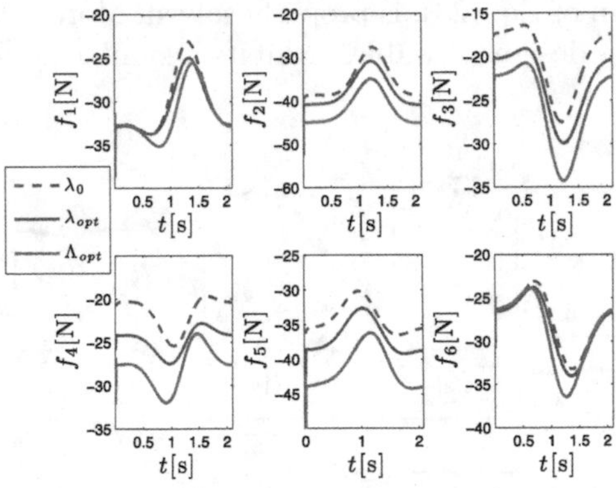

Fig. 11. Actuator force with $\boldsymbol{\lambda}_{opt}$.

6 Conclusion

This contribution presented an optimal design procedure to maximize the tracking accuracy over an imposed trajectory. In this way, the structure-control design variables of a Stewart-Gough robot are found simultaneously. By using this methodology, the optimal performance of the robot considering dynamic and kinematic properties simultaneously was achieved maximizing the positioning accuracy.

The sensitivity analysis demonstrated the influence in the variation of structure-control design variables on the position accuracy of robot. Specially, it was observed a great sensitivity of the structure variables of the mechanism in position accuracy of the robot.

Metaheuristic optimization algorithms have shown to be a straightforward optimization tool to find optimal structure-control design variables for this optimal design problem.

Further works will encompass the optimization of parallel robots considering advanced controllers such as predictive control.

Acknowledgments. The authors express their acknowledgements to the Graduate Program in Mechanical Engineering of the Federal University of Technology - Paraná funded by CAPES.

References

1. Lara-Molina, F.A., Rosario, J.M., Dumur, D.: Multi-objective optimization of Stewart-Gough manipulator using global indices. In: 2011 IEEE/ASME International Conference on Advanced Intelligent Mechatronics (AIM), pp. 79–85 (2011)
2. Liu, X.J., Jin, Z.L., Gao, F.: Optimum design of 3-DOF spherical parallel manipulators with respect to the conditioning and stiffness indices. Mech. Mach. Theory **35**(9), 1257–1267 (2000)
3. Dasgupta, B., Mruthyunjaya, T.S.: Closed-form dynamic equations of the general Stewart platform through the Newton-Euler approach. Mech. Mach. Theory **33**(7), 993–1012 (1998)
4. Khalil, W., Dombre, E.: Modeling, Identification and Control of Robots. Butterworth-Heinemann, Oxford (2004)
5. Lara-Molina, F.A., Rosário, J.M., Dumur, D.: Robust generalized predictive control of Stewart-Gough platform. In: Robotics Symposium, 2011 IEEE IX Latin American and IEEE Colombian Conference on Automatic Control and Industry Applications (LARC), pp. 1–6 (2011)
6. Lara-Molina, F.A., Rosário, J.M., Dumur, D., Wenger, P.: Robust generalized predictive control of the Orthoglide robot. Ind. Robot Int. J. **41**(3), 275–285 (2014)
7. Pietsch, I.T., Krefft, M., Becker, O.T., Bier, C.C., Hesselbach, J.: How to reach the dynamic limits of parallel robots? An autonomous control approach. IEEE Trans. Autom. Sci. Eng. **2**(4), 369–380 (2005)
8. Abdellatif, H., Heimann, B.: Advanced model-based control of a 6-DOF hexapod robot: a case study. IEEE/ASME Trans. Mechatron. **15**(2), 269–279 (2010)
9. Villarreal-Cervantes, M.G., Cruz-Villar, C.A., Alvarez-Gallegos, J.: Structure-control mechatronic design of the planar 5R 2DoF parallel robot. In: Proceedings of the 2009 IEEE International Conference on Mechatronics (2009)
10. da Silva, M.M., Gonçalves, L.A.M.: Mechatronic design concept and its application to pick-and-place robotic systems. J. Braz. Soc. Mech. Sci. Eng. **35**, 31–40 (2013)
11. Villarreal-Cervantes, M.G., Cruz-Villar, C.A., Alvarez-Gallegos, J., Portilla-Flores, E.A.: Robust structure-control design approach for mechatronic systems. IEEE/ASME Trans. Mechatron. **18**(5), 1592–1601 (2013)
12. Yu, W., Li, X., Carmona, R.: A novel PID tuning method for robot control. Ind. Robot Int. J. **40**(6), 574–582 (2013)
13. Gosselin, C., Angeles, J.: A global performance index for the kinematic optimization of robotic manipulators. J. Mech. Des. **113**(3), 220–226 (1991)
14. Melet, J.P.: Jacobian, manipulability, condition number, and accuracy of parallel robots. J. Mech. Des. **128**, 199–206 (2006)
15. Merlet, J.P.: Designing a parallel manipulator for a specific workspace. Int. J. Robot. Res. **16**(4), 545–556 (1997)
16. Zhang, D., Xi, F., Mechefske, C.M., Lang, S.Y.T.: Analysis of parallel kinematic machine with kinetostatic modelling method. Robot. Comput. Integr. Manuf. **20**(2), 151–165 (2004)

17. Nenchev, D.N., Bhattacharya, S., Uchiyama, M.: Dynamic analysis of parallel manipulators under the singularity-consistent parameterization. J. Robotica **15**(4), 375–384 (1997)
18. Saltelli, A., Ratto, M., Andres, T., Campolongo, F., Cariboni, J., Gatelli, D., Michaela, S., Tarantola, S.: Global Sensitivity Analysis: The Primer. Wiley, Chichester (2008)
19. Lara-Molina, F.A., Rosário, J.M., Dumur, D.: Architecture of predictive control for a Stewart platform manipulator. In: 2010 8th World Congress on Intelligent Control and Automation (WCICA), pp. 6584–6589 (2010)
20. Sobol', I.M.: On sensitivity estimation for nonlinear mathematical models. Matem. Mod. **2**(1), 112–118 (1990)
21. Florian, A.: An efficient sampling scheme: updated Latin hypercube sampling. Probab. Eng. Mech. **7**(2), 123–130 (1992)
22. Sivanandam, S.N., Deepa, S.N.: Genetic algorithm optimization problems. In: Sivanandam, S.N., Deepa, S.N. (eds.) Introduction to Genetic Algorithms, pp. 165–209. Springer, Heidelberg (2008)

A Genetic Algorithm Approach to the Automated System for Solving the Container Loading Problem

Rodrigo Nogueira Cardoso(✉), Marco Vinícius Muniz Ferreira, Alexandre Rodrigues de Sousa, and José Jean-Paul Zanlucchi Souza Tavares

Federal University of Uberlândia, Uberlândia, Brazil
{rodrigo.cardoso,marcomuniz,alexandre.sousa,jean.tavares}@ufu.br

Abstract. On the one hand the container loading problem has been widely studied in an effort to reduce logistical costs. On the other hand, automated planning research has as an objective assisting industrial processes by processing a system model and providing a list of actions that will enable the system to get from a given initial state to an objective. This works proposes an approach that combines CLP solving and automated planners to create a system that can execute the entire loading process. The CLP is solved by an improved genetic algorithm and its resulting packing pattern is converted to a format accepted by existing automated planners, whose output is a set of actions which can be executed to carry out the loading of the container.

Keywords: Genetic algorithm · Container loading problem · Automated system

1 Introduction

The container loading problem (CLP) as an optimization problem has been thoroughly studied in order to reduce costs and improve space utilization. Throughout the last decade a broad set of techniques have been developed to address such a complex and multivariate problem. The CLP is said to a NP complete problem, which means that an optimal solution cannot be achieved in a reasonable time, except for some particular problems. Such an optimal solution can only be estimated. For that reason, this estimation process has been addressed with several different technologies, such as, ant colony algorithms [1], bee colony algorithms [2], linear programming [3–7], tower building [8], tree search [9–12], genetic algorithms (GA) [13–18] and Branch and Bound technique [19].

The main concern of the previously mentioned techniques is almost always the packing pattern alone, leaving some of the practical aspects like packing procedure/technology aside. In previous works [20,21] the authors explored a hybrid approach between CLP techniques and Automated Planning (AP) aiming to include some of these missing practical aspects into the final CLP's solution.

F. Santos Osório and R. Sales Gonçalves (Eds.): LARS 2015/SBR 2015, CCIS 619, pp. 267–280, 2016.
DOI: 10.1007/978-3-319-47247-8_16

Toward the goal of building a fully functional and robust algorithm, a few modifications were proposed to the ongoing study in order to overcome some of the limitations that were faced during the previous developments. On both [20,21] the Linear Programming (LP) technique has been solely used to obtain the packing pattern, but as observed the LP suffered with the excessive growth on the number of variables when increasing the container size and/or the type of boxes, which led the authors to work with smaller size problems. With that on mind, within this works is proposed a new approach based on genetic algorithms to solve the CLP.

The GA technique has been used for several authors in different attempts to solve the CLP. Most of those approaches use the hybrid paradigm, combining some of the GA features along with various techniques. [13] were one of the precursive works on using GA for solving the CLP, in their development the GA is used to assist a tower building technique to place the generated towers on the container floor. [15] combined a heuristic to guide the GA searching process in a one-by-one box placement methodology. A more recent approach proposed by [16] used the global search features of GA to generate a decent individual used as input to an Ant Colony Optimization refinement algorithm. [17] have proposed a mixture of Greedy search and GA to solve the CLP by means of subdivision of the problem into three sub problems. A fewer works were actually relying on a pure GA, as shown by [18], but his results are yet to be refined.

In the next section, the CLP is briefly described, and its main features is given. The third section discusses on GA. In the fourth section, the automated system proposal is presented. The fifth section shows the results on the presented proposal.

2 Container Loading Problem

2.1 Characterization

The CLP describes the problem of packing a given set of three-dimensional rectangular boxes into a larger three-dimensional rectangular container [17] such that a given objective function is maximized or minimized.

The CLP belongs to a more general category of optimization problems called Cutting and Packing problems (C&P problems). The C&P problems consist of a combinatory optimization of small items into large objects.

The C&P problems can be defined by two sets of elements: the set of large objects, that correspond to the problem resource and the set of small items, which is the problem's demand. The C&P problems can be evaluated on one, two, three or even larger number of geometric dimensions. In order to obtain a solution one has to select a subset of small items, group them into one or more collections and assign each of the resulting collections to one of the large objects such that all small items of the subset lie entirely within the large object and the small items do not overlap. [22]

A typology for classifying the many variants that derive from the C&P concept is presented by [22]. According to the previously mentioned reference, in the present work, the authors focused on the tridimensional rectangular single large object placing problem (3D-R-SLOPP). The 3D-R-SLOPP states that a subset of a few types of boxes must be selected for packing, in order to maximize the usable volume of a single container. This problem was particularly addressed because it attends the development of a single machine able to carry out the entire loading process. In this problem, the container dimensions and the characteristics of the boxes are set before the optimization can start.

2.2 Problem Constraints

Before addressing the CLP it's necessary to evaluate the constraints which are used to guide the solution according to logistical demands. [23] was one of the first assessments that proposed a set of constraints to be used as guidelines to distinguish be-tween the different instances of CLP. More recently [24] proposed an update to [23] as to include newer logistical practices, technological particularities and those constraints that were already present in the literature but was not prescribed by the former set.

In [24] the prior restrictions were discretized into five categories. The first one relates to the container device itself. In this context, the container weight limit and weight distribution are stated. The second category discuss on the items, whether there are priority constrains, due to delivery deadlines and expiration dates, orientation constraints and pilling constraints. The third category is interested on the packing of the subset items. As a subset of items, it is possible to understand the composing parts of a machine or a set of products that have to be delivered to the same costumer or destination. The fourth category explains about the product positioning. The aim of this constraint is to facilitate the loading and unloading process of large products, multiple customers and multiple destinations. The fifth category regards on the packing pattern and considers the pattern stability when transporting the products, and pattern complexity as to the technology available for the loading and unloading processes.

In order to simplify the development of the fully automated system, this work is only focused on the stability or partial stability of the cargo.

3 Genetic Algorithm

The basic guidelines for evolutionary algorithms such as GA were presented by [25]. It's based on the natural selection (NS) mechanism proposed by Darwin in Origin of Species. In this mechanism the individuals are subjected to environmental pressure which ultimately select those who are best suited to survive. In nature every creature competes with the other species to meet their needs, through natural resources. The higher chance of survival means that an individual is more adapted to the environmental conditions and also are able to keep improving giving said conditions.

The GA methodology aims to simulate this NS in a computerized environment. In this environment the individual is defined as chromosome or a collection of genes, which is formulated according to the problem's needs, and comprehends one of the problem's solution. Each gene represents a parameter that will be adjusted in the course of the algorithm in order to improve the chance of survival. The measure of this environmental adjustment is given by a fitness function that, in turn, is evaluated through an expression whose variables are the individuals genes. The NS process is based on two natural events, which are the following: the reproductive process and spontaneous mutation. Both events have their own parallel on the simulated environment. The reproductive phase is comprised by two distinct steps: selection and crossing-over. The selection phase can be computationally implemented by means of different techniques, such as: roulette wheel, ranking selection, tournament selection, among others. The crossing-over phase can also be implement used techniques like: one point cross over, PMX, cyclic crossover, among others. At last the mutation algorithm is simulated by a random gene modification.

4 Automated Planning

Planning is the reasoning side of acting. It is an abstract, explicit deliberation process that chooses and organizes actions by anticipating their expected outcome. This deliberation aims at achieving as best as possible some pre-stated objectives. Automated planning is an area of artificial intelligence (AI) that studies this deliberation process computationally [26].

By using computers, AP aims to speed up the process of making choices, where there is no evident solution. Through tree search, AP tries to find a reasonably good route to the solution, optionally trying to minimize a given metric variable [21].

The AP consists on, given an initial state of the considered system, S_0, a final state S_f, it is possible to discover a group of action sequences $A_i = \{a_1, ..., a_n\}, n \in N$, where a_i is a subset of actions, such that

$$S_0 \xrightarrow{a_1} S_1 \xrightarrow{a_2} ... \xrightarrow{a_n} S_n \text{ [27].} \quad (1)$$

According to [26] the variety of applications for automated planning is becoming quite significant. For example, the web repository of PLANET, the European network on automated planning, refers to about 20 applications in aeronautics and space, agricultural, industrial, commercial and military domains, and this list is doubtless incomplete.

As previously said on the introduction section, this work aims to use automated planners, in particularly, domain independent planners. For using such planners, there is an entire paradigm for expressing the problem domain, the initial and final states, which is accomplished by using PDDL language.

The basic idea of the PDDL representation is that we have a domain and a problem. The domain contains the description of the manipulated objects, their features (computed as variables) and all the possible actions. The actions, which are parameterized according to the objects they need, are composed of preconditions for the action to take place and post conditions (results). The problem, on the other hand, specifies the initial and final states. It addresses all the variables and relations that are valid on the beginning of the planning procedure and on the desired state.

5 System's Proposal

This work's proposal is an attempt of developing an automated system to carry out the container loading process, by means of the integration between CLP solving techniques and automated planning. It's worth mentioning that the current approach is an update on previous attempts. Figure 1 shows the proposed architecture.

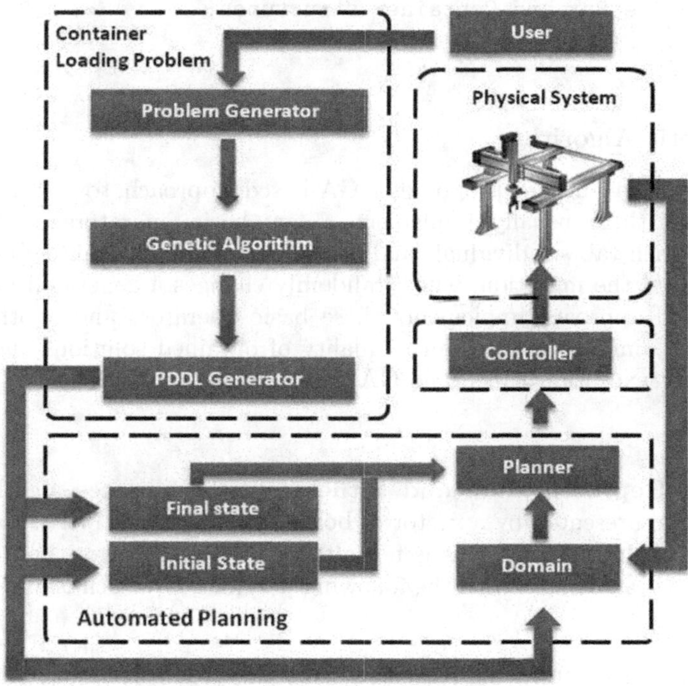

Fig. 1. Automated system's architecture.

The first part, represented by the container loading problem, is responsible for acquiring the information on the container and boxes, solving the stated problem by means of the GA and finally translating the solution into the PDDL language.

The second part takes the solution obtained with the GA and generates the steps to be sent to the controller which actuates over the physical system. In this work, particularly, the authors are focused on the systems intelligence rather than the controller and the physical plant.

5.1 Problem Generator

The problem generator module is a function that interacts with the user as to obtain all the information to feed the CLP. The steps carried out by this module is shown below.

```
1 - Obtain information about the containers dimensions and save
into a structure called Container.
2 - Obtain the number of box types and verify the numbers validity.
3 - Obtain information about the boxes dimensions and verify their
integrity.
4 - Save the box dimensions into a structure called Box.
5 - Print the generated problem.
6 - Return the Box and Container structures.
```

5.2 Genetic Algorithm

In this work the authors proposed a GA-based approach to solve the CLP. Genetic algorithms usually implement a few basic operators, such as the crossover, which takes individuals and generate descendants mixing their characteristics, and the mutation, which randomly changes a generated individual. The proposed approach implements these basic operators and another set of custom ones, aiming to improve the quality of obtained solutions and shorten the necessary execution time. The GA structure is described in the following sections.

Individual Representation and Fitness. In the proposed algorithm, the individual is represented by a vector of boxes. This vector is ordered according to the type of the box, and this is not altered within the algorithm. Each box in the vector is associated with the following parameters, which is also shown in Fig. 2:

- Position: its position in the XY plane from which it will be allocated at the lowest possible position regarding the Z axis.
- Rotation: a number from 1 to 6 representing how the box is rotated when being allocated.
- Index: the boxes are organized as a queue representing the order in which they are allocated in the container; the index parameter represents a boxs position in the queue.
- Type: a parameter that indicates the boxs type - at the beginning of the program the user must inform a set of box types describing their dimensions.

For an individual to be evaluated the first thing that needs to be done is the allocation process. First a tridimensional boolean matrix, called grid, is created matching the size of the container, and each element is initialized as false.

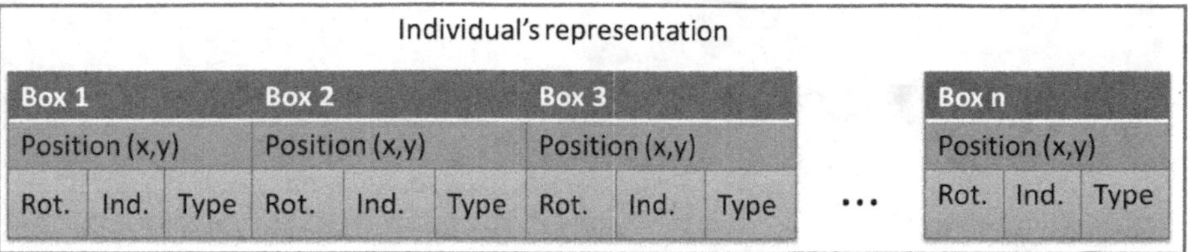

Fig. 2. Representation of a set of individuals.

Then, in the order specified by the index parameter, each box is allocated to the lowest Z position that fits two criteria: the submatrix of the grid representing all the places the box will occupy should have all elements set to false; and the submatrix representing the plane directly below the box should have at least a predetermined percentage of its elements set to true. Once a suitable position has been found, the calculated Z position for the box is stored and the respective submatrix of the grid is set to true. That way, if an element is set to false it represents empty spaces if not it represents occupied space. If no suitable position is found for a certain box, it is considered that the box could not be allocated and thus is out of the container.

Once all the boxes are allocated calculating the fitness consists on counting the number of empty spaces on the container and using that as the fitness value. There-fore, when the container is completely filled the fitness value is 0 and if it does reach that value the algorithm will terminate immediately as an optimal solution has been found.

Initial Population: The initial population is mostly generated by setting the boxes to random positions and rotations, and then randomizing the order of allocation. There are four individuals that are generated using a seeding technique. The seeds are generated by filling layers sequentially with boxes either in the order they were given to the algorithm, as to allow users to choose the priority in which they will be allocated in those individuals, or sorted by volume with three seeded individuals filling layers on the XY plane and the other on the XZ plane. This filling method is very simplistic, simply allocating boxes one after another, respecting the geometric constraints, until filling the layer. This process is repeated until no more boxes can be allocated without exceeding the containers dimensions, with no special treatment for rows and layers with mismatched boxes.

Parents Selection: Possible parents for the crossover operator are selected through a tournament-based approach. A set of individuals are randomly sampled from the population and the one with the lowest fitness value is chosen. This ensures that more suitable individuals have a greater probability of propagating their genes, while preserving variability.

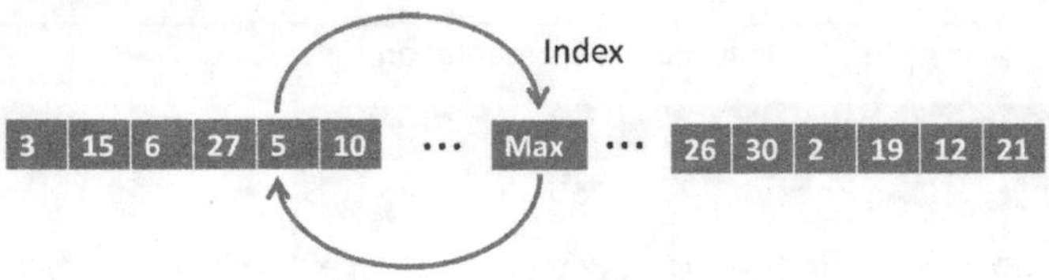

Fig. 3. Push back operator.

Crossover Operator: The crossover operator starts by generating a random crossover point based on the index parameter. The first child is generated by replicating all the boxes from the first parent whose index attribute is less or equal than the generated point, and the other boxes are taken from the second parent, and their indexes are recalculated so that these boxes are allocated after the ones from the first parent. The second child is generated analogously by reversing the rule for the parents genes.

Mutation Operator: The mutation operator in the proposed algorithm is biased towards moving boxes to XY positions related to vertical grid columns with more empty spaces. This is done by calculating a bidimentional matrix in which each elements position represents a XY position on the container, and their value is the number of vacant spaces in the grid column. The probability that a certain XY point will be selected is proportional to its respective value in the aforementioned matrix. A randomly chosen box is then set to the selected XY position.

"Push Back" Operator: This is a simple operator which selects a random box and swaps its index attribute with the box which has the highest index value in the individual.

"Improve" Operator: The improve operator chooses a random number of boxes and try to improve their positioning. For each of these boxes the algorithm evaluates if moving it one space along the X and Y axis or changing the rotation would improve the fitness of the individual. If it does, the algorithm commits the change, if not, the possible changes are discarded.

"Compact" Operator: This operator evaluates the individual after the occupation grid is generated and moves all boxes towards the origin of the X and Y axis until any geometric constraints are violated. After all boxes are evaluated, this process is repeated until no boxes can be moved further.

"Immigration/Emigration" Operator: This operator aims to improve variability whenever the algorithm detects the population isnt evolving after a predetermined number of generations. Whenever this condition is satisfied a percentage of the population is discarded. The individuals to be removed are selected

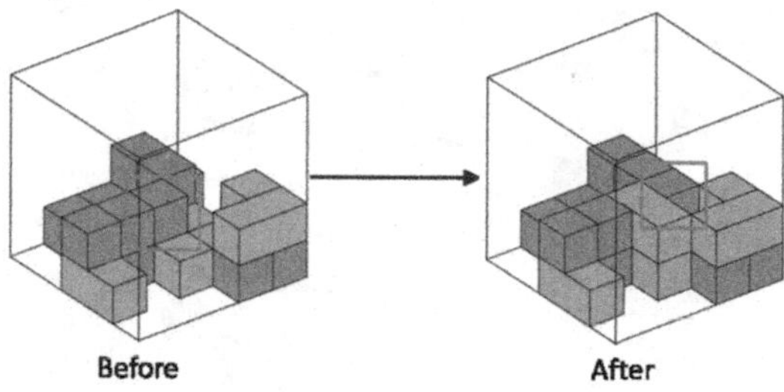

Fig. 4. Improve operator.

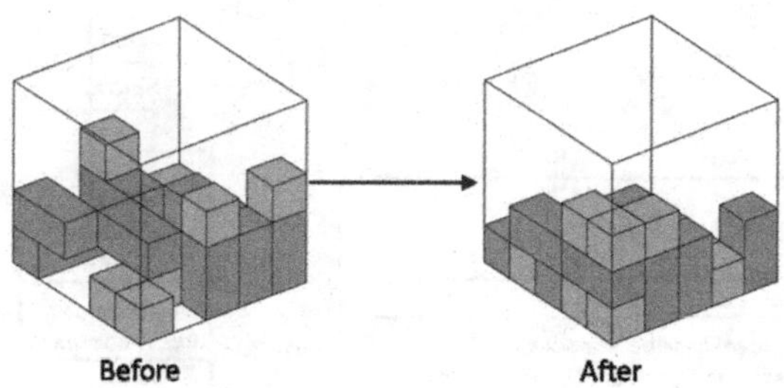

Fig. 5. Compact operator.

according to their fitness values, preserving the fitter part of the population. New elements are then randomly generated to replace those that were discarded and the algorithm resumes its normal operation.

5.3 PDDL Generator

After solving the CLP, the information on the problem's solution has to be passed on to the automated planner. The automated planner in turn is responsible for generating the list of actions, the so called plan, to achieve the previously obtained packing pattern. To this end the PDDL generator creates both problem and domain files to be run by the automated planners. The AP model is shown below by means of the UML standard.

Figure 6 represents the use case diagram for the system.

In order to achieve a reasonable performance with the automated planners, the resultant PDDL model had to be a very simplistic approach, in which there are only two possible actions: the *moveFloor* action that picks an unloaded product and places it on the container floor, and the stack action that picks an unloaded product and places it on top of a box or a group of boxes.

The class diagram provides a more detailed description on the PDDL model, as shown in Fig. 7.

For this problem, we have two classes: the products and places. The product represents the boxes to be packed in the container. It has a *pFree* boolean attribute which tells if a specific product is at the loading dock. The actions

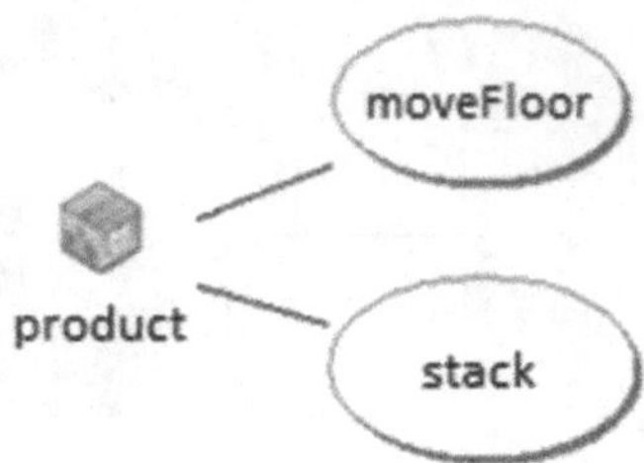

Fig. 6. Use case diagram.

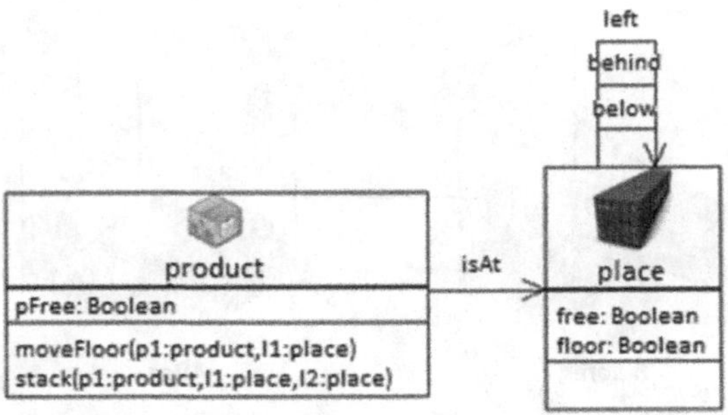

Fig. 7. Class diagram.

moveFloor and stack were already addressed previously. Connecting both product and place classes there is a relation called *isAt*, which can be null if the product is at the loading dock, or it can tell in which place a certain box is packed. The place class has two boolean at-tributes: free and floor. The free attribute tells if a certain place has been occupied or not, and the floor attribute tells if a certain place is located on the floor. The relations left, behind and below are responsible for organizing a set of places into a 3d manner.

6 Testing Methodology

In order to assess the algorithms ability to solve different CLP scenarios several tests were performed. For those tests the following values were assigned to different parameters of the algorithm:

- Population size: 30;
- Number of generations: 50;
- Crossover rate: 80 %;
- Mutation rate: 10 %;
- Improve rate: 5 %;
- Push back rate: 10 %;
- Compact rate: 10 %;
- Number of individuals in the tournament-based selection: 3;
- Number of generations without improvements for immigration/emigration: 5.

Table 1. Configurations used for testing.

Problem	Container			Box type 1				Box type 2				Box type 3			
	length	width	height	quantity	length	width	height	quantity	length	width	height	quantity	length	width	height
1	35	10	10	100	3	3	3	200	3	2	2	-	-	-	-
2	50	15	15	300	4	4	1	300	3	3	3	-	-	-	-
3	30	15	20	97	3	3	3	167	2	2	2	67	1	2	3
4	25	25	15	200	4	4	4	300	4	2	2	200	3	3	1
5	10	10	10	50	3	3	3	50	2	2	2	200	2	1	1
6	100	40	30	300	8	4	4	200	10	2	3	200	8	8	6

Keeping these parameters constant, ten different configuration of boxes were defined as described in Table 1, with up to three different types of boxes each. The algorithm was then executed five times for each of these setups and the results were used for statistical analysis.

In order to evaluate the algorithms performance under adverse conditions, one extra setup (problem 7) was proposed containing 10 different boxs types.

All test scenarios were devised so that the combined volume of available boxes would exceed the containers own volume and the boxes geometry would not allow an optimum scenario to be achieved by the seeding in the initial population.

Both the occupancy rate of the best element and the algorithms execution time were measured in each run. Execution time is heavily dependent of the computer setup in which it was run, and the machine used for testing has the following relevant specifications:

- Processor: Intel Core i7 4790 K operating at 4 GHz, raising up to 4.4 GHz on demand;
- Memory: 16 GB of DDR3 RAM operating at 1866 MHz;
- Operating system: Windows 10 Professional 64-bits
- Software: MATLAB 2012a

During all tests no heavy background tasks were allowed to run so as to not impair the processing power available to the algorithm.

7 Results

The results for the previously stated problems are shown in Table 2. For each problem the statistical analysis consists of finding the mean and the confidence interval of the occupied volume and execution time. It was adopted 0.05 as a significance level and the confidence interval was calculated by Students t-distribution.

Table 2. Results for the configurations.

Problem	Occupied volume (%)	Execution time (s)
1	79.39 ± 0.63	84.01 ± 10.32
2	90.29 ± 0.84	207.75 ± 15.38
3	93.39 ± 0.51	102.06 ± 11.55
4	93.20 ± 0.17	269.60 ± 24.89
5	94.44 ± 0.92	82.94 ± 6.79
6	92.23 ± 0.34	399.15 ± 65.22
7	75.43 ± 0.06	997.15 ± 122.75

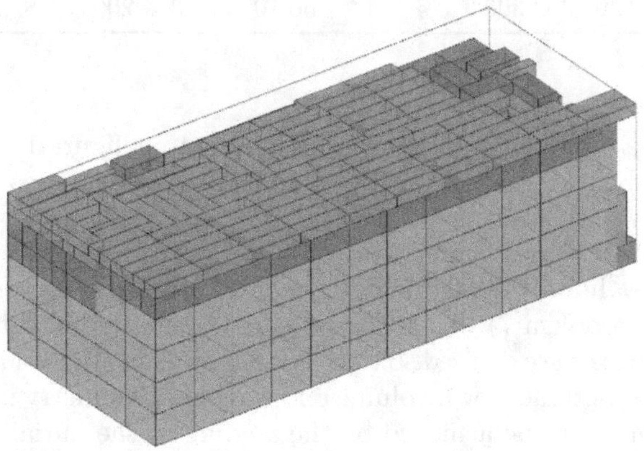

Fig. 8. Problem 6's result.

According to the results obtained its was observed that the algorithm manages to achieve a decent volume occupation for most of the problems analyzed. When analysing problem 1 no obvious reasons were found as to explain the obtained occupation. The algorithm was primarily intended to solve problems with only a few different types of boxes which was successfully accomplished regarding the results. The growth on execution time is directly related with the problems dimensions, which can still be increased once the maximum execution time was almost seven minutes (except for problem 7). Concerning to problem 7, it was observed that the algorithm was not robust enough to deal with so many different types of boxes as it wasnt capable of correctly assigning the box in order to obtain a compact and stable pattern. Its the authors opinion that a problem with such quantity of box types requires a greater population, so that diversity related to different box arrangements can be achieved.

8 Conclusions

This works is an attempt to improve the previous approaches on developing a fully automated system to carry out the entire container loading process. Despite the fact that, in this work, the authors focused only on the CLP part of the proposal, there was already a considerable advance on the capabilities to solve the CLP and to move forward to a more realistic approach. The results obtained with the GA revealed a reasonable performance when compared to the linear programming technique that was been used. In terms of dimensionality the GA approach allowed the authors to work with greater dimensions, meaning greater packing possibilities and the results were still obtained in a matter of minutes. It is worth noting that the entire work was developed using MATLAB, which is not a development environment that can be optimized for the specific problem that was hereby addressed.

While implementing the genetic algorithm it was noticed that a pure implementation yielded poor results. The addition of the seeding process and the additional operators greatly improved the obtained results, by applying domain-specific knowledge to the algorithm.

Still, its necessary to invest some time in optimizing the algorithm so that even larger problems can be addressed. After the tests were carried out the authors were able to identify certain difficulties towards solving the problem by means of the GA. It still lacks some guidance in order to correctly fill some void spaces that were generated in the course of the algorithms execution.

References

1. Yap, C.N., Lee, L.S., Majid, Z.A., Seow, H.V.: Ant colony optimization for container loading problem. J. Math. Stat. **8**(2), 169–175 (2012)
2. Dereli, T., Das, G.S.: A hybrid bee(s) algorithm for solving container loading problems. Appl. Soft Comput. **11**(2011), 2854–2862 (2010)
3. Carvalho, J.M.V.: LP models for bin packing and cutting stock problems. Eur. J. Oper. Res. **141**(2002), 253–273 (2002)
4. Chien, C.F., Lee, C.Y., Huang, Y.C., Wu, W.T.: An efficient computational procedure for determining the container-loading pattern. Comput. Ind. Eng. **56**, 965–978 (2008)
5. Junqueira, L., Morabito, R., Yamashita, D.S.: Three-dimensional container loading models with cargo stability and load bearing constraints. Comput. Oper. Res. **39**(2012), 74–85 (2010)
6. Junqueira, L., Morabito, R., Yamashita, D.S.: MIP-based approaches for the container loading problem with multi-drop constraints. Ann. Oper. Res. **199**(1), 51–75 (2011)
7. Che, C.H., Huang, W., Lim, A., Zhu, W.: The multiple container loading cost minimization problem. Eur. J. Oper. Res. **214**, 501–511 (2011)
8. Bischoff, E.E., Janetz, F., Ratcliff, M.S.W.: Loading pallets with non-identical items. Eur. J. Oper. Res. **84**, 681–692 (1995)
9. Wang, Z., Li, K.W., Levy, J.K.: A heuristic for the container loading problem: a tertiary-tree-based dynamic space decomposition approach. Eur. J. Oper. Res. **191**(2008), 86–99 (2007)

10. Ren, J., Tian, Y., Sawaragi, T.: A tree search method for the container loading problem with shipment priority. Eur. J. Oper. Res. **214**(2011), 526–535 (2011)
11. Lim, A., Ma, H., Xu, J., Zhang, X.: An iterated construction approach with dynamic prioritization for solving the container loading problems. Expert Syst. Appl. **39**(2012), 4292–4305 (2012)
12. Zhang, D., Peng, Y., Leung, S.C.H.: A heuristic block-loading algorithm based on multi-layer search for the container loading problem. Comput. Oper. Res. **39**(10), 2267–2276 (2011)
13. Gehring, H., Bortfeldt, A.: A genetic algorithm for solving the container loading problem. Int. Trans. Oper. Res. **4**(5/6), 401–418 (1997)
14. Soak, S.M., Lee, S.W.: A memetic algorithm for the quadratic multiple container packing problem. Appl. Intell. **36**(1), 119–135 (2010)
15. Yeung, L.H.W., Tang, W.K.S.: A hybrid genetic approach for container loading in logistics industry. IEEE Trans. Ind. Eletronics. **52**(2), 617–627 (2005)
16. Zhang, D., Du, L.: Hybrid ant colony optimization based on genetic algorithm for container loading problem. In: IEEE International Conference of Soft Computing and Pattern Recognition, Dalian, China (2011)
17. Remi-Omosowon, A., Cant, R., Langensiepen, C.: Hybridization and the collaborative combination of algorithms. In: IEEE 16th International Conference on Computer Modelling and Simulation. IEEE, Cambridge (2014)
18. Erdem, H. A.: Solving container loading problem with genetic algorithm. In: IEEE 15th International Symposium on Computational Intelligence and Informatics. IEEE, Budapest (2014)
19. Bortfeldt, A., Mack, D.: A heuristic for the three-dimensional strip packing problem. Eur. J. Oper. Res. **183**(3), 1267–1279 (2006)
20. Cardoso, R.N., Pereira B.L., Fonseca, J.P.S., Ferreira, M.V.M, Tavares, J.J.P.Z.S.: Automated planning integrated with linear programming applied in the container loading problem. In: IFAC International Conference on Management and Control of Production and Logistics. Fortaleza, Brazil (2013)
21. Cardoso, R.N., Ferreira, M.V.M, Souza, A.R., Tavares, J.J.P.Z.S.: Automated system for the container loading problem integrating linear programming and automated planning. In: 12th Latin America Robotics Symposium. Uberlndia, Brazil (2015)
22. Wäscher, G., Haußner, H., Schumann, H.: An improved typology of cutting and packing problems. Eur. J. Oper. Res. **183**(3), 1109–1130 (2006)
23. Dyckhoff, H.: A typology of cutting and packing problems. Eur. J. Oper. Res. **44**(2), 145–159 (1990)
24. Bortfeldt, A., Wäscher, G.: Constraints in container loading a state-of-the-art review. Eur. J. Oper. Res. **229**(1), 1–20 (2012)
25. Holland, J.H.: Adaptation in Natural and Artificial Systems. MIT Press, Cambridge (1975)
26. Ghallab, M., Nau, D., Traverso, P.: Automated Planning: Theory and Practice. Morgan Kaufmann Publishers, Burlington (2004)
27. Fonseca, J.P.S., Cardoso, R.N., Guimares, W.H.P., Ribeiro, K.S., Sousa, A.R., Tavares, J., Carvalho, J.C.M.: Automated planning and real systems based on plc: a practical application in a didactic bench of manufacturing automation. In: Proceedings of the Tampra Workshop at 22nd International Conference on Automated Planning and Scheduling (ICAPS), Atibaia, Brazil, pp. 37–44 (2012)

Trajectory Planning for UGV Using Clothoids

Lucas P.N. Matias[1(✉)], Tiago C. Santos[1], Denis F. Wolf[1], and Jefferson R. Souza[2]

[1] Institute of Mathematics and Computer Science, University of São Paulo, São Paulo, Brazil
lucas.matias@usp.br, {tiagocs,denis}@icmc.usp.br
[2] Faculty of Computing, Federal University of Uberlândia, Uberlândia, Brazil
jrsouza@ufu.br

Abstract. Path planning and autonomous navigation are the important challenges in mobile robotics. These are difficult tasks because the robot has to accurately and safely perform autonomous maneuverings. This work presents a methodology to plan the trajectory of a robot in dynamic and complex environments. Also, the changing lanes of one simulated car, which it traverse autonomously. A planner based in the AD* algorithm is used to plan a less costly trajectory to the destination for the task of automatic parking. For the changing lanes, we use the clothoid creation method, which is useful for avoid a vehicle in front of it. The methodology enables the robot to reach the goal, which is applied to determine the speed and steering of the robot. The results show that the methodology can create smooth clothoid trajectories to the vehicle follow.

Keywords: Path planning · UGV · clothoids

1 Introduction

Autonomous Intelligent Vehicles (Unmanned Ground Vehicle - UGV) are growing with several auto companies (Mercedes, Audi, Ford, Toyota) aiming the reduction of the number of accidents on streets and highways, reduction in the spending on accidents, increased efficiency in traffic of big capital and decrease in fuel consumption. Furthermore, allow mobility for the disabled and elderly.

Changing lanes is fundamental for UGV on streets, roads and highways all around the world. An interesting and relevant aspect of the lane change is that it can allow a vehicle to move from an unsafe lane for a safe lane while the vehicle is traveling along the path [5]. Consequently, the lane change being carried out in real time with UGV is an important issue for the researchers today.

In this paper we developed a methodology based on trajectory planning for UGV using Clothoids, AMCL (Adaptive Monte Carlo Localization) and Anytime Dynamic A* (AD*). The robots (Pioneer 3-AT - Fig. 1a and CaRINA 2 - Fig. 1b) are able to plan and follow a trajectory, avoiding obstacles and updating the trajectory when necessary. One of the problems found in the path planning is the reaction speed, because the planner should behave appropriately when

F. Santos Osório and R. Sales Gonçalves (Eds.): LARS 2015/SBR 2015, CCIS 619, pp. 281–298, 2016.
DOI: 10.1007/978-3-319-47247-8_17

obstacles appear blocking the robot's course. Furthermore, outdoor environments are dynamic, and can change quickly. The planner should be prepared to find a new path to achieve the goal and to avoid possible obstacles in the robots way.

(a) Pioneer 3-AT (b) CaRINA2

Fig. 1. The test platforms used in this work.

Our goal is the environment understanding, while keeping the safe robot. We employ a methodology to accomplish automatic path planning, reaching a goal and avoiding obstacles in the trajectory of the robot. Also, the methodology generates a clothoid from one UGV for performing the change of lanes.

The proposed methodology is applied for solving two relevant issues, which are automatic parking and changing lanes. Automatic parking consists of two steps. First, the robot localizes itself in an unknown environment and creates a map of the area. Secondly, a trajectory is generated using the AD* path planning algorithm that trades off automatically between the final goal and the obstacles visible to the robot. For problem with changing lanes, a mathematical formulation is presented for generation of a clothoid in order to move an autonomous vehicle of unsafe lanes into safe lanes while keeping the vehicle on the road.

The main contribution of this work is the integration of localization, mapping and planning in a full robotic application (Fig. 1a). This helps in tasks such as avoiding obstacles and autonomous parking. Furthermore, other contribution is the generation of clothoids, which are used for performing the trajectory of autonomous vehicles (Fig. 1b) in order to the vehicles changing lanes with safe.

The remainder of this book chapter is organized as follows. Section 2 reviews the state of the art in motion planning and automatic parking. Section 3 presents the proposed methodology, detailing the approaches to solve the two relevant issues of this work. Experimental setup, results and analysis are shown in Sect. 4. Finally, Sect. 5 discusses conclusions and suggestions for the future works.

2 Related Work

2.1 Motion Planning

A path planning algorithm [2] suggests directions for a vehicle to move in unknown environments using data from 3D LIDAR. In this surrounding, the vehicle should be following some traffic conventions and generate maps of the area. MRF (Markov Random Field) is used to estimate the probabilities that the vehicle is moving in different directions, and those probabilities are passed to a planner. This updates the trajectories over time. Lastly, it is possible to observe that the MRF is useful to determine the direction of the vehicle and planning trajectories.

A real-time planning [10] is employed for on-road autonomous driving. This method is divided in two steps, which are on-road behavior planning and online path generation to traverse appropriate paths in dynamic environment using UGV. The method is successfully applied in several on-road traffic scenarios.

In [12], an anytime method is built for interval path planning, which is a variation of the A* algorithm for dynamic environments. This method uses intervals instead of goal points to achieve a solution. Results demonstrate that the method is fast to find a solution and is able to execute in real-time for several applications. However, the experiments in this study were simulated.

A path planning system based on modified potential field is proposed in [15]. This system combines artificial neural networks and potential fields to generate the appropriate trajectory of the vehicle. The planning method finds a global solution for the simulated experiment. The results demonstrate that the system is efficient in finding a solution quickly, but the vehicle has not been tested in a real environment to verify its effectiveness.

2.2 Autonomous Car Parking

A path planning [3] for semi-structured outdoor environments is presented. First, a map based on static objects of the surrounding is constructed. Secondly, a topological lane-network of the best fit of the map is created. The path planning is based on the A* algorithm, the solution yielded being optimized to create a non-linear trajectory. Results show that using topological graphs is a good option, because they incur in lower computational cost and respect the car restrictions.

A methodology is proposed in [8] for autonomous navigation of vehicles in parking garages. This methodology uses maps and laser sensors data to calculate the trajectory of the vehicle and to localize it when the GPS data is not available. The trajectory is created using a planner based on the A* algorithm, which sends the commands of speed and steering for the vehicle to accomplish this trajectory. The results prove that the car performed the parking without problems.

An autonomous parking system [7] considers the free parking spaces and the cars receives this information. The car gets close to a parking space and verifies if fits a vehicle. Having found the parking space, the system uses a implementation of the Rapidly-exploring Random Trees (RRT*) algorithm to correctly park the vehicle. In the results presented, the driver left the car and the system could find a parking space and park the vehicle in this free space. When the driver called back the car using a wireless application, the car went back to the driver.

A parking structure with connected intelligent vehicles is proposed in [4]. The author explains that the current parking structure is inefficient because too much space is used. The work shows a new parking structure looking to minimize the space used in parking lots. Using intelligent vehicles, such as a vehicular ad hoc networking, the results shows that the current space used in parking lots could be reduced to nearly half with the vehicles working in collaborative ways.

3 Proposed Methodology

Our goal is to solve two issues, which are automatic parking and changing lanes. For automatic parking, we propose an approach (Fig. 2) to plan a trajectory for the robot in simulated and real (indoor and outdoor) environments. For changing lanes, we develop a method using simple clothoids as shown in Fig. 4.

3.1 Automatic Parking

We propose a methodology (Fig. 2) composed of two steps described below:

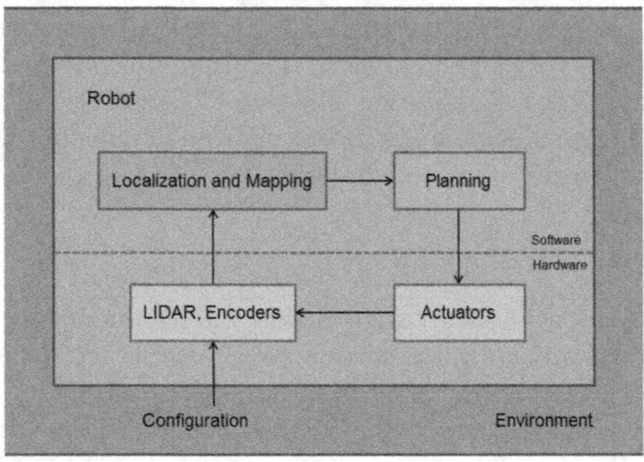

Fig. 2. Methodology to solve automatic parking and obstacles avoidance.

- Localization and Mapping: Localizes the robot in an unknown environment and creates a map of the region. Localization is used by the planning step to plan the robot. We used an Adaptive Monte Carlo Localization (AMCL)[1] package built in ROS[2] to localize a robot moving in a 2D space. Also, we used a Simultaneous Localization and Mapping (SLAM) package built in ROS called Gmapping[3] to create a 2D occupancy grid map.
- Planning: AD* planner is employed to plan the trajectory to be traversed, where a user defines the goal in the ROS framework and the planner provides the lower cost trajectory given the obstacles in front of the robot. After that, the mobile robot is able to follow the trajectory built properly.

[1] AMCL http://wiki.ros.org/amcl.
[2] Robot Operating System http://wiki.ros.org/.
[3] Gmapping http://wiki.ros.org/gmapping.

Localization and Mapping. To know where to go, the robot needs knowledge about the place it is in. ROS framework provides an AMCL package that estimates the self-localization of a mobile robot. More specifically, AMCL [14] is a probabilistic localization system that allows a robot moving in the 2D space. It uses a particle filter to estimate the pose of a robot based on wheel odometry and LIDAR sensor data. Also, AMCL needs a known map of the environment.

Using the SICK laser positioned on the robot, the ROS framework and the MORSE[4] simulator, we control the robot through a simulated environment built in the Blender[5]. Using the Gmapping package from ROS we built a map with the SICK laser sensor data. As the robot moves, data acquired from laser is integrated to the map. Thus, the map is built during the robot's navigation. In the simulation, the robot was driven through the environment and the map was built. This map is used by the AD* planner to plan the trajectory that the robot will follow to the given destination. The map built has information about obstacles on the environment. It shows where the robot can move unimpeded and where there are obstacles that the robot must avoid. The map of the environment is shown in Fig. 3, built by the SICK laser sensor data.

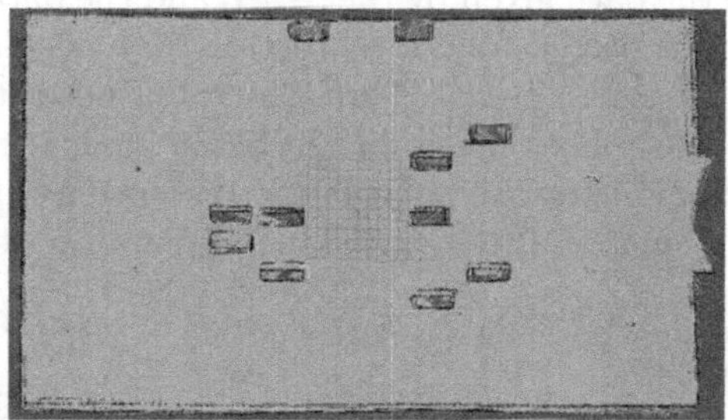

Fig. 3. Map generated using SICK laser sensor data and ROS packages, built by driving the robot through the simulated environment.

Planning. The planning algorithm is divided into two parts, which are the global and local planners. Global planner uses the map of the environment to plan the trajectory that the robot has to follow. The local planner is responsible to use the sensor data and the path given by the global planner to decide what is the less costly movement to perform, sending the movement command to the robot.

To make the global planning, we use the SBPL (Search-Based Planning Library) [1], which uses the AD* algorithm [9] to find the best path (less costly). AD* is a combination of A* that finds a solution quickly, and D* that improves the first solution found. AD* finds a fast solution, and improves on the solution found. To reach the solution the global planner uses cost function where the

[4] MORSE https://www.openrobots.org/wiki/morse/.

[5] Blender https://www.blender.org/.

costm is the cost to use the motion founded, *costa* is the actual position, *costn* is the next position and *dist* is the distance from initial and final position:

$$costm = max(costa, costn) \tag{1}$$

$$cost = (1 + costm) * dist \tag{2}$$

For the path calculation, we provide a goal point and its orientation. Using the given goal, the planner tries to compute the best solution to this point considering the limitations of the robot. These physical limitations are given to the planner as parameters. These parameters are primitive motions of the robot. Based on these primitive motions, the robot calculates a path that is possible for it to follow. As the robot moves along the path, the planner periodically checks if the robot moves away from the trajectory. If the path followed diverges from the planned path, the global planner recalculates a new lowest cost trajectory to update the trajectory. To visualize the computed path and the robot's sensor data through the environment, we use the Rviz[6], which presents the analyzed environment and the robot movements from of the robot's perspective.

The local planner used is the provided by ROS, Base Local Planner (BLP)[7]. This planner receives the path given by the SBPL and the data received from the laser. This information is passed to a cost function that calculates the movement searching the lower cost where *pscale* is how the robot needs to keep in the path planned, *gscale* how the robots needs to reach the goal, *oscale* how the robot needs to keep away from obstacles, *pdist* is the distance from the path, *gdist* is the distance from goal and *odist* is the distance from an obstacle.

$$cost = (pscale * pdist) + (gscale * gdist) + (oscale * odist) \tag{3}$$

The solution is computed according the proximity to the route, proximity to the goal and proximity to an obstacle. These three variables of proximity are passed to the planner as parameters. This solution is passed to the robot as the speed and steering commands. Local planner takes charge of avoiding obstacles while the robot navigates, trying to follow the path passed by global planner, but always being careful with obstacles along the way. The robot can not find a viable solution to avoid obstacles in the way, the local planner makes the robot stop moving, to prevent a collision.

3.2 Changing Lanes

We develop a methodology (Fig. 4) composed of three steps described below: Clothoid Creation, Clothoid Division and Trajectory Creation.

We developed a method to changing lanes, which uses clothoids to create a trajectory between two points and it's orientation. For creating the trajectory, we first develop a code which create one clothoid between two points. In this case,

[6] 3D visualization tool for ROS http://wiki.ros.org/rviz.

[7] Base Local Planner http://wiki.ros.org/base_local_planner.

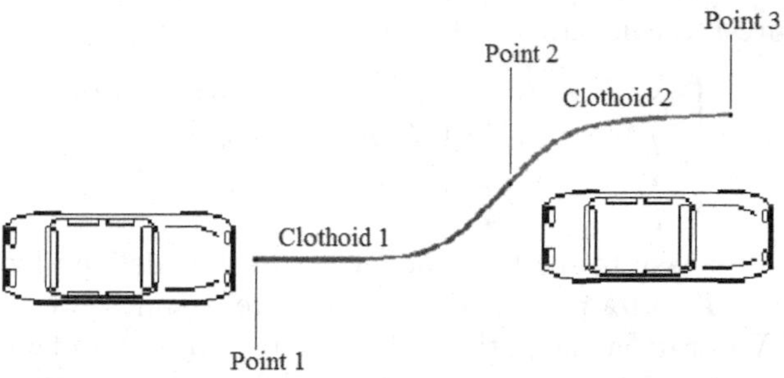

Fig. 4. Proposed methodology to solve changing lanes.

we estimate a regular clothoid with the maximum orientation variation of $\frac{\pi}{2}$. To create a complete trajectory we set the initial point and the goal point, we look the orientation and distance between points and calculate intermediate points between the setting points, if necessary. With intermediate points we created clothoids in every couple of points and linking the created clothoids, so we can describe a trajectory from the initial point to the destination point.

Clothoid Creation. Each clothoid segment is defined by a s, which is the distance along the segment and a k that is the curvature in every point in the segment. A clothoid is defined by a linear variation curvature along the segment, $k(s) = \sigma s$, and the σ is the rate of change of the curvature. In each s in the path, the state (x, y, ψ) is provided by the equations below:

$$\psi(s) = \int_0^s k(z)dz + \psi_0 \tag{4}$$

$$x(s) = \int_0^s \cos\psi(z)dz + x_0 \tag{5}$$

$$y(s) = \int_0^s \sin\psi(z)dz + y_0 \tag{6}$$

where s is defined by the interval $[0,\ L]$ and L is the length of the segment.

Clothoid Division. A clothoid can be divided by three parts, which are an entry clothoid, an arc and an exit clothoid equal to the entry clothoid [13]. On the entry clothoid the curvature is increased, the arc can give an degree of freedom of the path and the exit clothoid start with the final curvature of the entry clothoid and end with zero curvature. To add an arc, we have the λ parameter which define the length of the arc as a fraction of the total length of the path.

$$\lambda = \frac{L_{arc}}{L} \tag{7}$$

The curvature k is calculated as follow:

$$k(s) = \begin{cases} \sigma s & \forall s \in [0, L\frac{1-\lambda}{2}] \quad \textit{Entry Clothoid} \\ \sigma L\frac{1-\lambda}{2} & \forall s \in (L\frac{1-\lambda}{2}, L\frac{1+\lambda}{2}) \quad \textit{Arc} \\ \sigma(L-s) & \forall s \in [L\frac{1+\lambda}{2}, L] \quad \textit{Exit Clothoid} \end{cases} \tag{8}$$

In the Eq. 5 it's easy to see that the curvature k is defined by sharpness σ, length of the path L and λ parameter. To calculate σ and L, we used the initial and final state. We start at the initial state $q_1 = (x_1, y_1, \psi_1)$ and goes to the final state $q_2 = (x_2, y_2, \psi_2)$. The clothoid can be calculated for any two state q_1 and q_2, where the angle α between ψ_1 and a line created connecting the two points (x_1, y_1) and (x_2, y_2) has less than $\frac{\pi}{2}$ degrees. Then, we used the absolute value of the angle α of the orientation variation along the path:

$$\alpha = 2\tan^{-1}(\frac{y_2 - y_1}{x_2 - x_1}) = \psi_2 - \psi_1 \tag{9}$$

From substituting the Eq. 5 in Eq. 1 we have the equation

$$\alpha = \psi_2 - \psi_1 = \int_0^L k(z)dz = \frac{\sigma L^2}{4}(1 - \lambda^2) \tag{10}$$

The $D(\alpha, \lambda)$ is the ratio of the euclidean distance d and the length L from the path connecting them, $D(\alpha, \lambda) = \frac{d}{L}$. From the Eq. 7 and the definition of D we can reach the equations of L and σ

$$L = \frac{d}{D(\alpha, \lambda)} \tag{11}$$

$$\sigma = \frac{4\alpha}{L^2(1 - \lambda^2)} \tag{12}$$

where d is the euclidean distance between the two points and α is given by Eq. 6. $D(\alpha, \lambda)$ can be found from a the geometry of a symmetric path given by α and λ. We followed the approach given by Kanayama and Hartman [6], and for our case $D(\alpha, \lambda)$ is calculated by

$$D(\alpha, \lambda) = 2\int_0^{0.5} \cos\psi(z)dz \tag{13}$$

where $\psi(z)$ is calculated by.

$$\begin{cases} \frac{2\alpha}{1+\gamma} z & \forall z \in [0, \frac{\gamma}{2}] \\ \frac{2\alpha}{1-\gamma^2}(-z^2 + z - \frac{\gamma^2}{4}) & \forall z \in (\frac{\gamma}{2}, \frac{1}{2}] \end{cases} \tag{14}$$

In [5], Eq. 11 considers a lane change creating two connected clothoids, in this work we proposed to create only one clothoid and use this method to create many clothoids as necessary. The γ parameter is the fraction of distance between the two clothoids created divided by an third state q_i, in our case we have only one clothoid and $\gamma = 0.5$ always once the q_i is at the middle of the clothoid.

Trajectory Creation. To create a safe trajectory we use the clothoid creation method and many points as necessary. With the points the planner create the clothoids for each couple of points and connect them. With a fast clothoid creation method it's easy to create trajectories and pass them to the vehicle follow. To create a lane change trajectory we set three points, with these points we create one clothoid connecting the first and the second point and another clothoid connecting the second and the third point, as shown in Fig. 4.

4 Experimental Results

To evaluate the capabilities and performance of the methodology, we implemented the described system (Sect. 3) and tested its performance on a simulated wheeled vehicle (Fig. 8), Pioneer 3-AT robot (Fig. 1a) and an autonomous vehicle (CaRINA 2 - Fig. 1b). The robots are equipped with a laser scanner and odometry, and were deployed in the simulated (the tasks of automatic parking and changing lanes) and real (automatic parking task) scenarios.

Our methodology runs on a low end PC and is implemented in C++. We use ROS to interface with the robot and our method, to provide localization, to sensor data processing and send commands of speed and steering of the robot.

4.1 Obstacle Avoidance

This task was made in two distinct scenarios. First, indoor environment (Fig. 5), more specifically a hallway of a building. Secondly, outside area of the university library. We used a pioneer 3-AT robot with a HOKUYO laser sensor at the front of the robot and a laptop above it for controlling it. The laser sensor received the same range of the SICK laser used in the simulated experiments.

Fig. 5. Indoor environment adapted to make the tests with the robot.

The indoor scenario was built with some boxes for the robot to execute the obstacle avoidance. The goal was the robot to traverse the environment avoiding the obstacles. Using laser data and odometry, the planner created a trajectory and could cross the environment avoiding obstacles and trying make the trajectory with the most proximity to make the less costly trajectory. We have a video achieved with the our methodology employed in the obstacle avoidance on the real indoor environment. VIDEO: **Obstacle Avoidance Indoor**[8].

We built an outdoor scenario (Fig. 6) with some boxes on the floor close of an university library. In this outdoor experiment, we tried to make the robot to traverse the environment avoiding the obstacles using the path planner. First, we built the map of the environment, then using this map we set a goal in the map that the robot need to reach. Using some maneuvers, the robot could reach the goal avoiding the obstacles (Fig. 7). This test was repeated five times, in the five tests the robot reached the goal. In some of these tests the robot took more time to reach the proposed goal, but also reached it. We have a video showing the robot traversing the environment. VIDEO: **Obstacle Avoidance Outdoor**[9].

Fig. 6. Outdoor scenario adapted to make the obstacle avoidance using robot.

4.2 Automatic Parking

The local planner had a limitation, the simulated vehicle could not navigate backwards. BLP was developed to holonomic vehicles that can move to any direction on the 2D plane rotating on its center. Because of this, the BLP can not send the negative speed commands, this makes the vehicle can not navigate backwards, because the lower speed sent to the robot (zero). A solution found was change the source code of the BLP to look backwards too, so the planner can keep navigating even when a reverse movement is needed. Originally, BLP search for best forward solutions only, therefore, only positive speed commands are calculated. In the source code, we changed the code and made the planner to

[8] Obstacle Avoidance Indoor https://youtu.be/D4ayQQOwNr4.

[9] Obstacle Avoidance Outdoorhttps://youtu.be/FsBoip6NwJQ.

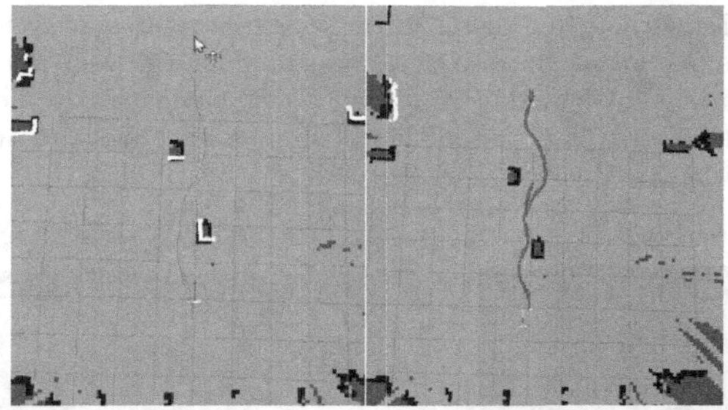

Fig. 7. At the left side, in green, we have the path calculated by the global planner, in the right side we have the path that the robot made. (Color figure online)

look for forward solutions, after that, the planner searched for backward solutions keeping with best of these solutions sending its commands for the robot to follow.

Before we start to test the simulated on the vehicle, we made some evaluations on the simulator to verify how the planners would work. After that, some tests and parameters adjustments, the planner on the simulation could plan a good trajectory with the lower cost (Fig. 8) and following the trajectory appropriate, always looking to the obstacles on the way and avoiding them when necessary.

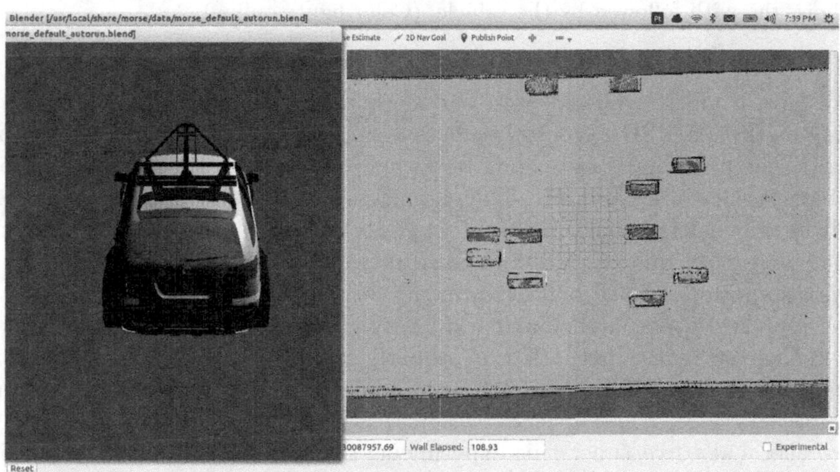

Fig. 8. The car model on the MORSE simulator and path calculated by the SBPL (global planner), this path will be followed by the car and, if necessary, the SBPL will find a better path and pass it to the local planner to follow.

The results obtained (Fig. 9) were good for the simulated environment, because the planning system can to plan a trajectory and to follow it, approaching the obstacles without hit them. If the vehicle navigates away from the trajectory, the planner recalculates a new trajectory for the vehicle to follow, so, the vehicle always will be looking for a way to keep navigating to the destination (goal) and, in the worst case, the robot will stop and not move to any direction.

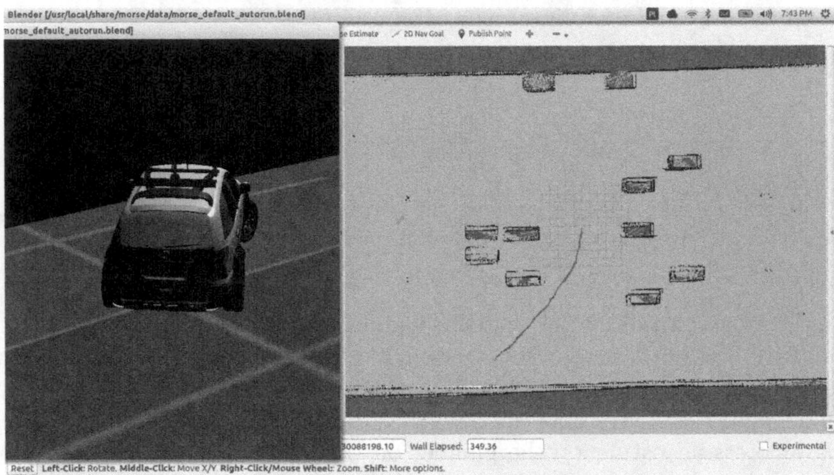

Fig. 9. The car model on the MORSE simulator in the final trajectory performed. The green line presents the trajectory provided by the AD* planning algorithm and the red line shows the path followed by the vehicle. (Color figure online)

We have a video achieved with our methodology applied in the automatic parking on the MORSE. We propose here a parking system with only a forward sensor, so the planner avoid making backwards movements since it can not look the obstacles that are behind it. The planner sends the steering and speed commands to the vehicle, which instantly set these commands making the wheels of simulated car turn quickly. VIDEO: **Simulation CaRINA Parking**[10].

We also performed the real experiments, which is shown in Fig. 10.

We proposed a perpendicular parking, which the planner needs to calculate some maneuver to try to park the properly robot (Fig. 11). The AD* planning algorithm successfully completed the trajectory autonomously and appropriately.

We also made five times this experiment, the robot performed the parking correctly. Sometimes, the robot took a time to park it, but in all the demonstrated tests the planner made the parking successfully. We made a video of the robot parking in this scenario. VIDEO: **Real Pioneer Parking**[11].

[10] Simulation CaRINA Parking https://youtu.be/WySNAvHiG10.
[11] Real Pioneer Parking https://youtu.be/XZTgjl59EJU.

Fig. 10. Outdoor environment adapted to make the parking tests with the robot.

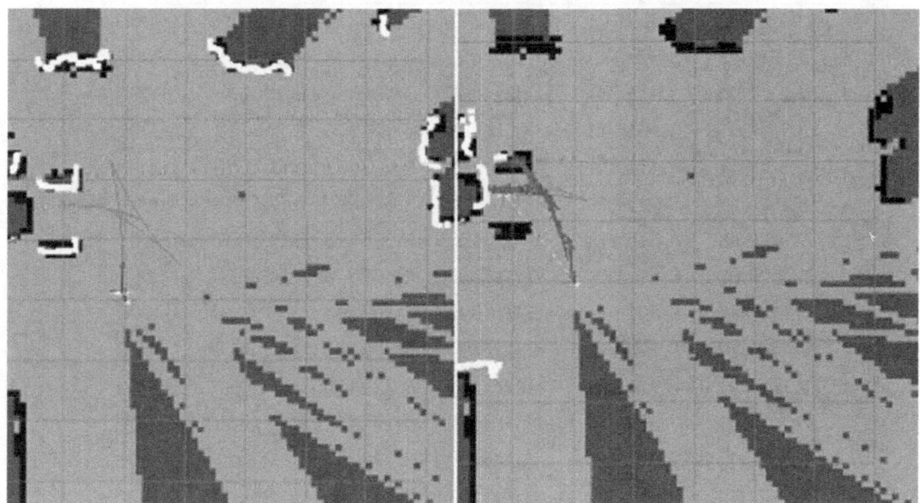

Fig. 11. At the left side, in green, we have the path calculated by the global planner, in the right side we have the path that the robot made.

4.3 Changing Lanes

To test the clothoid creation and changing lanes method, we use the MORSE simulator to validate the clothoid created. We used a model of the CaRINA on the simulation, the CaRINA so in the work of Massera [11] is proposed a controller to the vehicle, which will follow the points passed trying minimize the error as the vehicle follow the points. In the simulation, we passed the clothoid to the controller and it follow the passed points (as shown in Fig. 12). We started the vehicle in the origin point and setted the goal point at the point (−25, −5). The initial state of the vehicle was $q_1 = (0, 0, 0)$ and the final state was

$q_2 = (-25, -5, 0)$. To create this trajectory, we calculated the intermediate point between the two states creating an intermediate state $q_i = (-12.5, -2.5, \pi/6)$. We created two clothoids, which can be seen all the points from the Table 1.

Table 1. Clothoid creation results.

X	Y	Curvature	Psi	Length	Velocity
0.0	0.0	0.0	0.0	0.0	0.0
−1.4687720	−0.0058950	0.00817571	3.152950	0.025617	1.00000
−2.9235300	−0.043871	0.01635143	3.185205	0.051235	1.00000
−4.3752020	−0.144330	0.02452714	3.238357	0.076852	1.00000
−5.8173290	−0.337306	0.03270285	3.312406	0.102469	1.00000
−7.2377120	−0.651733	0.04087857	3.407353	0.128086	1.00000
−8.6202080	−1.104586	0.03270285	3.502532	0.153704	1.00000
−9.9597330	−1.672612	0.02452714	3.576813	0.179321	1.00000
−11.260647	−2.324589	0.01635143	3.630198	0.204938	1.00000
−12.532709	−3.031429	0.00817571	3.662686	0.230555	1.00000
−13.789207	−3.765690	0.00000000	3.674276	0.256173	1.00000
−15.390137	−4.415179	−0.0093993	3.662601	0.022669	1.00000
−16.700126	−5.044748	−0.0187987	3.629667	0.045338	1.00000
−17.965523	−5.618262	−0.0281980	3.575473	0.068007	1.00000
−19.273008	−6.115198	−0.0375973	3.500018	0.090676	1.00000
−20.503277	−6.507356	−0.0469967	3.403304	0.113345	1.00000
−21.762572	−6.774535	−0.0375973	3.306353	0.136014	1.00000
−22.988701	−6.928396	−0.0284591	3.231476	0.158683	1.00000
−24.176451	−7.012948	−0.0190598	3.177457	0.181352	1.00000
−25.329343	−7.027786	−0.0096604	3.143255	0.204021	1.00000
−26.466373	−7.025806	−0.0002611	3.131195	0.226690	1.00000

Table 1 shows the results of the clothoid creation method, this tries to reach the final state but obeying the geometric limitations. According with the variation of the orientation during the trajectory, the vehicle will increase more the x or y in the path created. Despite of the difference between the goal point and the final calculated point, the method approximates to reach the final state and in the simulation the vehicle could follow the created clothoid without problem. Figure 13 presents the trajectory created in green by the clothoid creation method and the trajectory executed in red by the controller of the vehicle (Massera [11]).

Figures 15 and 14 shows the results of the orientation and lateral error from the vehicle while it tried to follow the trajectory created. The controller tries to keep in the trajectory reducing the orientation (psi) and the lateral (x, y) error.

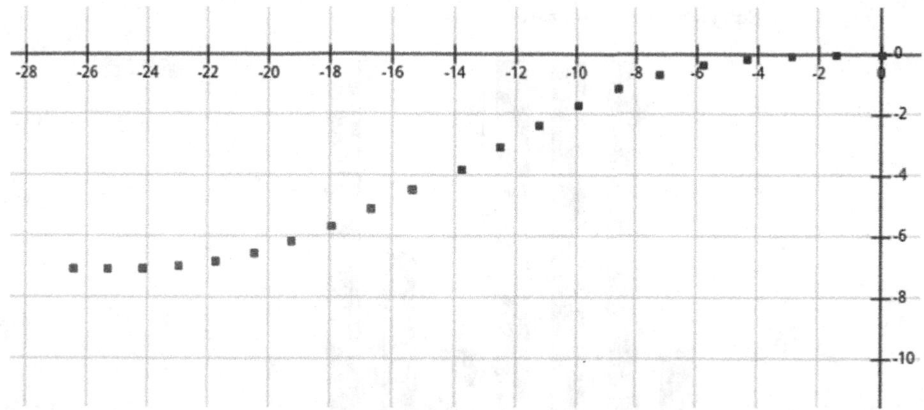

Fig. 12. The two clothoids created from the clothoid creation method. The first clothoid is shown in blue and the second clothoid is presented in red. (Color figure online)

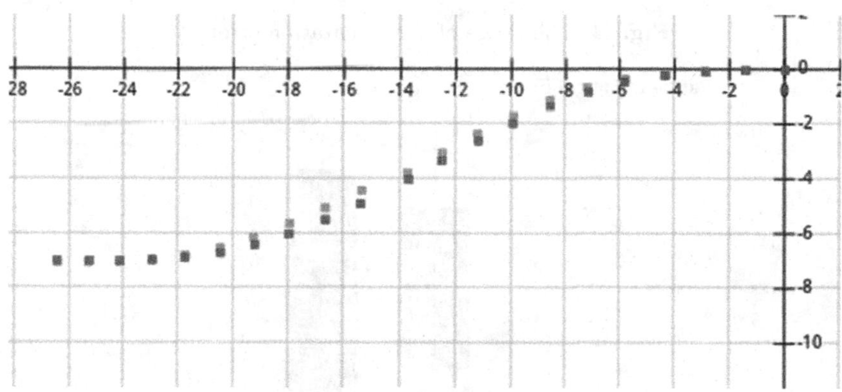

Fig. 13. The trajectory created in green and the executed by controller in red. (Color figure online)

The lateral error is pretty small and it does not affect the trajectory. The orientation error seems big but does not inflict any considerable change on the distance of the vehicle and the proposed trajectory. The method created shows a smooth clothoid trajectory, which has no trouble to the vehicle follow. We recorded the vehicle following the trajectory created. The video shows the path executed by the car guided by the clothoid trajectory only to test if the trajectory is able to be followed by the vehicle. VIDEO: **Simulated Lane Change**[12].

[12] Simulated Lane Change https://goo.gl/9svDgL.

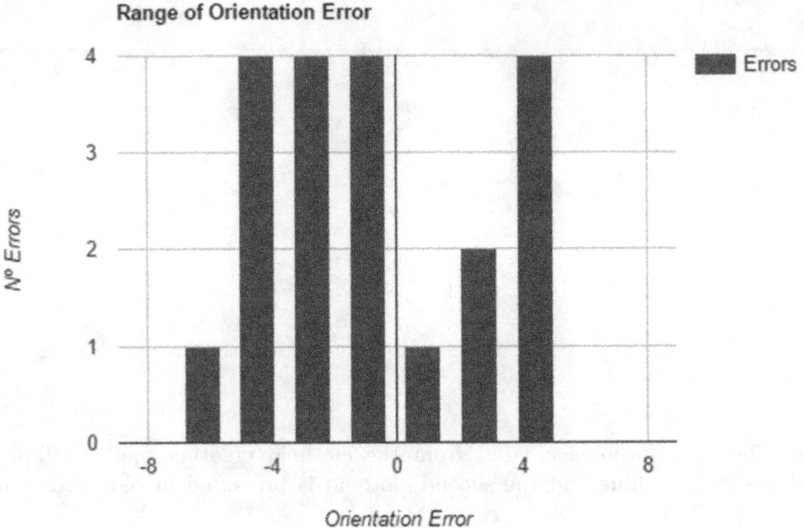

Fig. 14. The range of the orientation error.

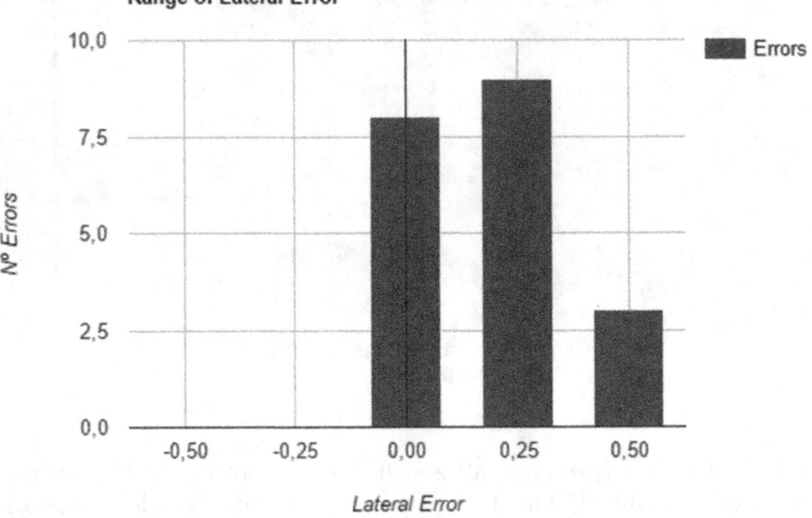

Fig. 15. The range of the lateral error.

5 Conclusions

We propose a methodology to solve the tasks of automatic parking and changing lanes using robots in simulated, indoor and outdoor scenarios. The methodology uses the map of the place, the LIDAR data and robot's odometry. The area map is built, which the global planner calculates a trajectory to the goal and the local planner sends the speed and steering commands for the vehicle to follow. Also, we implement a method to create clothoids for the task of changing lanes.

We demonstrate four tests on the scenarios. Two of them were evaluated qualitatively (obstacle avoidance indoor and simulation CaRINA parking). Other two (obstacle avoidance outdoor and real pioneer parking) were analyzed quantitatively, all the executions the robot successfully navigated autonomously. The planner calculated a path through the environment avoiding the obstacles, while the vehicle follows the path trying to reach the goal. If it distanced the path, the planner calculates a new route to the goal. A limitation on the local planner was found, it was not developed for Ackermann robots, so, when the vehicle needed to go backwards on the simulated environment, the planners made the vehicle just stop. This limitation can be solved by changing the source code of the local planner, making the planner consider the negative speed commands. Furthermore, we tested the clothoid creation method and execute the trajectory in the simulation tests, which tried to approximate to the goal state respecting the geometric limitations. With the clothoid created was possible to pass the trajectory points to the vehicle controller execute the lane change without issues.

As future work, we will perform overtaking tests on Brazilian highways to verify if the global and local planners will work appropriately using CaRINA 2.

Acknowledgment. The authors would like to acknowledge the support by CNPq, USP and UFU. Also, would like to acknowledge to Carlos Massera by support with the controller of the vehicle that makes possible the tests with the clothoid creation method.

References

1. Butzke, J., Sapkota, K., Prasad, K., MacAllister, B., Likhachev, M.: State lattice with controllers: augmenting lattice-based path planning with controller-based motion primitives. In: International Conference on Intelligent Robots and Systems (IROS) (2014)
2. Dolgov, D., Thrun, S.: Detection of principal directions in unknown environments for autonomous navigation. In: Proceedings of the Robotics: Science and Systems (RSS) (2008)
3. Dolgov, D., Thrun, S.: Autonomous driving in semi-structured environments: mapping and planning. In: International Conference on Robotics and Automation (ICRA) (2009)
4. Ferreira, M., Damas, L., Conceicao, H., dOrey, P. M., Fernandes, R., Steenkiste, P.: Self-automated parking lots for autonomous vehicles based on vehicular ad hoc networking. In: International Vehicles (IV) (2014)
5. Funke, J., Gerdes, J. C.: Simple clothoid paths for autonomous vehicle lane changes at the limits of handling. In: Proceedings of the ASME Dynamic Systems and Control Conference (2013)
6. Kanayama, Y.J., Hartman, B.I.: Smoth local-path planning for autonomous vehicle. Int. J. Robot. Res. **16**(3), 263–284 (1997)
7. Kim, S.W., Liu, W., Marczuk, K.A.: Autonomous parking from a random drop point. In: Intelligent Vehicles Symposium (IV) (2014)

8. Kummerle, R., Hahnel, D., Dolgov, D., Thrun, S., Burgard, W.: Autonomous driving in a multi-level parking structure. In: International Conference on Robotics and Automation (ICRA) (2009)
9. Likhachev, M., Ferguson, D., Gordon, G., Stentz, A.T., Thrun, S.: Anytime dynamic a* : an anytime, replanning algorithm. In: International Conference on Automated Planning and Scheduling (ICAPS) (2005)
10. Ma, L., Jiaotong, X., China, X., Yang, J., Zhang, M.: A two-level path planning method for on-road autonomous driving. In: International Conference on Intelligent Systems Design and Engineering Application (ISDEA) (2012)
11. Massera Filho, C., Wolf, D.F.: Dynamic inversion-based control for front wheel drive autonomous ground vehicles near the limits of handling. In: Intelligent Transportation System (2014)
12. Narayanan, V., Phillips, M., Likhachev, M.: Anytime safe interval path planning for dynamic environments. In: International Conference on Intelligent Robot Systems (IROS) (2012)
13. Scheuer, A., Fraichard, T.: Planning continuous-curvature paths for car-like robots. In: International Conference on Intelligent Robots and Systems (1996)
14. Thrun, S., Fox, D., Burgard, W., Dellaert, F.: Robust monte carlo localization for mobile robots. Artif. Intell. **128**(1), 99–141 (2000)
15. iaoming, T.: Local obstacle avoidance planning of logistics system AGV based vector field. In: Management Science and Industrial Engineering (MSIE) (2011)

Author Index

Printed in the United States
By Bookmasters

Printed in the United States
By Bookmasters